Instrumentation and Control Systems

Instrumentation and Control Systems

W. Bolton

AMSTERDAM • BOSTON • HEIDELBERG • LONDON • NEW YORK • OXFORD
PARIS • SAN DIEGO • SAN FRANCISCO • SINGAPORE • SYDNEY • TOKYO
Newnes is an imprint of Elsevier

Newnes
An imprint of Elsevier
Linacre House, Jordan Hill, Oxford OX2 8DP
200 Wheeler Road, Burlington, MA 01803

British Library Cataloguing in Publication Data
A catalogue record for this book is available from the British Library

Library of Congress Cataloguing in Publication Data
A catalogue record for this book is available from the Library of Congress

ISBN 0 7506 6432 0

For information on all Newnes publications
visit our website at www.newnespress.com

Printed and bound in Great Britain by Biddles Ltd, King's Lynn, Norfolk

Contents

Preface

Aims

This book has the aims of covering the new specification of the Edexcel Level 4 BTEC units of *Instrumentation and Control Principles* and *Control Systems and Automation* for the Higher National Certificates and Diplomas in Engineering and also providing a basic introduction to instrumentation and control systems for undergraduates. The book aims to give an appreciation of the principles of industrial instrumentation and an insight into the principles involved in control engineering.

Structure of the book

The book has been designed to give a clear exposition and guide readers through the principles involved in the design and use of instrumentation and control systems, reviewing background principles where necessary. Each chapter includes worked examples, multiple-choice questions and problems; answers are supplied to all questions and problems. There are numerous case studies in the text and application notes indicating applications of the principles.

Coverage of Edexcel units

Basically, the Edexcel unit *Instrumentation and Control Principles* is covered by chapters 1 to 6 with the unit *Control Systems and Automation* being covered by chapters 8 to 13 with chapter 5 including the overlap between the two units. Chapter 7 on PLCs is included to broaden the coverage of the book from these units.

Performance outcomes

The following indicate the outcomes for which each chapter has been planned. At the end of the chapters the reader should be able to:

Chapter 1: Measurement systems
    Read and interpret performance terminology used in the specifications of instrumentation.

Chapter 2: Instrumentation system elements
    Describe and evaluate sensors, signal processing and display elements commonly used with instrumentation used   in the

measurement of position, rotational speed, pressure, flow, liquid level and temperature.

Chapter 3: Instrumentation case studies
Explain how system elements are combined in instrumentation for some commonly encountered measurements.

Chapter 4: Control systems
Explain what is meant by open and closed-loop control systems, the differences in performance between such systems and explain the principles involved in some simple examples of such systems.

Chapter 5: Process controllers
Describe the function and terminology of a process controller and the use of proportional, derivative and integral control laws.
Explain PID control and how such a controller can be tuned.

Chapter 6: Correction elements
Describe common forms of correction/regulating elements used in control systems.
Describe the forms of commonly used pneumatic/hydraulic and electric correction elements.

Chapter 7: PLC systems
Describe the functions of logic gates and the use of truth tables.
Describe the basic elements involved with PLC systems and devise programs for them to carry out simple control tasks.

Chapter 8: System models
Explain how models for physical systems can be constructed in terms of simple building blocks.

Chapter 9: Transfer function
Define the term transfer function and explain how it used to relate outputs to inputs for systems.
Use block diagram simplification techniques to aid in the evaluation of the overall transfer function of a number of system elements.

Chapter 10: System response
Use Laplace transforms to determine the response of systems to common forms of inputs.
Use system parameters to describe the performance of systems when subject to a step input.
Analyse systems and obtain values for system parameters.
Explain the properties determining the stability of systems.

Chapter 11: Frequency response
Explain how the frequency response function can be obtained for a system from its transfer function.
Construct Bode plots from a knowledge of the transfer function.
Use Bode plots for first and second-order systems to describe their frequency response.
Use practically obtained Bode plots to deduce the form of the transfer function of a system.

Compare compensation techniques.

Chapter 12: Nyquist diagrams
Draw and interpret Nyquist diagrams.

Chapter 13: Controllers
Explain the reasons for the choices of P, PI or PID controllers.
Explain the effect of dead time on the behaviour of a control system.
Explain the uses of cascade control and feedforward control.

W. Bolton

1 Measurement systems

1.1 Introduction

This chapter is an introduction to the instrumentation systems used for making measurements and deals with the basic elements of such systems and the terminology used to describe their performance in use.

1.1.1 Systems

The term *system* will be freely used throughout this book and so here is a brief explanation of what is meant by a system and how we can represent systems.

If you want to use an amplifier then you might not be interested in the internal working of the amplifier but what output you can obtain for a particular input. In such a situation we can talk of the amplifier being a system and describe it by means of specifying how the output is related to the input. With an engineering system an engineer is more interested in the inputs and outputs of a system than the internal workings of the component elements of that system.

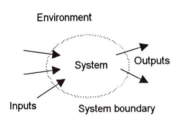

Figure 1.1 *A system*

> A *system* can be defined as an arrangement of parts within some boundary which work together to provide some form of output from a specified input or inputs. The boundary divides the system from the environment and the system interacts with the environment by means of signals crossing the boundary from the environment to the system, i.e. inputs, and signals crossing the boundary from the system to the environment, i.e. outputs (Figure 1.1).

A useful way of representing a system is as a *block diagram*. Within the boundary described by the box outline is the system and inputs to the system are shown by arrows entering the box and outputs by arrows leaving the box. Figure 1.2 illustrates this for an electric motor system; there is an input of electrical energy and an output of mechanical energy, though you might consider there is also an output of waste heat. The interest is in the relationship between the output and the input rather than the internal science of the motor and how it operates. It is convenient to think of the system in the box operating on the input to produce the output. Thus, in the case of an amplifier system (Figure 1.3) we can think of the system multiplying the input V by some factor G, i.e. the amplifier gain, to give the output GV.

Often we are concerned with a number of linked systems. For example we might have a CD player system linked to an amplifier system which,

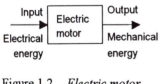

Figure 1.2 *Electric motor system*

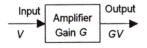

Figure 1.3 *Amplifier system*

in turn, is linked to a loudspeaker system. We can then draw this as three interconnected boxes (Figure 1.4) with the output from one system becoming the input to the next system. In drawing a system as a series of interconnected blocks, it is necessary to recognise that the lines drawn to connect boxes indicate a flow of information in the direction indicated by the arrow and not necessarily physical connections.

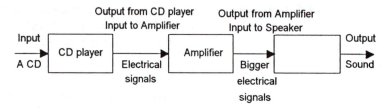

Figure 1.4 *Interconnected systems*

1.2 Instrumentation systems

The purpose of an *instrumentation system* used for making measurements is to give the user a numerical value corresponding to the variable being measured. Thus a thermometer may be used to give a numerical value for the temperature of a liquid. We must, however, recognise that, for a variety of reasons, this numerical value may not actually be the true value of the variable. Thus, in the case of the thermometer, there may be errors due to the limited accuracy in the scale calibration, or reading errors due to the reading falling between two scale markings, or perhaps errors due to the insertion of a cold thermometer into a hot liquid, lowering the temperature of the liquid and so altering the temperature being measured. We thus consider a measurement system to have an input of the true value of the variable being measured and an output of the measured value of that variable (Figure 1.5). Figure 1.6 shows some examples of such instrumentation systems.

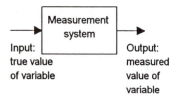

Figure 1.5 *An instrumentation/ measurement system*

An *instrumentation system* for making measurements has an input of the true value of the variable being measured and an output of the measured value.

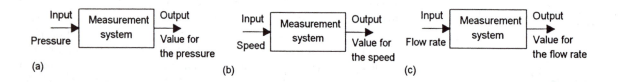

Figure 1.6 *Example of instrumentation systems: (a) pressure measurement, (c) speedometer, (c) flow rate measurement*

1.2.1 The constituent elements of an instrumentation system

An instrumentation system for making measurements consists of several elements which are used to carry out particular functions. These functional elements are:

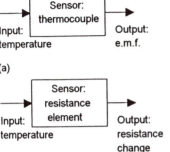

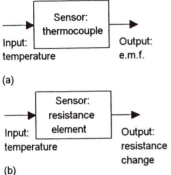

Figure 1.7 *Sensors: (a) thermo-couple, (b) resistance thermometer element*

1   *Sensor*
This is the element of the system which is effectively in contact with the process for which a variable is being measured and gives an output which depends in some way on the value of the variable and which can be used by the rest of the measurement system to give a value to it. For example, a thermocouple is a sensor which has an input of temperature and an output of a small e.m.f. (Figure 1.7(a)) which in the rest of the measurement system might be amplified to give a reading on a meter. Another example of a sensor is a resistance thermometer element which has an input of temperature and an output of a resistance change (Figure 1.7(b)).

2   *Signal processor*
This element takes the output from the sensor and converts it into a form which is suitable for display or onward transmission in some control system. In the case of the thermocouple this may be an amplifier to make the e.m.f. big enough to register on a meter (Figure 1.8(a)). There often may be more than item, perhaps an element which puts the output from the sensor into a suitable condition for further processing and then an element which processes the signal so that it can be displayed. The term *signal conditioner* is used for an element which converts the output of a sensor into a suitable form for further processing. Thus in the case of the resistance thermometer there might be a signal conditioner, a Wheatstone bridge, which transforms the resistance change into a voltage change, then an amplifier to make the voltage big enough for display (Figure 1.8(b)).

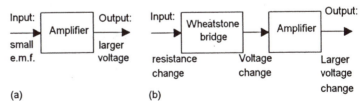

Figure 1.8 *Examples of signal processing*

3   *Data presentation*
This presents the measured value in a form which enables an observer to recognise it (Figure 1.9). This may be via a display, e.g. a pointer moving across the scale of a meter or perhaps information on a visual display unit (VDU). Alternatively, or additionally, the signal may be recorded, e.g. on the paper of a chart recorder or perhaps on magnetic disc, or transmitted to some other system such as a control system.

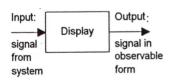

Figure 1.9 *A data presentation element*

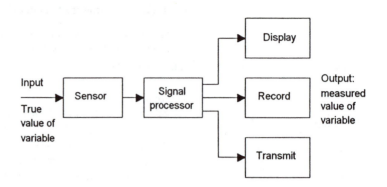

Figure 1.10 *Measurement system elements*

Figure 1.10 shows how these basic functional elements form a measurement system.

The term *transducer* is often used in relation to measurement systems. Transducers are defined as an element that converts a change in some physical variable into a related change in some other physical variable. It is generally used for an element that converts a change in some physical variable into an electrical signal change. Thus sensors can be transducers. However, a measurement system may use transducers, in addition to the sensor, in other parts of the system to convert signals in one form to another form.

Example

With a resistance thermometer, element A takes the temperature signal and transforms it into resistance signal, element B transforms the resistance signal into a current signal, element C transforms the current signal into a display of a movement of a pointer across a scale. Which of these elements is (a) the sensor, (b) the signal processor, (c) the data presentation?

The sensor is element A, the signal processor element B and the data presentation element is C. The system can be represented by Figure 1.11.

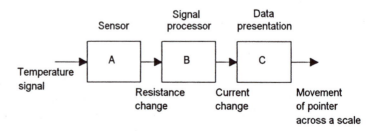

Figure 1.11 *Example*

1.3 Performance terms

The following are some of the more common terms used to define the performance of measurement systems and functional elements.

1.3.1 Accuracy and error

Accuracy is the extent to which the value indicated by a measurement system or element might be wrong. For example, a thermometer may have an accuracy of ±0.1°C. Accuracy is often expressed as a percentage of the full range output or full-scale deflection (f.s.d). For example, a system might have an accuracy of ±1% of f.s.d. If the full-scale deflection is, say, 10 A, then the accuracy is ±0.1 A. The accuracy is a summation of all the possible errors that are likely to occur, as well as the accuracy to which the system or element has been calibrated.

The term *error* is used for the difference between the result of the measurement and the true value of the quantity being measured, i.e.

error = measured value − true value

Thus if the measured value is 10.1 when the true value is 10.0, the error is +0.1. If the measured value is 9.9 when the true value is 10.0, the error is −0.1.

> *Accuracy* is the indicator of how close the value given by a measurement system can be expected to be to the true value. The *error* of a measurement is the difference between the result of the measurement and the true value of the quantity being measured.

Application
The accuracy of a digital thermometer is quoted in its specification as:

Full scale accuracy - better than 2%

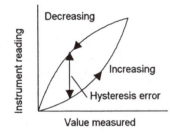

Figure 1.12 *Hysteresis error*

Errors can arise in a number of ways and the following describes some of the errors that are encountered in specifications of instrumentation systems.

1 *Hysteresis error*

The term *hysteresis error* (Figure 1.12) is used for the difference in outputs given from the same value of quantity being measured according to whether that value has been reached by a continuously increasing change or a continuously decreasing change. Thus, you might obtain a different value from a thermometer used to measure the same temperature of a liquid if it is reached by the liquid warming up to the measured temperature or it is reached by the liquid cooling down to the measured temperature.

2 *Non-linearity error*

The term *non-linearity error* (Figure 1.13) is used for the error that occurs as a result of assuming a linear relationship between the input and output over the working range, i.e. a graph of output plotted against input is assumed to give a straight line. Few systems or elements, however, have a truly linear relationship and thus errors occur as a result of the assumption of linearity. Linearity error

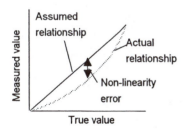

Figure 1.13 *Non-linearity error*

Application
A load cell is quoted in its specification as having:

Non-linearity error ±0.03% of full range
Hysteresis error ±0.02% of full range

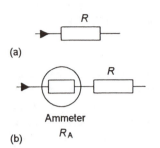

Figure 1.14 *Loading with an ammeter: (a) circuit before meter introduced, (b) extra resistance introduced by meter*

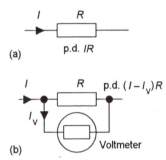

Figure 1.15 *Loading with a voltmeter: (a) before meter, (b) with meter present*

Application
See Appendix A for a discussion of how the accuracy of a value determined for some quantity can be computed from values obtained from a number of measurements, e.g. the accuracy of the value of the density of some material when computed from measurements of its mass and volume, both the mass and volume measurements having errors.

is usually expressed as a percentage error of full range or full scale output.

3   *Insertion error*
When a cold thermometer is put in to a hot liquid to measure its temperature, the presence of the cold thermometer in the hot liquid changes the temperature of the liquid. The liquid cools and so the thermometer ends up measuring a lower temperature than that which existed before the thermometer was introduced. The act of attempting to make the measurement has modified the temperature being measured. This effect is called *loading* and the consequence as an *insertion error*. If we want this modification to be small, then the thermometer should have a small heat capacity compared with that of the liquid. A small heat capacity means that very little heat is needed to change its temperature. Thus the heat taken from the liquid is minimised and so its temperature little affected.

Loading is a problem that is often encountered when measurements are being made. For example, when an ammeter is inserted into a circuit to make a measurement of the circuit current, it changes the resistance of the circuit and so changes the current being measured (Figure 1.14). The act of attempting to make such a measurement has modified the current that was being measured. If the effect of inserting the ammeter is to be as small as possible and for the ammeter to indicate the original current, the resistance of the ammeter must be very small when compared with that of the circuit.

When a voltmeter is connected across a resistor to measure the voltage across it, then what we have done is connected a resistance, that of the voltmeter, in parallel with the resistance across which the voltage is to be measured. If the resistance of the voltmeter is not considerably higher than that of the resistor, the current through the resistor is markedly changed by the current passing through the meter resistance and so the voltage being measured is changed (Figure 1.15). The act of attempting to make the measurement has modified the voltage that was being measured. If the effect of inserting the voltmeter in the circuit is to be as small as possible, the resistance of the voltmeter must be much larger than that of the resistance across which it is connected. Only then will the current bypassing the resistor and passing through the voltmeter be very small and so the voltage not significantly changed.

Example

Two voltmeters are available, one with a resistance of 1 kΩ and the other 1 MΩ. Which instrument should be selected if the indicated value is to be closest to the voltage value that existed across a 2 kΩ resistor before the voltmeter was connected across it?

The 1 MΩ voltmeter should be chosen. This is because when it is in parallel with 2 kΩ, less current will flow through it than if the 1 kΩ voltmeter had been used and so the current through the resistor will

Figure 1.16 *Multi-range meter*

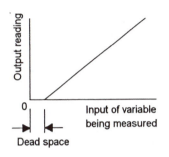

Figure 1.17 *Dead space*

be closer to its original value. Hence the indicated voltage will be closer to the value that existed before the voltmeter was connected into the circuit.

1.3.2 Range

The *range* of variable of system is the limits between which the input can vary. For example, a resistance thermometer sensor might be quoted as having a range of −200 to +800°C. The meter shown in Figure 1.16 has the dual ranges 0 to 4 and 0 to 20. The range of variable of an instrument is also sometimes called its *span*.

The term *dead band* or *dead space* is used if there is a range of input values for which there is no output. Figure 1.17 illustrates this. For example, bearing friction in a flow meter using a rotor might mean that there is no output until the input has reached a particular flow rate threshold.

1.3.3 Precision, repeatability and reproducibility

The term *precision* is used to describe the degree of freedom of a measurement system from random errors. Thus, a high precision measurement instrument will give only a small spread of readings if repeated readings are taken of the same quantity. A low precision measurement system will give a large spread of readings. For example, consider the following two sets of readings obtained for repeated measurements of the same quantity by two different instruments:

20.1 mm, 20.2 mm, 20.1 mm, 20.0 mm, 20.1 mm, 20.1 mm, 20.0 mm

19.9 mm, 20.3 mm, 20.0 mm, 20.5 mm, 20.2 mm, 19.8 mm, 20.3 mm

The results of the measurement give values scattered about some value. The first set of results shows a smaller spread of readings than the second and indicates a higher degree of precision for the instrument used for the first set.

The terms repeatability and reproducibility are ways of talking about precision in specific contexts. The term *repeatability* is used for the ability of a measurement system to give the same value for repeated measurements of the same value of a variable. Common cause of lack of repeatability are random fluctuations in the environment, e.g. changes in temperature and humidity. The error arising from repeatability is usually expressed as a percentage of the full range output. For example, a pressure sensor might be quoted as having a repeatability of ±0.1% of full range. Thus with a range of 20 kPa this would be an error of ±20 Pa.

The term *reproducibility* is used to describe the ability of a system to give the same output when used with a constant input with the system or elements of the system being disconnected from its input and then reinstalled. The resulting error is usually expressed as a percentage of the full range output.

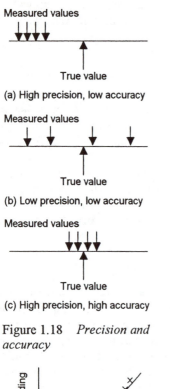

Figure 1.18 *Precision and accuracy*

(a) High precision, low accuracy

(b) Low precision, low accuracy

(c) High precision, high accuracy

Figure 1.19 *Sensitivity as slope of input–output graph*

Application
An iron–constantan thermocouple is quoted as having a sensitivity at 0°C of 0.05 mV/°C.

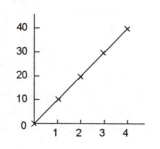

Figure 1.20 *Example*

Note that precision should not be confused with accuracy. High precision does not mean high accuracy. A high precision instrument could have low accuracy. Figure 1.18 illustrates this:

> The term *precision* is used to describe the degree of freedom of a measurement system from random errors. The *repeatability* of a system is its ability to give the same output for repeated applications of the same input value, without the system or element being disconnected from its input or any change in the environment in which the test is carried out. The *reproducibility* of a system is its ability to give the same output when it and/or elements of the system are disconnected from the input and then reinstalled.

1.3.4 Sensitivity

The *sensitivity* indicates how much the output of an instrument system or system element changes when the quantity being measured changes by a given amount, i.e. the ratio ouput/input. For example, a thermocouple might have a sensitivity of 20 µV/°C and so give an output of 20 µV for each 1°C change in temperature. Thus, if we take a series of readings of the output of an instrument for a number of different inputs and plot a graph of output against input (Figure 1.19), the sensitivity is the slope of the graph.

The term is also frequently used to indicate the sensitivity to inputs other than that being measured, i.e. environmental changes. For example, the sensitivity of a system or element might be quoted to changes in temperature or perhaps fluctuations in the mains voltage supply. Thus a pressure measurement sensor might be quoted as having a temperature sensitivity of ±0.1% of the reading per °C change in temperature.

Example

A spring balance has its deflection measured for a number of loads and gave the following results. Determine its sensitivity.

Load in kg	0	1	2	3	4
Deflection in mm	0	10	20	30	40

Figure 1.20 shows the graph of output against input. The graph has a slope of 10 mm/kg and so this is the sensitivity.

Example

A pressure measurement system (a diaphragm sensor giving a capacitance change with output processed by a bridge circuit and displayed on a digital display) is stated as having the following characteristics. Explain the significance of the terms:

Application
A commercial pressure measurement system is quoted in the manufacturer's specification as having:

Range 0 to 10 kPa
Supply Voltage ±15 V dc
Linearity error ±0.5% FS
Hysteresis error ±0.15% FS
Sensitivity 5 V dc for full range
Thermal sensitivity ±0.02%/°C
Thermal zero drift 0.02%/°C FS
Temperature range 0 to 50°C

Range: 0 to 125 kPa and 0 to 2500 kPa
Accuracy: ±1% of the displayed reading
Temperature sensitivity: ±0.1% of the reading per °C

The range indicates that the system can be used to measure pressures from 0 to 125 kPa or 0 to 2500 kPa. The accuracy is expressed as a percentage of the displayed reading, thus if the instrument indicates a pressure of, say, 100 kPa then the error will be ±1 kPa. The temperature sensitivity indicates that if the temperature changes by 1°C that displayed reading will be in error by ±0.1% of the value. Thus for a pressure of, say, 100 kPa the error will be ±0.1 kPa for a 1°C temperature change.

1.3.5 Stability

The stability of a system is its ability to give the same output when used to measure a constant input over a period of time. The term *drift* is often used to describe the change in output that occurs over time. The drift may be expressed as a percentage of the full range output. The term *zero drift* is used for the changes that occur in output when there is zero input.

1.3.6 Dynamic characteristics

The terms given above refer to what can be termed the *static characteristics*. These are the values given when steady-state conditions occur, i.e. the values given when the system or element has settled down after having received some input. The *dynamic characteristics* refer to the behaviour between the time that the input value changes and the time that the value given by the system or element settles down to the steady-state value. For example, Figure 1.21 shows how the reading of an ammeter might change when the current is switched on. The meter pointer oscillates before settling down to give the steady-state reading. The following are terms commonly used for dynamic characteristics.

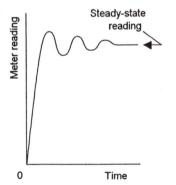

Figure 1.21 *Oscillations of a meter reading*

1 *Response time*
   This is the time which elapses after the input to a system or element is abruptly increased from zero to a constant value up to the point at which the system or element gives an output corresponding to some specified percentage, e.g. 95%, of the value of the input.

2 *Rise time*
   This is the time taken for the output to rise to some specified percentage of the steady-state output. Often the rise time refers to the time taken for the output to rise from 10% of the steady-state value to 90 or 95% of the steady-state value.

3 *Settling time*
   This is the time taken for the output to settle to within some percentage, e.g. 2%, of the steady-state value.

1.4 Reliability

If you toss a coin ten times you might find, for example, that it lands heads uppermost six times out of the ten. If, however, you toss the coin for a very large number of times then it is likely that it will land heads uppermost half of the times. The probability of it landing heads uppermost is said to be half. The *probability* of a particular event occurring is defined as being

$$\text{probability} = \frac{\text{number of occurrences of the event}}{\text{total number of trials}}$$

when the total number of trials is very large. The probability of the coin landing with either a heads or tails uppermost is likely to be 1, since every time the coin is tossed this event will occur. A probability of 1 means a certainty that the event will take place every time. The probability of the coin landing standing on edge can be considered to be zero, since the number of occurrences of such an event is zero. The closer the probability is to 1 the more frequent an event will occur; the closer it is to zero the less frequent it will occur.

Reliability is an important requirement of a measurement system. The *reliability* of a measurement system, or element in such a system, is defined as being the probability that it will operate to an agreed level of performance, for a specified period, subject to specified environmental conditions. The agreed level of performance might be that the measurement system gives a particular accuracy. The reliability of a measurement system is likely to change with time as a result of perhaps springs slowly stretching with time, resistance values changing as a result of moisture absorption, wear on contacts and general damage due to environmental conditions. For example, just after a measurement system has been calibrated, the reliability should be 1. However, after perhaps six months the reliability might have dropped to 0.7. Thus the system cannot then be relied on to always give the required accuracy of measurement, it typically only giving the required accuracy seven times in ten measurements, seventy times in a hundred measurements.

A high reliability system will have a low failure rate. *Failure rate* is the number of times during some period of time that the system fails to meet the required level of performance, i.e.:

$$\text{Failure rate} = \frac{\text{number of failures}}{\text{number of systems observed} \times \text{time observed}}$$

A failure rate of 0.4 per year means that in one year, if ten systems are observed, 4 will fail to meet the required level of performance. If 100 systems are observed, 40 will fail to meet the required level of performance. Failure rate is affected by environmental conditions. For example, the failure rate for a temperature measurement system used in hot, dusty, humid, corrosive conditions might be 1.2 per year, while for the same system used in dry, cool, non-corrosive environment it might be 0.3 per year.

With a measurement system consisting of a number of elements, failure occurs when just one of the elements fails to reach the required performance. Thus in a system for the measurement of the temperature of a fluid in some plant we might have a thermocouple, an amplifier and a meter. The failure rate is likely to be highest for the thermocouple since that is the element in contact with the fluid while the other elements are likely to be in the controlled atmosphere of a control room. The reliability of the system might thus be markedly improved by choosing materials for the thermocouple which resist attack by the fluid. Thus it might be in a stainless steel sheath to prevent fluid coming into direct contact with the thermocouple wires.

Example

The failure rate for a pressure measurement system used in factory A is found to be 1.0 per year while the system used in factory B is 3.0 per year. Which factory has the most reliable pressure measurement system?

The higher the reliability the lower the failure rate. Thus factory A has the more reliable system. The failure rate of 1.0 per year means that if 100 instruments are checked over a period of a year, 100 failures will be found, i.e. on average each instrument is failing once. The failure rate of 3.0 means that if 100 instruments are checked over a period of a year, 300 failures will be found, i.e. instruments are failing more than once in the year.

1.5 Requirements

The main requirement of a measurement system is *fitness for purpose*. This means that if, for example, a length of a product has to be measured to a certain accuracy that the measurement system is able to be used to carry out such a measurement to that accuracy. For example, a length measurement system might be quoted as having an accuracy of ±1 mm. This would mean that all the length values it gives are only guaranteed to this accuracy, e.g. for a measurement which gave a length of 120 mm the actual value could only be guaranteed to be between 119 and 121 mm. If the requirement is that the length can be measured to an accuracy of ±1 mm then the system is fit for that purpose. If, however, the criterion is for a system with an accuracy of ±0.5 mm then the system is not fit for that purpose.

In order to deliver the required accuracy, the measurement system must have been calibrated to give that accuracy. *Calibration* is the process of comparing the output of a measurement system against standards of known accuracy. The standards may be other measurement systems which are kept specially for calibration duties or some means of defining standard values. In many companies some instruments and items such as standard resistors and cells are kept in a company standards department and used solely for calibration purposes.

1.5.1 Calibration

Calibration should be carried out using equipment which can be traceable back to national standards with a separate calibration record kept for each measurement instrument. This record is likely to contain a description of the instrument and its reference number, the calibration date, the calibration results, how frequently the instrument is to be calibrated and probably details of the calibration procedure to be used, details of any repairs or modifications made to the instrument, and any limitations on its use.

The *national standards* are defined by international agreement and are maintained by national establishments, e.g. the National Physical Laboratory in Great Britain and the National Bureau of Standards in the United States. There are seven such *primary standards*, and two *supplementary* ones, the primary ones being:

1 *Mass*
   The mass standard, the kilogram, is defined as being the mass of an alloy cylinder (90% platinum–10% iridium) of equal height and diameter, held at the International Bureau of Weights and Measures at Sèvres in France. Duplicates of this standard are held in other countries.

2 *Length*
   The length standard, the metre, is defined as the length of the path travelled by light in a vacuum during a time interval of duration 1/299 792 458 of a second.

3 *Time*
   The time standard, the second, is defined as a time duration of 9 192 631 770 periods of oscillation of the radiation emitted by the caesium–133 atom under precisely defined conditions of resonance.

4 *Current*
   The current standard, the ampere, is defined as that constant current which, if maintained in two straight parallel conductors of infinite length, of negligible circular cross-section, and placed one metre apart in a vacuum, would produce between these conductors a force equal to 2×10^{-7} N per metre of length.

5 *Temperature*
   The kelvin (K) is the unit of thermodynamic temperature and is defined so that the temperature at which liquid water, water vapour and ice are in equilibrium (known as the triple point) is 273.16 K. A temperature scale devised by Lord Kelvin forms the basis of the absolute practical temperature scale that is used and is based on a number of fixed temperature points, e.g. the freezing point of gold at 1337.58 K.

6 *Luminous intensity*
   The candela is defined as the luminous intensity, in a given direction, of a specified source that emits monochromatic radiation of frequency 540×10^{12} Hz and that has a radiant intensity of 1/683 watt per unit steradian (a unit solid angle, see below).

7   *Amount of substance*
The mole is defined as the amount of a substance which contains as many elementary entities as there are atoms in 0.012 kg of the carbon 12 isotope.

The *supplementary standards* are:

1   *Plane angle*
The radian is the plane angle between two radii of a circle which cuts off on the circumference an arc with a length equal to the radius (Figure 1.22).

2   *Solid angle*
The steradian is the solid angle of a cone which, having its vertex in the centre of the sphere, cuts off an area of the surface of the sphere equal to the square of the radius (Figure 1.23).

Primary standards are used to define national standards, not only in the primary quantities but also in other quantities which can be derived from them. For example, a resistance standard of a coil of manganin wire is defined in terms of the primary quantities of length, mass, time and current. Typically these national standards in turn are used to define reference standards which can be used by national bodies for the calibration of standards which are held in calibration centres.

The equipment used in the calibration of an instrument in everyday company use is likely to be *traceable* back to national standards in the following way:

1   National standards are used to calibrate standards for calibration centres.

2   Calibration centre standards are used to calibrate standards for instrument manufacturers.

3   Standardised instruments from instrument manufacturers are used to provide in-company standards.

4   In-company standards are used to calibrate process instruments.

There is a simple traceability chain from the instrument used in a process back to national standards (Figure 1.24). In the case of, say, a glass bulb thermometer, the traceability might be:

1   National standard of fixed thermodynamic temperature points

2   Calibration centre standard of a platinum resistance thermometer with an accuracy of ±0.005°C

3   An in-company standard of a platinum resistance thermometer with an accuracy of ±0.01°C

4   The process instrument of a glass bulb thermometer with an accuracy of ±0.1°C

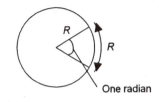

Figure 1.22   *The radian*

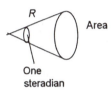

Figure 1.23   *The steradian*

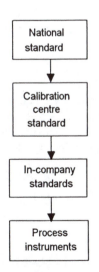

Figure 1.24   *Traceability chain*

1.5.2 Safety systems

Statutory safety regulations lay down the responsibilities of employers and employees for safety in the workplace. These include for employers the duty to:

- Ensure that process plant is operated and maintained in a safe way so that the health and safety of employees is protected.

- Provide a monitoring and shutdown system for processes that might result in hazardous conditions.

Employees also have duties to:

- Take reasonable care of their own safety and for the safety of others.

- Avoid misusing or damaging equipment that is designed to protect people's safety.

Thus, in the design of measurement systems, due regard has to be paid to safety both in their installation and operation. Thus:

- The failure of any single component in a system should not create a dangerous situation.

- A failure which results in cable open or short circuits or short circuiting to ground should not create a dangerous situation.

- Foreseeable modes of failure should be considered for fail-safe design so that, in the event of failure, the system perhaps switches off into a safe condition.

- Systems should be easily checked and readily understood.

The main risks from electrical instrumentation are electrocution and the possibility of causing a fire or explosion as a consequence of perhaps cables or components overheating or arcing sparks occurring in an explosive atmosphere. Thus it is necessary to ensure that an individual cannot become connected between two points with a potential difference greater than about 30 V and this requires the careful design of earthing so that there is always an adequate earthing return path to operate any protective device in the event of a fault occurring.

Problems

Questions 1 to 5 have four answer options: A. B, C and D. Choose the correct answer from the answer options.

1 Decide whether each of these statements is True (T) or False (F).

Sensors in a measurement system have:
(i) An input of the variable being measured.
(ii) An output of a signal in a form suitable for further processing in the measurement system.

Which option BEST describes the two statements?

A   (i) T (ii) T
B   (i) T (ii) F
C   (i) F (ii) T
D   (i) F (ii) F

2 The following lists the types of signals that occur in sequence at the various stages in a particular measurement system:

(i)   Temperature
(ii) Voltage
(iii) Bigger voltage
(iv) Movement of pointer across a scale

The signal processor is the functional element in the measurement system that changes the signal from:

A   (i) to (ii)
B   (ii) to (iii)
C   (iii) to (iv)
D   (ii) to (iv)

3 Decide whether each of these statements is True (T) or False (F).

The discrepancy between the measured value of the current in an electrical circuit and the value before the measurement system, an ammeter, was inserted in the circuit is bigger the larger:
(i)   The resistance of the meter.
(ii) The resistance of the circuit.

Which option BEST describes the two statements?

A   (i) T (ii) T
B   (i) T (ii) F
C   (i) F (ii) T
D   (i) F (ii) F

4 Decide whether each of these statements is True (T) or False (F).

A highly reliable measurement system is one where there is a high chance that the system will:
(i)   Require frequent calibration.
(ii) Operate to the specified level of performance.
Which option BEST describes the two statements?

A   (i) T (ii) T
B   (i) T (ii) F
C   (i) F (ii) T
D   (i) F (ii) F

5 Decide whether each of these statements is True (T) or False (F).

A measurement system which has a lack of repeatability is one where there could be:
(i) Random fluctuations in the values given by repeated measurements of the same variable.
(ii) Fluctuations in the values obtained by repeating measurements over a number of samples.

Which option BEST describes the two statements?

A   (i) T   (ii) T
B   (i) T   (ii) F
C   (i) F   (ii) T
D   (i) F   (ii) F

6   List and explain the functional elements of a measurement system.
7   Explain the terms (a) reliability and (b) repeatability when applied to a measurement system.
8   Explain what is meant by calibration standards having to be traceable to national standards.
9   Explain what is meant by 'fitness for purpose' when applied to a measurement system.
10 The reliability of a measurement system is said to be 0.6. What does this mean?
11 The measurement instruments used in the tool room of a company are found to have a failure rate of 0.01 per year. What does this mean?
12 Determine the sensitivity of the instruments that gave the following readings:

(a)

Load kg	0	2	4	6	8
Deflection mm	0	18	36	54	72

(b)

Temperature °C	0	10	20	30	40
Voltage mV	0	0.59	1.19	1.80	2.42

(c)

Load N	0	1	2	3	4
Charge pC	0	3	6	9	12

13 Calibration of a voltmeter gave the following data. Determine the maximum hysteresis error as a percentage of the full-scale range.

Increasing input:

Standard mV	0	1.0	2.0	3.0	4.0
Voltmeter mV	0	1.0	1.9	2.9	4.0

Decreasing input:

Standard mV	4.0	3.0	2.0	1.0	0
Voltmeter mV	4.0	3.0	2.1	1.1	0

2 Instrumentation system elements

2.1 Introduction

This chapter discusses the sensors, signal processors and data presentation elements commonly used in engineering. The term *sensor* is used for an element which produces a signal relating to the quantity being measured. The term *signal processor* is used for the element that takes the output from the sensor and converts it into a form which is suitable for data presentation. *Data presentation* is where the data is displayed, recorded or transmitted to some control system.

2.2 Displacement sensors

A displacement sensor is here considered to be one that can be used to:

1   Measure a linear displacement, i.e. a change in linear position. This might, for example, be the change in linear displacement of a sensor as a result of a change in the thickness of sheet metal emerging from rollers.

2   Measure an angular displacement, i.e. a change in angular position. This might, for example, be the change in angular displacement of a drive shaft.

3   Detect motion, e.g. this might be as part of an alarm or automatic light system, whereby an alarm is sounded or a light switched on when there is some movement of an object within the 'view' of the sensor.

4   Detect the presence of some object, i.e. a proximity sensor. This might be in an automatic machining system where a tool is activated when the presence of a work piece is sensed as being in position.

Displacement sensors fall into two groups: those that make direct contact with the object being monitored, by spring loading or mechanical connection with the object, and those which are non-contacting. For those linear displacement methods involving contact, there is usually a sensing shaft which is in direct contact with the object being monitored, the displacement of this shaft is then being monitored by a sensor. This shaft movement may be used to cause changes in electrical voltage, resistance, capacitance, or mutual inductance. For angular displacement methods involving mechanical connection, the rotation of a shaft might directly drive, through gears, the rotation of the sensor element, this perhaps generating an e.m.f. Non-contacting proximity sensors might consist of a beam of infrared light being broken by the presence of the

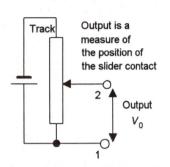

Figure 2.1 *Potentiometer*

Application
The following is an example of part of the specification of a commercially available displacement sensor using a plastic conducting potentiometer track:

Ranges from 0 to 10 mm to 0 to 2 m
Non-linearity error ±0.1% of full range
Resolution ±0.02% of full range
Temperature sensitivity ±120 parts per million/°C
Resolution ±0.02% of full range

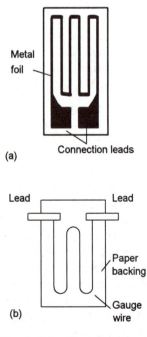

Figure 2.2 *Strain gauges*

object being monitored, the sensor then giving a voltage signal indicating the breaking of the beam, or perhaps the beam being reflected from the object being monitored, the sensor giving a voltage indicating that the reflected beam has been detected. Contacting proximity sensors might be just mechanical switches which are tripped by the presence of the object. The following are examples of displacement sensors.

2.2.1 Potentiometer

A *potentiometer* consists of a resistance element with a sliding contact which can be moved over the length of the element and connected as shown in Figure 2.1. With a constant supply voltage V_s, the output voltage V_o between terminals 1 and 2 is a fraction of the input voltage, the fraction depending on the ratio of the resistance R_{12} between terminals 1 and 2 compared with the total resistance R of the entire length of the track across which the supply voltage is connected. Thus $V_o/V_s = R_{12}/R$. If the track has a constant resistance per unit length, the output is proportional to the displacement of the slider from position 1. A rotary potentiometer consists of a coil of wire wrapped round into a circular track, or a circular film of conductive plastic or a ceramic-metal mix termed a cermet, over which a rotatable sliding contact can be rotated. Hence an angular displacement can be converted into a potential difference. Linear tracks can be used for linear displacements.

With a wire wound track the output voltage does not continuously vary as the slider is moved over the track but goes in small jumps as the slider moves from one turn of wire to the next. This problem does not occur with a conductive plastic or the cermet track. Thus, the smallest change in displacement which will give rise to a change in output, i.e. the resolution, tends to be much smaller for plastic tracks than wire-wound tracks. Errors due to non-linearity of the track for wire tracks tend to range from less than 0.1% to about 1% of the full range output and for conductive plastics can be as low as about 0.05%. The track resistance for wire-wound potentiometers tends to range from about 20 Ω to 200 kΩ and for conductive plastic from about 500 Ω to 80 kΩ. Conductive plastic has a higher temperature coefficient of resistance than wire and so temperature changes have a greater effect on accuracy.

2.2.2 Strain-gauged element

Strain gauges consist of a metal foil strip (Figure 2.2(a)), flat length of metal wire (Figure 2.2(b)) or a strip of semiconductor material which can be stuck onto surfaces like a postage stamp. When the wire, foil, strip or semiconductor is stretched, its resistance R changes. The fractional change in resistance $\Delta R/R$ is proportional to the strain ε, i.e.:

$$\frac{\Delta R}{R} = G\varepsilon$$

where G, the constant of proportionality, is termed the *gauge factor*. Metal strain gauges typically have gauge factors of the order of 2.0. When such a strain gauge is stretched its resistance increases, when

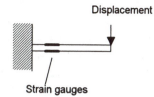

Displacement

Strain gauges

Figure 2.3 *Strain-gauged cantilever*

Application
A commercially available displacement sensor, based on the arrangement shown in Figure 2.3, has the following in its specification:

Range 0 to 100 mm
Non-linearity error ±0.1% of full range
Temperature sensitivity ±0.01% of full range/°C

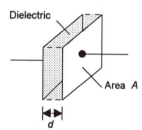

Dielectric

Area *A*

d

Figure 2.4 *Parallel plate capacitor*

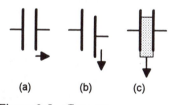

(a)   (b)   (c)

Figure 2.5 *Capacitor sensors*

Application
A commercially available capacitor displacement sensor based on the use of the sliding capacitor plate (Figure 2.5 (b)) includes in its specification:

Ranges available from 0 to 5 mm to 0 to 250 mm
Non-linearity and hysteresis error ±0.01% of full range

compressed its resistance decreases. Strain is (change in length/original length) and so the resistance change of a strain gauge is a measurement of the change in length of the gauge and hence the surface to which the strain gauge is attached. Thus a displacement sensor might be constructed by attaching strain gauges to a cantilever (Figure 2.3), the free end of the cantilever being moved as a result of the linear displacement being monitored. When the cantilever is bent, the electrical resistance strain gauges mounted on the element are strained and so give a resistance change which can be monitored and which is a measure of the displacement. With strain gauges mounted as shown in Figure 2.3, when the cantilever is deflected downwards the gauge on the upper surface is stretched and the gauge on the lower surface compressed. Thus the gauge on the upper surface increases in resistance while that on the lower surface decreases. Typically, this type of sensor is used for linear displacements of the order of 1 mm to 30 mm, having a non-linearity error of about ± 1% of full range.

A problem that has to be overcome with strain gauges is that the resistance of the gauge changes when the temperature changes and so methods have to be used to compensate for such changes in order that the effects of temperature can be eliminated. This is discussed later in this chapter when the circuits used for signal processing are discussed.

2.2.3 Capacitive element

The capacitance C of a parallel plate capacitor (Figure 2.4) is given by:

$$C = \frac{\varepsilon_r \varepsilon_0 A}{d}$$

where ε_r is the relative permittivity of the dielectric between the plates, ε_0 a constant called the permittivity of free space, A the area of overlap between the two plates and d the plate separation. The capacitance will change if the plate separation d changes, the area A of overlap of the plates changes, or a slab of dielectric is moved into or out of the plates, so varying the effective value of ε_r (Figure 2.5). All these methods can be used to give linear displacement sensors.

One form that is often used is shown in Figure 2.6 and is referred to as a *push–pull* displacement sensor. It consists of two capacitors, one between the movable central plate and the upper plate and one between the central movable plate and the lower plate. The displacement x moves the central plate between the two other plates. Thus when the central plate moves upwards it decreases the plate separation of the upper capacitor and increases the separation of the lower capacitor. Thus the capacitance of the upper capacitor is increased and that of the lower capacitor decreased. When the two capacitors are incorporated in opposite arms of an alternating current bridge, the output voltage from the bridge is proportional to the displacement. Such a sensor has good long-term stability and is typically used for monitoring displacements from a few millimetres to hundreds of millimetres. Non-linearity and hysteresis errors are about ± 0.01% of full range.

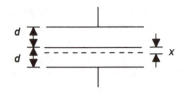

Figure 2.6   *Push-pull displacement sensor*

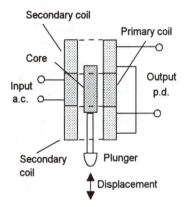

Figure 2.7   *LVDT*

Application
A commercially available displacement sensor using a LVDT has the following in its specification:

Ranges ±0.125 mm to ±470 mm
Non-linearity error ±0.25% of full range
Temperature sensitivity ±0.01% of full range
Signal conditioning incorporated within the housing of the LVDT

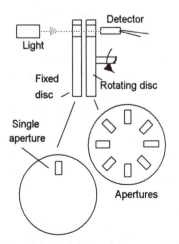

Figure 2.8   *Optical incremental encoder*

2.2.4 Linear variable differential transformer

The *linear variable differential transformer*, generally referred to by the abbreviation *LVDT*, is a transformer with a primary coil and two secondary coils. Figure 2.7 shows the arrangement, there being three coils symmetrically spaced along an insulated tube. The central coil is the primary coil and the other two are identical secondary coils which are connected in series in such a way that their outputs oppose each other. A magnetic core is moved through the central tube as a result of the displacement being monitored. When there is an alternating voltage input to the primary coil, alternating e.m.f.s are induced in the secondary coils. With the magnetic core in a central position, the amount of magnetic material in each of the secondary coils is the same and so the e.m.f.s induced in each coil are the same. Since they are so connected that their outputs oppose each other, the net result is zero output. However, when the core is displaced from the central position there is a greater amount of magnetic core in one coil than the other. The result is that a greater e.m.f. is induced in one coil than the other and then there is a net output from the two coils. The bigger the displacement the more of the core there is in one coil than the other, thus the difference between the two e.m.f.s increases the greater the displacement of the core. Typically, LVDTs have operating ranges from about ±2 mm to ±400 mm. Non-linearity errors are typically about ±0.25%. LVDTs are very widely used for monitoring displacements.

2.2.5 Optical encoders

An *encoder* is a device that provides a digital output as a result of an angular or linear displacement. Position encoders can be grouped into two categories: incremental encoders, which detect changes in displacement from some datum position, and absolute encoders, which give the actual position. Figure 2.8 shows the basic form of an *incremental encoder* for the measurement of angular displacement of a shaft. It consists of a disc which rotates along with the shaft. In the form shown, the rotatable disc has a number of windows through which a beam of light can pass and be detected by a suitable light sensor. When the shaft rotates and disc rotates, a pulsed output is produced by the sensor with the number of pulses being proportional to the angle through which the disc rotates. The angular displacement of the disc, and hence the shaft rotating it, can thus be determined by the number of pulses produced in the angular displacement from some datum position. Typically the number of windows on the disc varies from 60 to over a thousand with multi-tracks having slightly offset slots in each track. With 60 slots occurring with 1 revolution then, since 1 revolution is a rotation of 360°, the minimum angular displacement, i.e. the resolution, that can be detected is 360/60 = 6°. The resolution thus typically varies from about 6° to 0.3° or better.

With the incremental encoder, the number of pulses counted gives the angular displacement, a displacement of, say, 50° giving the same number of pulses whatever angular position the shaft starts its rotation from. However, the *absolute encoder* gives an output in the form of a binary number of several digits, each such number representing a particular angular position. Figure 2.9 shows the basic form of an absolute encoder for the measurement of angular position.

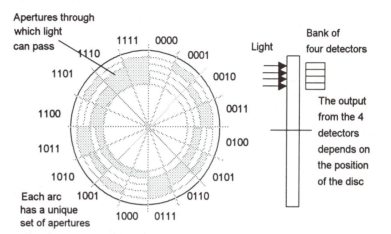

Figure 2.9 *The rotating wheel of the absolute encoder. Note that though the normal form of binary code is shown in the figure, in practice a modified form of binary code called the Gray code is generally used. This code, unlike normal binary, has only one bit changing in moving from one number to the next. Thus we have the sequence 0000, 0001, 0011, 0010, 0011, 0111, 0101, 0100, 1100, 1101, 1111.*

With the form shown in the figure, the rotating disc has four concentric circles of slots and four sensors to detect the light pulses. The slots are arranged in such a way that the sequential output from the sensors is a number in the binary code, each such number corresponding to a particular angular position. A number of forms of binary code are used. Typical encoders tend to have up to 10 or 12 tracks. The number of bits in the binary number will be equal to the number of tracks. Thus with 10 tracks there will be 10 bits and so the number of positions that can be detected is 2^{10}, i.e. 1024, a resolution of $360/1024 = 0.35°$.

The incremental encoder and the absolute encoder can be used with linear displacements if the linear displacement is first converted to a rotary motion by means of a tracking wheel (Figure 2.10).

2.2.6 Moiré fringes

Moiré fringes are produced when light passes through two gratings which have rulings inclined at a slight angle to each other. Movement of one grating relative to the other causes the fringes to move. Figure 2.11(a) illustrates this. Figure 2.11(b) shows a transmission form of instrument using Moiré fringes and Figure 2.11(c) a reflection form. With both, a long grating is fixed to the object being displaced. With the transmission form, light passes through the long grating and then a smaller fixed grating, the transmitted light being detected by a photocell. With the reflection form, light is reflected from the long grating through a smaller fixed grating and onto a photocell.

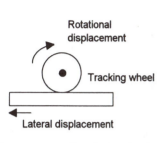

Figure 2.10 *Tracking wheel*

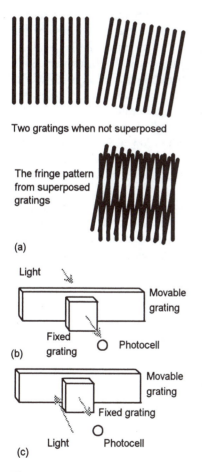

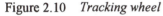

Figure 2.11 *(a) Moiré fringes, (b) transmission and (c) reflection forms of instruments*

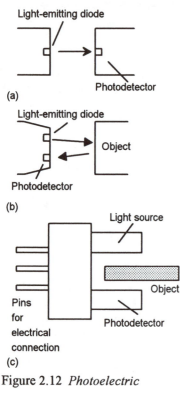

Figure 2.12 *Photoelectric proximitry sensors*

Coarse grating instruments might have 10 to 40 lines per millimetre, fine gratings as many as 400 per millimetre. Movement of the long grating relative to the fixed short grating results in fringes moving across the view of the photocell and thus the output of the cell is a sequence of pulses which can be counted. The displacement is thus proportional to the number of pulses counted. Displacements as small as 1 μm can be detected by this means. Such methods have high reliability and are often used for the control of machine tools.

2.2.7 Optical proximity sensors

There are a variety of optical sensors that can be used to determine whether an object is present or not. Photoelectric switch devices can either operate as *transmissive types* where the object being detected breaks a beam of light, usually infrared radiation, and stops it reaching the detector (Figure 2.12(a)) or *reflective types* where the object being detected reflects a beam of light onto the detector (Figure 2.12(b)).

In both types the radiation emitter is generally a *light-emitting diode (LED)*. The radiation detector might be a *phototransistor*, often a pair of transistors, known as a *Darlington pair*; using the pair increases the sensitivity. Depending on the circuit used, the output can be made to switch to either high or low when light strikes the transistor. Such sensors are supplied as packages for sensing the presence of objects at close range, typically at less than about 5 mm. Figure 2.12(c) shows a U-shaped form where the object breaks the light beam.

Another possibility is a *photodiode*. Depending on the circuit used, the output can be made to switch to either high or low when light strikes the diode. Yet another possibility is a *photoconductive cell*. The resistance of the photoconductive cell, often cadmium sulphide, depends on the intensity of the light falling on it.

Figure 2.13 illustrates a proximity sensor based on reflection. A LED emits infrared radiation which is reflected by the object. The reflected radiation is then detected by a phototransistor. In the absence of the object there is no detected reflected radiation; when the object is in the proximity, there is.

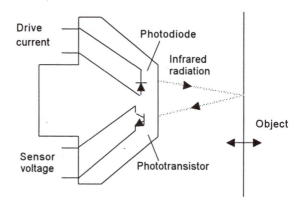

Figure 2.13 *Proximity sensor*

Another form of optical sensor is the *pyroelectric sensor*. Such sensors give a voltage signal when the infrared radiation falling on them changes, no signal being given for constant radiation levels. Lithium tantulate is a widely used pyroelectric material. Figure 2.14 shows an example of such a sensor. Such sensors can be used with burglar alarms or for the automatic switching on of a light when someone walks up the drive to a house. A special lens is put in front of the detector. When a object which emits infra-red radiation is in front of the detector, the radiation is focused by the lens onto the detector. But only for beams of radiation in certain directions will a focused beam fall on the detector and give a signal. Thus when the object moves then the focused beam of radiation is effectively switched on and off as the object cuts across the lines at which its radiation will be detected. Thus the pyroelectric detector gives a voltage output related to the changes in the signal.

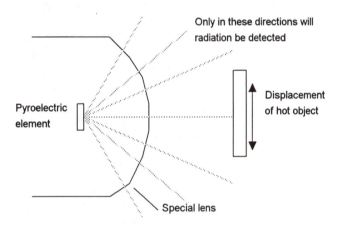

Figure 2.14 *Pyroelectric sensor*

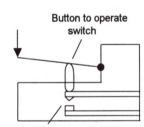

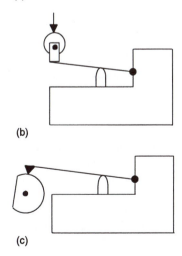

(a)

(b)

(c)

Figure 2.15 *Limit switches:
(a) Lever, (b) roller, (c) cam*

2.2.8 Mechanical switches

There are many situations where a sensor is required to detect the presence of some object. The sensor used in such situations can be a mechanical switch, giving an on–off output when the switch contacts are opened or closed by the presence of an object. Figure 2.15 illustrates the forms of a number of such switches. Switches are used for such applications as where a work piece closes the switch by pushing against it when it reaches the correct position on a work table, such a switch being referred to as a *limit switch*. The switch might then be used to switch on a machine tool to carry out some operation on the work piece. Another example is a light being required to come on when a door is opened, as in a refrigerator. The action of opening the door can be made to close the contacts in a switch and trigger an electrical circuit to switch on the lamp.

Figure 2.16 shows another form of a non-contact switch sensor, a *reed switch*. This consists of two overlapping, but not touching, strips of a

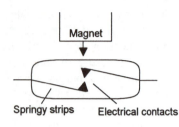

Figure 2.16   *Reed switch*

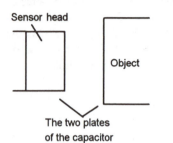

Figure 2.17   *Capacitive proximity switch*

2.3 Speed sensors

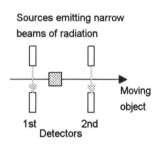

Figure 2.18   *Measurement of linear speed*

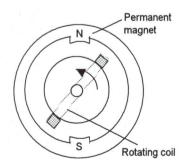

Figure 2.19   *The tachogenerator*

spring magnetic material sealed in a glass or plastic envelope. When a magnet or current carrying coil is brought close to the switch, the strips become magnetised and attract each other. The contacts then close. Typically a magnet closes the contacts when it is about 1 mm from the switch.

2.2.9 Capacitive proximity switch

A proximity switch that can be used with metallic and non-metallic objects is the *capacitive proximity switch*. The capacitance of a pair of plates separated by some distance depends on the separation, the smaller the separation the higher the capacitance. The sensor of the capacitive proximity switch is just one of the plates of the capacitor, the other plate being the metal object whose proximity is to be detected (Figure 2.17). Thus the proximity of the object is detected by a change in capacitance. The sensor can also be used to detect non-metallic objects since the capacitance of a capacitor depends on the dielectric between its plates. In this case the plates are the sensor and the earth and the non-metallic object is the dielectric. The change in capacitance can be used to activate an electronic switch circuit and so give an on–off device. Capacitive proximity switches can be used to detect objects when they are typically between 4 and 60 mm from the sensor head.

The following are examples of sensors that can be used to monitor linear and angular speeds.

2.3.1 Optical methods

Linear speeds can be measured by determining the time between when the moving object breaks one beam of radiation and when it breaks a second beam some measured distance away (Figure 2.18). Breaking the first beam can be used to start an electronic clock and breaking the second beam to stop the clock.

2.3.2 Incremental encoder

The incremental encoder described above in Section 2.2.5 can be used for a measurement of angular speed or a rotating shaft, the number of pulses produced per second being counted.

2.3.3 Tachogenerator

The basic tachogenerator consists of a coil mounted in a magnetic field (Figure 2.19). When the coil rotates electromagnetic induction results in an alternating e.m.f. being induced in the coil. The faster the coil rotates the greater the size of the alternating e.m.f. Thus the size of the alternating e.m.f. is a measure of the angular speed. Typically such a sensor can be used up to 10 000 revs per minute and has a non-linearity error of about ±1% of the full range.

2.4 Fluid pressure sensors

Deflection   Diaphragm

Diaphragm deflects because
pressure greater on other side
than this side

(a)

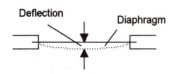

(b)

Figure 2.20 *Diaphragm sensors*

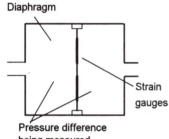

Diaphragm

Strain
gauges

Pressure difference
being measured

Figure 2.21 *Diaphragm pressure gauge using strain gauges*

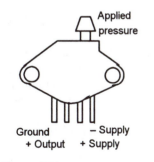

Applied
pressure

Ground   – Supply
  + Output   + Supply

Figure 2.22   *MPX100AP*

Application
The specification of a MPX pressure
sensor typically includes:

Pressure range 0 to 100 kPa
Supply voltage 10 V
Sensitivity 0.4 mV/kPa
Linearity ±0.25% of full scale
Pressure hysteresis ±0.1% of full scale
Response time (10% to 90%) 1.0 ms

Many of the devices used to monitor fluid pressure in industrial processes involve the monitoring of the elastic deformation of diaphragms, bellows and tubes. The following are some common examples of such sensors.

> The term *absolute pressure* is used for a pressure measured relative to a vacuum, *differential pressure* when the difference between two pressures is measured and *gauge pressure* for the pressure measured relative to some fixed pressure, usually the atmospheric pressure.

2.4.1 Diaphragm sensor

The movement of the centre of a circular diaphragm as a result of a pressure difference between its two sides is the basis of a pressure gauge (Figure 2.20(a)). For the measurement of the absolute pressure, the opposite side of the diaphragm is a vacuum, for the measurement of pressure difference the pressures are connected to each side of the diaphragm, for the gauge pressure, i.e. the pressure relative to the atmospheric pressure, the opposite side of the diaphragm is open to the atmosphere. The amount of movement with a plane diaphragm is fairly limited; greater movement can, however, be produced with a diaphragm with corrugations (Figure 2.20(b)).

The movement of the centre of a diaphragm can be monitored by some form of displacement sensor. Figure 2.21 shows the form that might be taken when strain gauges are used to monitor the displacement, the strain gauges being stuck to the diaphragm and changing resistance as a result of the diaphragm movement. Typically such sensors are used for pressures over the range 100 kPa to 100 MPa, with an accuracy up to about ±0.1%.

One form of diaphragm pressure gauge uses strain gauge elements integrated within a silicon diaphragm and supplied, together with a resistive network for signal processing, on a single silicon chip as the Motorola MPX pressure sensor (Figure 2.22). With a voltage supply connected to the sensor, it gives an output voltage directly proportional to the pressure. In one form it has a built-in vacuum on one side of the diaphragm and so the deflection of the diaphragm gives a measure of the absolute pressure applied to the other side of the diaphragm. The output is a voltage which is proportional to the applied pressure with a sensitivity of 0.6 mV/kPa. Other versions are available which have one side of the diaphragm open to the atmosphere and so can be used to measure gauge pressure, others allow pressures to be applied to both sides of the diaphragm and so can be used to measure differential pressures. Such sensors are available for use for the measurement of absolute pressure, differential pressure or gauge pressure, e.g. MPX2100 has a pressure range of 100 kPa and with a supply voltage of 16 V d.c. gives a voltage output over the full range of 40 mV.

Figure 2.23 shows the form that might be taken by a capacitance diaphragm pressure gauge. The diaphragm forms one plate of a

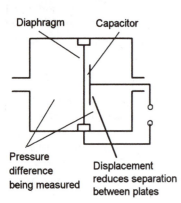

Figure 2.23 *Diaphragm gauges: using capacitance*

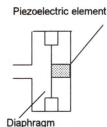

Figure 2.24 *Basic form of a piezoelectric sensor*

Application

A commercially available piezo-electric diaphragm pressure gauge has in its specification:

Ranges 0 to 20 MPa, 0 to 200 MPa, 0 to 500 MPa, 0 to 1000 MPa
Non-linearity error ±0.5%
Sensitivity –0.1 pC/kPa
Temperature sensitivity ±0.5% of full scale for use +20°C tp +100°C

capacitor, the other plate being fixed. Displacement of the diaphragm results in changes in capacitance. The range of such pressure gauges is about 1 kPa to 200 kPa with an accuracy of about ±0.1%.

Another form of diaphragm sensor uses a LVDT to monitor the displacement of the diaphragm. Such an arrangement is typically used for low pressure measures where high stability is required. The total error due to non-linearity, hysteresis and repeatability can be of the order of ±0.5% of full scale.

2.4.2 Piezoelectric sensor

When certain crystals are stretched or compressed, charges appear on their surfaces. This effect is called *piezo-electricity*. Examples of such crystals are quartz, tourmaline, and zirconate-titanate.

A piezoelectric pressure gauge consists essentially of a diaphragm which presses against a piezoelectric crystal (Figure 2.24). Movement of the diaphragm causes the crystal to be compressed and so charges produced on its surface. The crystal can be considered to be a capacitor which becomes charged as a result of the diaphragm movement and so a potential difference appears across it. The amount of charge produced and hence the potential difference depends on the extent to which the crystal is compressed and hence is a measure of the displacement of the diaphragm and so the pressure difference between the two sides of the diaphragm. If the pressure keeps the diaphragm at a particular displacement, the resulting electrical charge is not maintained but leaks away. Thus the sensor is not suitable for static pressure measurements. Typically such a sensor can be used for pressures up to about 1000 MPa with a non-linearity error of about ±1.0% of the full range value.

2.4.3 Bourdon tube

The *Bourdon tube* is an almost rectangular or elliptical cross-section tube made from materials such as stainless steel or phosphor bronze. With a C-shaped tube (Figure 2.25(a)), when the pressure inside the tube increases the closed end of the C opens out, thus the displacement of the closed end becomes a measure of the pressure. A C-shaped Bourdon tube can be used to rotate, via gearing, a shaft and cause a pointer to move across a scale. Such instruments are robust and typically used for pressures in the range 10 kPa to 100 MPa with an accuracy of about ±1% of full scale.

Another form of Bourdon instrument uses a helical-shaped tube (Figure 2.25(b)). When the pressure inside the tube increases, the closed end of the tube rotates and thus the rotation becomes a measure of the pressure. A helical-shaped Bourdon tube can be used to move the slider of a potentiometer and so give an electrical output related to the pressure. Helical tubes are more expensive but have greater sensitivity. Typically they are used for pressures up to about 50 MPa with an accuracy of about ±1% of full range.

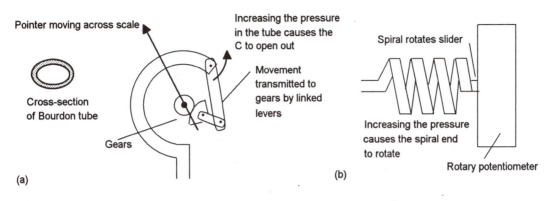

Figure 2.25 *Bourdon tube instruments: (a) geared form, (b) potentiometer form*

2.5 Fluid flow

For a fluid flowing through a pipe of cross-sectional area A_1 with a velocity v_1 (Figure 2.26), in 1 s the fluid advances a distance v_1 and so amount of fluid passing a particular point per second is A_1v_1 and the volume rate of flow is A_1v_1. If the fluid then flows through a constriction of cross-sectional area A_2 in the pipe then we must have:

$$A_2v_2 = A_1v_1$$

and so there must be an increase in velocity. An increase in velocity means an acceleration and therefore a force is required to move the fluid through the constriction. This force is provided by the pressure in the fluid dropping at the constriction. The traditional methods used for the measurement of fluid flow involve devices based on the measurement of the pressure difference occurring at a constriction and using it as a measure of the flow rate. The relationship between the pressure drop and the volume rate of flow is non-linear, i.e. the flow rate is not directly proportional to the pressure difference but to the square root of the pressure difference. The venturi tube and the orifice plate described below are common examples.

Other methods have, however, been developed which more rapidly and efficiently record the flow rate and often with less interference to the flow.

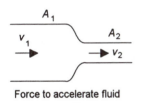

Figure 2.26 *Pressure drop at a constriction*

2.5.1 Differential pressure methods

There are a number of forms of differential pressure devices based on the above equation and involving constant size constrictions, e.g. the venturi tube, nozzles, Dall tube and orifice plate. In addition there are other devices involving variable size constrictions, e.g. the rotameter. The following are discussions of the characteristics of the above devices.

The *venturi tube* (Figure 2.27) has a gradual tapering of the pipe from the full diameter to the constricted diameter. The presence of the venturi tube results in a pressure loss occurring in the system of about 10 to 15%, a comparatively low value. The pressure difference between the flow prior to the constriction and the constriction can be measured with a

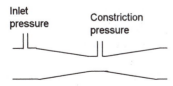

Figure 2.27 *Venturi tube*

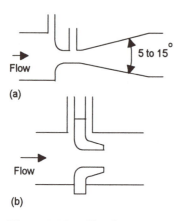

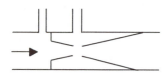

Figure 2.28   *Nozzles:*
(a) venturi, (b) flow

Figure 2.29   *Dall tube*

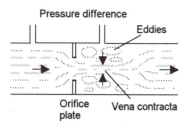

Figure 2.30   *Orifice plate*

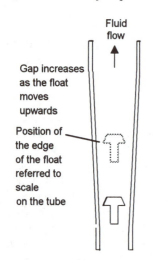

Figure 2.31   *Rotameter*

simple U-tube manometer or a differential diaphragm pressure cell. The instrument can be used with liquids containing particles, is simple in operation, capable of accuracy of about ±0.5%, has a long-term reliability, but is comparatively expensive and has a non-linear relationship between pressure and the volume rate of flow.

A cheaper form of venturi is provided by the *nozzle flow meter* (Figure 2.28). Two types of nozzle are used, the venturi nozzle and the flow nozzle. The venturi nozzle (Figure 2.28(a)) is effectively a venturi tube with an inlet which is considerably shortened. The flow nozzle (Figure 2.28(b)) is even shorter. Nozzles produce pressure losses of the order of 40 to 60%. Nozzles are cheaper than venturi tubes, give similar pressure differences, and have an accuracy of about ±0.5%. They have the same non-linear relationship between the pressure and the volume rate of flow.

The *Dall tube* (Figure 2.29) is another variation of the venturi tube. It gives a higher differential pressure and a lower pressure drop. The Dall tube is only about two pipe diameters long and is often used where space does not permit the use of a venturi tube.

The *orifice plate* (Figure 2.30) is simply a disc with a hole. The effect of introducing it is to constrict the flow to the orifice opening and the flow channel to an even narrower region downstream of the orifice. The narrowest section of the flow is not through the orifice but downstream of it and is referred to as the *vena contracta*. The pressure difference is measured between a point equal to the diameter of the tube upstream of the orifice and a point equal to half the diameter downstream. The orifice plate has the usual non-linear relationship between the pressure difference and the volume rate of flow. It is simple, reliable, produces a greater pressure difference than the venturi tube and is cheaper but less accurate, about ±1.5%. It also produces a greater pressure drop. Problems of silting and clogging can occur if particles are present in liquids.

The *rotameter* (Figure 2.31) is an example of a *variable area flow meter*; a constant pressure difference is maintained between the main flow and that at the constriction by changing the area of the constriction. The rotameter has a float in a tapered vertical tube with the fluid flow pushing the float upwards. The fluid has to flow through the constriction which is the gap between the float and the walls of the tube and so there is a pressure drop at that point. Since the gap between the float and the tube walls increases as the float moves upwards, the pressure drop decreases. The float moves up the tube until the fluid pressure is just sufficient to balance the weight of the float. The greater the flow rate the greater the pressure difference for a particular gap and so the higher up the tube the float moves. A scale alongside the tube can thus be calibrated to read directly the flow rate corresponding to a particular height of the float. The rotameter is cheap, reliable, has an accuracy of about ±1% and can be used to measure flow rates from about 30×10^{-6} m^3/s to 1 m^3/s.

The *Pitot tube* can be used to directly measure the velocity of flow of a fluid, rather than the volume rate of flow and consists essentially of just a small tube inserted into the fluid with an opening pointing directly

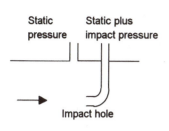

Figure 2.32 *Pitot tube*

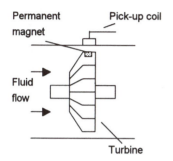

Figure 2.33 *Basic principle of the turbine flowmeter*

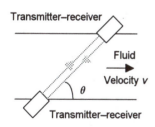

Figure 2.34 *Ultrasonic flow meter*

Application
A commercially available time of flight ultrasonic flow meter includes the following in its specification:

Accuracy ±1% of flow value
Non-linearity error ±1% of flow value
Repeatability ±0.5% of flow value

upstream (Figure 2.32). The fluid impinging on the open end of the tube is brought to rest and the pressure difference measured between this point and the pressure in the fluid at full flow. The difference in pressure between where the fluid is in full flow and the point where it is stopped is due to the kinetic energy of the fluid being transformed to potential energy, this showing up as an increase in pressure. Because kinetic energy is $\frac{1}{2}mv^2$, the velocity is proportional to the square root of the pressure difference.

2.5.2 Turbine meter

The *turbine flowmeter* (Figure 2.33) consists of a multi-bladed rotor that is supported centrally in the pipe along which the flow occurs. The rotor rotates as a result of the fluid flow, the angular velocity being approximately proportional to the flow rate. The rate of revolution of the rotor can be determined by attaching a small permanent magnet to one of the blades and using a pick-up coil. An induced e.m.f. pulse is produced in the coil every time the magnet passes it. The pulses are counted and so the number of revolutions of the rotor can be determined. The meter is expensive, with an accuracy of typically about ±0.1%. Another form uses helical screws which rotate as a result of the fluid flow.

2.5.3 Ultrasonic time of flight flow meter

Figure 2.34 shows one way ultrasonic waves can be used to determine the flow rate of a fluid. There are a pair of ultrasonic receiver–transmitters, one on each side of the pipe through which the fluid flows. If c is the velocity of the sound in still fluid, for the beam of sound going from left-to-right in the direction of the fluid flow the speed is $(c + v \cos \theta)$ while for the sound going from right-to-left in the opposite direction to the fluid flow the speed is $(c - v \cos \theta)$. If L is the distance between the two transmitter–receivers, then the times taken to go in the two directions are $L/(c + v \cos \theta)$ and $L/(c - v \cos \theta)$. The differences in these times is:

$$\Delta t = \frac{2Lv \cos \theta}{c^2 + v \cos^2 \theta} \approx \frac{2Lv \cos \theta}{c^2}$$

Thus measurement of the time can be used to determine the flow velocity. This method can be used for pipes from 75 mm to 1500 mm diameter, with fluid velocities from about 0.2 m/s to 12 m/s with an accuracy of about ±1% or better.

2.5.4 Vortex flow rate method

When a fluid flow encounters a body, the layers of fluid close to the surfaces of the body are slowed down. With a streamlined body, these boundary layers follow the contours of the body until virtually meeting at the rear of the object. This results in very little wake being produced. With a non-steamlined body, a so-called *bluff body*, the boundary layers detach from the body much earlier and a large wake is produced. When the boundary layer leaves the body surface it rolls up into vortices. These

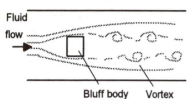

Figure 2.35 *Vortex shedding*

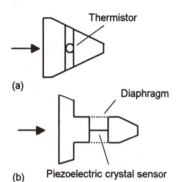

Figure 2.36   *Detection systems: (a) thermistor, (b) piezoelectric crystal*

are produced alternately from the top and bottom surfaces of the body (Figure 2.35). The result is two parallel rows of vortices moving downstream with the distance between successive vortices in each row being the same, a vortex in one row occurring half way between those in the other row.

For a particular bluff body, the number of vortices produced per second *f*, i.e. the frequency, is proportional to the flow rate. A number of methods are used for the measurement of the frequency. For example, a thermistor might be located behind the face of the bluff body (Figure 2.37(a)). The thermistor, heated as a result of a current passing through it, senses vortices due to the cooling effect caused by their breaking away. Another method uses a piezoelectric crystal mounted in the bluff body (Figure 2.36(b)). Flexible diaphragms react to the pressure disturbances produced by the vortices and are detected by the crystal.

Vortex flow meters are used for both liquids and gases, having an output which is independent of density, temperature or pressure, and having an accuracy of about ±1%. They are used at pressures up to about 10 MPa and temperatures of 200°C.

2.5.5 Coriolis flow meter

If a skater is spinning with arms outstretched and then pulls in his or her arms, they spin faster. As a consequence we can think of there being a torque acting on the skater's body to result in the increased angular velocity. This torque is considered to arise from a tangential force called the *Coriolis force*. When we move an object in a rotating system, it seems to be pushed sideways. For a body of mass M moving with constant linear radial velocity v and subject to an angular velocity ω the Coriolis force is $2M\omega v$.

The *Coriolis flow meter* consists basically of a C-shaped pipe (Figure 2.37) through which the fluid flows. The pipe, and fluid in the pipe, is given an angular acceleration by being set into vibration, this being done by means of a magnet mounted in a coil on the end of a tuning fork-like leaf spring. Oscillations of the spring then set the C-tube into oscillation. The result is an angular velocity that alternates in direction. At some instant the Coriolis force acting on the fluid in the upper limb is in one direction and in the lower limb in the opposite direction, this being because the velocity of the fluid is in opposite directions in the upper and lower limbs. The resulting Coriolis forces on the fluid in the two limbs are thus in opposite directions and cause the limbs of the C to become displaced. When the direction of the angular velocity is reversed then the forces reverse in direction and the limbs become displaced in the opposite direction. These displacements are proportional to the mass flow rate of fluid through the tube. The displacements are monitored by means of optical sensors, their outputs being a pulse with a width proportional to the mass flow rate. The flow meter can be used for liquids or gases and has an accuracy of ±0.5%. It is unaffected by changes in temperature or pressure.

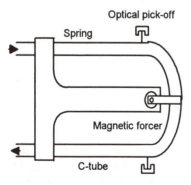

Figure 2.37 *Coriolis flow meter*

2.6 Liquid level

Methods used to measure the level of liquid in a vessel include those based on:

1   Floats whose position is directly related to the liquid level.

2   Archimedes' principle and a measurement of the upthrust acting on an object partially immersed in the liquid; the term displacer is used.

3   A measurement of the pressure at some point in the liquid, the pressure due to a column of liquid of height h being $h\rho g$, where ρ is the liquid density and g the acceleration due to gravity.

4   A measurement of the weight of the vessel containing the liquid plus liquid. The weight of the liquid is $Ah\rho g$, where A is the cross-sectional area of the vessel, h the height of liquid, ρ its density and g the acceleration due to gravity and thus changes in the height of liquid give weight changes.

5   A change in electrical conductivity when the liquid rises between two probes.

6   A change in capacitance as the liquid rises up between the plates of a capacitor.

7   Ultrasonic and nuclear radiation methods.

The following give examples of the above methods used for liquid level measurements.

2.6.1 Floats

Figure 2.38 shows a simple *float system*. The float is at one end of a pivoted rod with the other end connected to the slider of a potentiometer. Changes in level cause the float to move and hence move the slider over the potentiometer resistance track and so give a potential difference output related to the liquid level.

2.6.2 Displacer gauge

When an object is partially or wholly immersed in a fluid it experiences an upthrust force equal to the weight of fluid displaced by the object. This is known as *Archimedes' principle*. Thus a change in the amount of an object below the surface of a liquid will result in a change in the upthrust. The resultant force acting on such an object is then its weight minus the upthrust and thus depends on the depth to which the object is immersed. For a vertical cylinder of cross-sectional area A in a liquid of density ρ, if a height h of the cylinder is below the surface then the upthrust is $hA\rho g$, where g is the acceleration due to gravity, and so the apparent weight of the cylinder is $(mg - hA\pi g)$, where m is the mass of the cylinder. Such *displacer gauges* need calibrating for liquid level determinations for particular liquids since the upthrust depends on the liquid density. Figure 2.39 shows a simple version of a displacer gauge.

Application

A problem with floats and displacers is that such instruments tend to incorporate seals which require frequent maintenance in corrosive liquid applications, also there is the problem of fluids coating the floats and apparently changing the buoyancy.

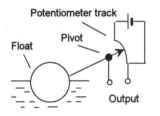

Figure 2.38   *Potentiometer float gauge*

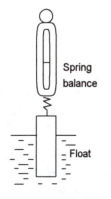

Figure 2.39   *Displacer gauge*

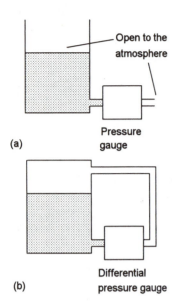

(a)

(b)

Figure 2.40 *Pressure level gauges*

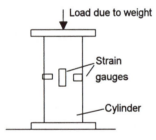

Figure 2.41 *Load cell*

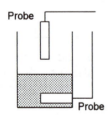

Figure 2.42 *Conductivity level indicator*

Application
An integrated circuit LM1830N can be used for signal processing with conductivity probes so that an output is given which can be used to activate a loudspeaker or a LED. The circuit compares the resistance of the liquid with the IC's internal reference resistance.

2.6.3 Differential pressure

The pressure due to a height h of liquid above some level is $h\rho g$, where ρ is the liquid density and g the acceleration due to gravity. With a tank of liquid open to the atmosphere, the pressure difference can be measured between a point near the base of the tank and the atmosphere. The result is then proportional to the height of liquid above the pressure measurement point (Figure 2.40(a)). With a closed tank, the pressure difference has to be measured between a point near the bottom of the tank and in the gases above the liquid surface (Figure 2.40(b)). The pressure gauges used for such measurements tend to be diaphragm instruments.

2.6.4 Load cell

The weight of a tank of liquid can be used as a measure of the height of liquid in the tank. Load cells are commonly used for such weight measurements. Typically, a *load cell* consists of a strain gauged cylinder (Figure 2.41) which is included in the supports for the tank of liquid. When the level of the liquid changes, the weight changes and so the load on the load cell changes and the resistances of the strain gauges change. The resistance changes of the strain gauges are thus a measure of the level of the liquid. Since the load cells are completely isolated from the liquid, the method is useful for corrosive liquids.

2.6.5 Electrical conductivity level indicator

Conductivity methods can be used to indicate when the level of a high electrical conductivity liquid reaches a critical level. One form has two probes, one probe mounted in the liquid and the other either horizontally at the required level or vertically with its lower end at the critical level (Figure 2.42). When the liquid is short of the required level, the resistance between the two probes is high since part of the electrical path between the two probes is air. However, when the liquid level reaches the critical level, there is a path entirely through the liquid and so the conductivity drops. Foaming, splashing and turbulence can affect the results.

2.6.6 Capacitive level indicator

A common form of *capacitive level gauge* consists of two concentric conducting cylinders, or a circular rod inside a cylinder, acting as capacitor plates with the liquid between them acting as the dielectric of a capacitor (Figure 2.43). If the liquid is an electrical insulator then the capacitor plates can be bare metal, if the liquid is conducting then they are metal coated with an insulator, e.g. Teflon. The arrangement consists essentially of two capacitors in parallel, one formed between the plates inside the liquid and the other from that part of the plates in the air above the liquid. A change in the liquid level changes the total capacitance of the arrangement. Errors can arise as a result of temperature changes since they will produce a change in capacitance

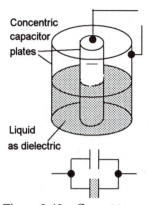

Figure 2.43 *Capacitive gauge*

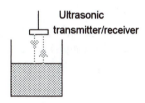

Figure 2.44 *Ultrasonic gauge*

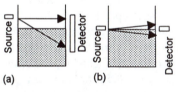

Figure 2.45 *Radionic gauges*

without any change in level. Errors can also arise if, when the liquid level drops, the electrodes remain coated with liquid. The system can be used, with suitable choice of electrode material, for corrosive liquids and is capable of reasonable accuracy.

2.6.7 Ultrasonic level gauge

In one version of an *ultrasonic level indicator*, an ultrasonic transmitter/ receiver is placed above the surface of the liquid (Figure 2.44). Ultrasonic pulses are produced, travel down to the liquid surface and are then reflected back to the receiver. The time taken from emission to reception of the pulses can be used as a measure of the position of the liquid surface. Because the receiver/transmitter can be mounted outside the liquid, it is particularly useful for corrosive liquids. Errors are produced by temperature changes since they affect the speed of the sound wave. Such errors are typically about 0.18% per °C.

2.6.8 Nucleonic level indicators

One form of level indicator uses gamma radiation from a radioactive source, generally cobalt–60, caesium–137 or radium–226. A detector is placed on one side of the container and the source on the other. The intensity of the radiation depends on the amount of liquid between the source and detector and can be used to determine the level of the liquid. Figure 2.45 shows two possible arrangements. With a compact source and extended detector, level changes over the length of the detector can be determined. A compact source and a compact detector can be used where small changes in a small range of level are to be detected. Such methods can be used for liquids, slurries and solids, and, since no elements of the system are in the liquid, for corrosive and high temperature liquids.

2.7 Temperature sensors

The expansion or contraction of solids, liquids or gases, the change in electrical resistance of conductors and semiconductors, thermoelectric e.m.f.s and the change in the current across the junction of semiconductor diodes and transistors are all examples of properties that change when the temperature changes and can be used as basis of temperature sensors. The following are some of the more commonly used temperature sensors.

2.7.1 Bimetallic strips

A *bimetallic strip* consists of two different metal strips of the same length bonded together (Figure 2.46). Because the metals have different coefficients of expansion, when the temperature increases the composite strip bends into a curved strip, with the higher coefficient metal on the outside of the curve. The amount by which the strip curves depends on the two metals used, the length of the composite strip and the change in temperature. If one end of a bimetallic strip is fixed, the amount by which the free end moves is a measure of the temperature. This

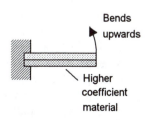

Figure 2.46 *Bimetallic strip*

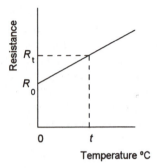

Figure 2.47 *Resistance variation with temperature for metals*

Application
A commercially available platinum resistance thermometer includes the following in its specification:

Range −200°C to +800°C
Accuracy ±0.01°C
Sensitivity 0.4 Ω/°C for 100 Ω

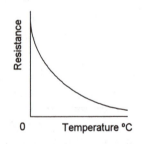

Figure 2.48 *Variation of resistance with temperature for thermistors*

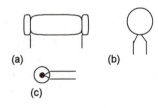

Figure 2.49 *Thermistors: (a) rod, (b) disc, (c) bead*

movement may be used to open or close electric circuits, as in the simple thermostat commonly used with domestic heating systems. Bimetallic strip devices are robust, relatively cheap, have an accuracy of the order of ±1% and are fairly slow reacting to changes in temperature.

2.7.2 Liquid in glass thermometers

The *liquid in glass thermometer* involves a liquid expanding up a capillary tube. The height to which the liquid expands is a measure of the temperature. With mercury as the liquid, the range possible is −35°C to +600°C, with alcohol −80°C to +70°C, with pentane −200°C to +30°C. Such thermometers are direct reading, fragile, capable of reasonable accuracy under standardised conditions, fairly slow reacting to temperature changes, and cheap.

2.7.3 Resistance temperature detectors (RTDs)

The resistance of most metals increases in a reasonably linear way with temperature (Figure 2.47) and can be represented by the equation:

$$R_t = R_0(1 + at)$$

where R_t is the resistance at a temperature t °C, R_0 the resistance at 0°C and a a constant for the metal, termed the temperature coefficient of resistance. *Resistance temperature detectors* (RTDs) are simple resistive elements in the form of coils of metal wire, e.g. platinum, nickel or copper alloys. Platinum detectors have high linearity, good repeatability, high long term stability, can give an accuracy of ±0.5% or better, a range of about −200°C to +850°C, can be used in a wide range of environments without deterioration, but are more expensive than the other metals. They are, however, very widely used. Nickel and copper alloys are cheaper but have less stability, are more prone to interaction with the environment and cannot be used over such large temperature ranges.

2.7.4 Thermistors

Thermistors are semiconductor temperature sensors made from mixtures of metal oxides, such as those of chromium, cobalt, iron, manganese and nickel. The resistance of thermistors decreases in a very non-linear manner with an increase in temperature, Figure 2.48 illustrating this. The change in resistance per degree change in temperature is considerably larger than that which occurs with metals. For example, a thermistor might have a resistance of 29 kΩ at −20°C, 9.8 kΩ at 0°C, 3.75 kΩ at 20°C, 1.6 kΩ at 40°C, 0.75 kΩ at 60°C. The material is formed into various forms of element, such as beads, discs and rods (Figure 2.49). Thermistors are rugged and can be very small, so enabling temperatures to be monitored at virtually a point. Because of their small size they have small thermal capacity and so respond very rapidly to changes in temperature. The temperature range over which they can be used will depend on the thermistor concerned, ranges within about

Application
The following is part of the specification for a bead thermistor temperature sensor:

Accuracy ±5%
Maximum power 250 mW
Dissipation factor 7 mW/°C
Response time 1.2 s
Thermal time constant 11 s
Temperature range –40°C to +125°C

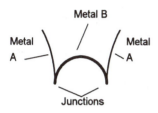

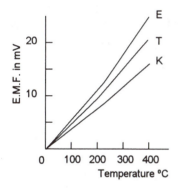

Figure 2.50   *Thermocouple*

Figure 2.51   *Thermocouples: chromel-constantan (E), chromel-alumel (K), copper-constantan (T)*

Table 2.1 *Thermocouples*

	Materials	Range °C	Sensitivity mV/°C
E	Chromel-constantan	0 to 980	63
J	Iron-constantan	-180 to 760	53
K	Chromel-alumel	-180 to 1260	41
R	Platinum-platinum/rhodium 13%	0 to 1750	8
T	Copper-constantan	-180 to 370	43

–100°C to +300°C being possible. They give very large changes in resistance per degree change in temperature and so are capable, over a small range, of being calibrated to give an accuracy of the order of 0.1°C or better. However, their characteristics tend to drift with time. Their main disadvantage is their non-linearity.

2.7.5 Thermocouples

When two different metals are joined together, a potential difference occurs across the junction. The potential difference depends on the two metals used and the temperature of the junction. A *thermocouple* involves two such junctions, as illustrated in Figure 2.50. If both junctions are at the same temperature, the potential differences across the two junctions cancel each other out and there is no net e.m.f. If, however, there is a difference in temperature between the two junctions, there is an e.m.f. The value of this e.m.f. E depends on the two metals concerned and the temperatures t of both junctions. Usually one junction is held at 0°C and then, to a reasonable extent, the following relationship holds:

$$E = at + bt^2$$

where a and b are constants for the metals concerned. Figure 2.51 shows how the e.m.f. varies with temperature for a number of commonly used pairs of metals. Standard tables giving the e.m.f.s at different temperatures are available for the metals usually used for thermocouples. Commonly used thermocouples are listed in Table 2.1, with the temperature ranges over which they are generally used and typical sensitivities. These commonly used thermocouples are given reference letters. The base-metal thermocouples, E, J, K and T, are relatively cheap but deteriorate with age. They have accuracies which are typically about ±1 to 3%. Noble-metal thermocouples, e.g. R, are more expensive but are more stable with longer life. They have accuracies of the order of ±1% or better. Thermocouples are generally mounted in a sheath to give them mechanical and chemical protection. The response time of an unsheathed thermocouple is very fast. With a sheath this may be increased to as much as a few seconds if a large sheath is used.

A thermocouple can be used with the reference junction at a temperature other than 0°C. However, the standard tables assume that the junction is at 0°C junction and hence a correction has to be applied before the tables can be used. The correction is applied using what is known as the *law of intermediate temperatures*, namely:

$$E_{t,0} = E_{t,I} + E_{I,0}$$

The e.m.f. $E_{t,0}$ at temperature t when the cold junction is at 0°C equals the e.m.f. $E_{t,I}$ at the intermediate temperature I plus the e.m.f. $E_{I,0}$ at temperature I when the cold junction is at 0°C. Consider a type E thermocouple. The following is data from standard tables.

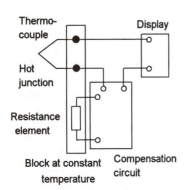

Figure 2.52 *Cold junction compensation*

Application
Integrated circuits are available which combine amplification with cold junction compensation for thermocouples, e.g. the Analog Devices AD594 (Figure 2.53). This, when used with a +5 V supply and a constantan–iron thermocouple, gives an output of 10 mV/°C.

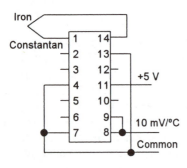

Figure 2.53 *AD594*

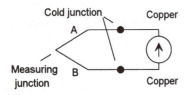

Figure 2.54 *The junctions with a measurement instrument*

Application
The specification for an integrated LM35 temperature sensor includes:

Accuracy at 25°C ±0.4%
Non-linearity 0.2°C
Sensitivity 10 mV/°C

Temp. (°C)	0	20	200
e.m.f. (mV)	0	1.192	13.419

Thus, using the law of intermediate temperatures, the thermoelectric e.m.f. at 200° C with the cold junction at 20°C is:

$$E_{200,20} = E_{200,0} - E_{20,0} = 13.419 - 1.192 = 12.227 \text{ mV}$$

Note that this is not the e.m.f. given by the tables for a temperature of 180°C with a cold junction at 0°C, namely 11.949 mV.

To maintain one junction of a thermocouple at 0°C, it needs to be immersed in a mixture of ice and water. This, however, is often not convenient and a *compensation circuit* (Figure 2.52) is used to provide an e.m.f. which varies with the temperature of the 'cold' junction in such a way that when it is added to the thermocouple e.m.f. it generates a combined e.m.f. which is the same as would have been generated if the cold junction had been at 0°C (see Section 2.9.3).

When a thermocouple is connected to a measuring circuit, other metals are involved (Figure 2.54). Thus we can have as the 'hot' junction that between metals A and B and the 'cold' junction effectively extended by the introduction of copper leads and the measurement instrument. Provided the junctions with the intermediate materials are at the same temperature, there is no extra e.m.f. involved and we still have the e.m.f. as due to the junction between metals A and B.

2.7.6 Thermodiodes and transistors

When the temperature of doped semiconductors changes, the mobility of their charge carriers change. As a consequence, when a p-n junction has a potential difference across it, the current through the junction is a function of the temperature. Such junctions for use as temperature sensors are supplied, together with the necessary signal processing circuitry as integrated circuits, e.g. LM3911 which gives an output voltage proportional to temperature. In a similar manner, transistors can be used as temperature sensors. An integrated circuit temperature sensor using transistors is LM35. This gives an output, which is a linear function of temperature, of 10 mV/°C when the supply voltage is 5 V.

2.7.7 Pyrometers

Methods used for the measurement of temperature which involve the radiation emitted by the body include:

1   *Optical pyrometer*
    This is based on comparing the brightness of the light emitted by the hot body with that from a known standard.

2   *Total radiation pyrometer*
    This involves the measurement of the total amount of radiation emitted by the hot body by a resistance element or a thermopile.

The *optical pyrometer*, known generally as the *disappearing filament pyrometer*, involves just the visible part of the radiation emitted by a hot object. The radiation is focused onto a filament so that the radiation and the filament can both be viewed in focus through an eyepiece (Figure 2.55). The filament is heated by an electrical current until the filament and the hot object seem to be the same colour, the filament image then disappearing into the background of the hot object. The filament current is then a measure of the temperature. A red filter between the eyepiece and the filament is generally used to make the matching of the colours of the filament and the hot object easier. Another red filter may be introduced between the hot object and the filament with the effect of making the object seem less hot and so extending the range of the instrument.

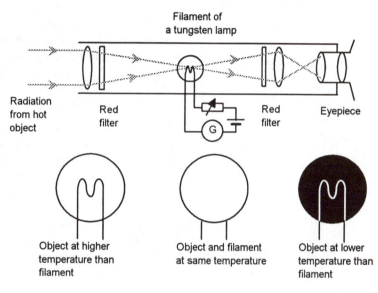

Figure 2.55 *Disappearing filament pyrometer*

The disappearing filament pyrometer has a range of about 600°C to 3000°C, an accuracy of about ±0.5% of the reading and involves no physical contact with the hot object. It can thus be used for moving or distant objects.

The *total radiation pyrometer* involves the radiation from the hot object being focused onto a radiation detector. Figure 2.56 shows the basic form of an instrument which uses a mirror to focus the radiation onto the detector. Some forms use a lens to focus the radiation. The detector is typically a thermopile with often up to 20 or 30 thermocouple junctions, a resistance element or a thermistor. The detector is said to be *broad band* since it detects radiation over a wide band of frequencies and so the output is the summation of the power emitted at every wavelength. It is proportional to the fourth power of the temperature (the Stefan–Boltzmann law). The accuracy of broad band total radiation pyrometers is typically about ±0.5% and ranges are available within the region 0°C to 3000°C. The time constant (a measure of how fast the

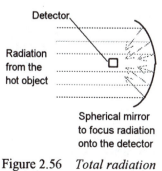

Figure 2.56 *Total radiation pyrometer*

system responds to a change in temperature and is the time taken to reach about 63% of the final value) for the instrument varies from about 0.1 s when the detector is just one thermocouple or small bead thermistor to a few seconds with a thermopile involving many thermocouples. Some instruments use a rotating mechanical chopper to chop the radiation before it impinges on the detector. The aim is to obtain an alternating output from the detector, since amplification is easier with an alternating voltage. It is thus of particular benefit when the level of radiation is low. However, choppers can only be used with detectors which have a very small time constant and thus tend to be mainly used with small bead thermistor detectors.

2.8 Sensor selection

The selection of a sensor for a particular application requires a consideration of:

1   The nature of the measurement required, i.e. the sensor input. This means considering the variable to be measured, its nominal value, the range of values, the accuracy required, the required speed of measurement, the reliability required and the environmental conditions under which the measurement is to be made.

2   The nature of the output required from the sensor, this determining the signal processing required. The selection of sensors cannot be taken in isolation from a consideration of the form of output that is required from the system after signal processing, and thus there has to be a suitable marriage between sensor and signal processing.

Then possible sensors can be identified, taking into account such factors as their range, accuracy, linearity, speed of response, reliability, life, power supply requirements, ruggedness, availability and cost.

Example

Select a sensor which can be used to monitor the temperature of a liquid in the range 10°C to 80°C to an accuracy of about 1°C and which will give an output which can be used to change the current in an electrical circuit.

There are a number of forms of sensor that can be used to monitor such a temperature in the range, and to the accuracy, indicated. The choice is, however, limited by the requirement for an output which can change the current in an electrical circuit. This would suggest a resistance thermometer. In view of the limited accuracy and range required, a thermistor might thus be considered.

Example

Select a sensor which can be used for the measurement of the level of a corrosive acid in a circular vessel of diameter 1 m and will give an electrical output. The acid level can vary from 0 to 3 m and the minimum change in level to be detected is 0.1 m. The empty vessel has a weight of 50 kg. The acid has a density of 1050 kg/m³.

Because of the corrosive nature of the acid there could be problems in using a sensor which is inserted in the liquid. Thus a possibility is to use a load cell, or load cells, to monitor the weight of the vessel. Such cells would give an electrical output. The weight of the liquid changes from 0 when empty to, when full, $1050 \times 3 \times \pi(1^2/4) \times 9.8$ = 24.3 kN. Adding this to the weight of the empty vessel gives a weight that varies from about 0.5 kN to 4.9 kN. A change of level of 0.1 m gives a change in weight of $0.10 \times 1050 \times \pi(1^2/4) \times 9.8 =$ 0.8 kN. If the load of the vessel is spread between three load cells, each will require a range of about 0 to 5 kN with a resolution of about 0.3 kN.

2.9 Signal processing

The output signal from the sensor of a measurement system has generally to be processed in some way to make it suitable for display or use in some control system. For example, the signal may be too small and have to be amplified, be analogue and have to be made digital, be digital and have to be made analogue, be a resistance change and have to be made into a current change, be a voltage change and have to be made into a suitable size current change, be a pressure change and have to be made into a current change, etc. All these changes can be referred to as *signal processing*. For example, the output from a thermocouple is a very small voltage, a few millivolts. A signal processing module might then be used to convert this into a larger voltage and provide cold junction compensation (i.e. allow for the cold junction not being at 0°C). Note that the term *signal conditioning* is sometimes used for the conversion of the output from a sensor into a suitable form for signal processing. The following gives some of the elements that are used in signal processing.

2.9.1 Resistance to voltage converter

Consider how the resistance change produced by a thermistor when subject to a temperature change can be converted into a voltage change. Figure 2.57 shows how a *potential divider circuit* can be used. A constant voltage, of perhaps 6 V, is applied across the thermistor and another resistor in series. With a thermistor with a resistance of 4.7 kΩ, the series resistor might be 10 kΩ. The output signal is the voltage across the 10 kΩ resistor. When the resistance of the thermistor changes, the fraction of the 6 V across the 10 kΩ resistor changes.

The output voltage is proportional to the fraction of the total resistance which is between the output terminals. Thus:

$$\text{output} = \frac{R}{R + R_t}V$$

where V is the total voltage applied, in Figure 2.57 this is shown as 6 V, R the value of the resistance between the output terminals (10 kΩ) and R_t

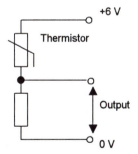

+6 V

Thermistor

Output

0 V

Figure 2.57   *Resistance to voltage conversion for a thermistor*

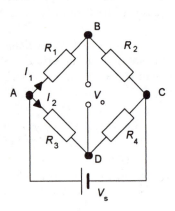

Figure 2.58   *Wheatstone bridge*

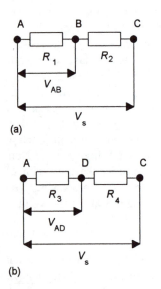

(a)

(b)

Figure 2.59   *Potential drops across AB and AD*

the resistance of the thermistor at the temperature concerned. The potential divider circuit is thus an example of a simple resistance to voltage converter.

Another example of such a converter is the *Wheatstone bridge*, Figure 2.58 shows the basic form of the bridge. The resistance element being monitored forms one of the arms of the bridge. When the output voltage V_o is zero, then there is no potential difference between B and D and so the potential at B must equal that at D. The potential difference across R_1, i.e. V_{AB}, must then equal that across R_3, i.e. V_{AD}. Thus:

$$I_1 R_1 = I_2 R_2$$

We also must have the potential difference across R_2, i.e. V_{BC}, equal to that across R_4, i.e. V_{DC}. Since there is no current through BD then the current through R_2 must be the same as that through R_1 and the current through R_4 the same as that through R_3. Thus:

$$I_1 R_2 = I_2 R_4$$

Dividing these two equations gives:

$$\frac{R_1}{R_2} = \frac{R_3}{R_4}$$

The bridge is said to be *balanced*.

Now consider what happens when one of the elements has a resistance which changes from this balanced condition. The supply voltage V_s is connected between points A and C and thus the potential drop across the resistor R_1 is the fraction $R_1/(R_1 + R_2)$ of the supply voltage (Figure 2.59(a)). Hence:

$$V_{AB} = \frac{V_s R_1}{R_1 + R_2}$$

Similarly, the potential difference across R_3 (Figure 2.59(b)) is:

$$V_{AD} = \frac{V_s R_3}{R_3 + R_4}$$

Thus the difference in potential between B and D, i.e. the output potential difference V_o, is:

$$V_o = V_{AB} - V_{AD} = V_s\left(\frac{R_1}{R_1 + R_2} - \frac{R_3}{R_3 + R_4}\right)$$

This equation gives the balanced condition when $V_o = 0$.

Consider resistance R_1 to be a sensor which has a resistance change, e.g. a strain gauge which has a resistance change when strained. A change in resistance from R_1 to $R_1 + \delta R_1$ gives a change in output from V_o to $V_o + \delta V_o$, where:

$$V_o + \delta V_o = V_s\left(\frac{R_1 + \delta R_1}{R_1 + \delta R_1 + R_2} - \frac{R_3}{R_3 + R_4}\right)$$

Hence:

$$(V_o + \delta V_o) - V_o = V_s\left(\frac{R_1 + \delta R_1}{R_1 + \delta R_1 + R_2} - \frac{R_1}{R_1 + R_2}\right)$$

If δR_1 is much smaller than R_1 then the denominator $R_1 + \delta R_1 + R_2$ approximates to $R_1 + R_2$ and so the above equation approximates to:

$$\delta V_o \approx V_s\left(\frac{\delta R_1}{R_1 + R_2}\right)$$

With this approximation, the change in output voltage is thus proportional to the change in the resistance of the sensor. We thus have a resistance to voltage converter. Note that the above equation only gives the output voltage when there is no load resistance across the output. If there is such a resistance then the loading effect has to be considered (see Section 1.3.1).

As the output voltage is proportional to the bridge excitation voltage, voltage drops along the cables from the voltage supply and the cable resistance between resistors in the bridge can affect the output, this being a particular problem if temperature changes cause resistance changes in these cables. Three-wire compensation (Figure 2.60(a)) can be used to help overcome the problem of cable resistance between, say, a temperature sensor and the bridge. Figure 2.60(b) shows a four-wire form of compensation, there being two parallel, dummy, leads.

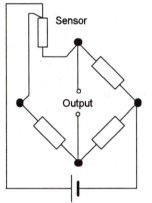

(a) Three-wire compensation

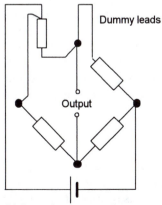

(b) Four-wire compensation

Figure 2.60 *Temperature compensation*

Example

A platinum resistance coil is to be used as a temperature sensor and has a resistance at 0°C of 100 Ω. It forms one arm of a Wheatstone bridge with the bridge being balanced at this temperature and each of the other arms also being 100 Ω. If the temperature coefficient of resistance of platinum is 0.0039 K^{-1}, what will be the output voltage from the bridge per degree change in temperature if the supply voltage is 6.0 V?

The variation of the resistance of the platinum with temperature can be represented by:

$$R_t = R_0(1 + at)$$

where R_t is the resistance at t °C, R_0 the resistance at 0°C and a the temperature coefficient of resistance. Hence:

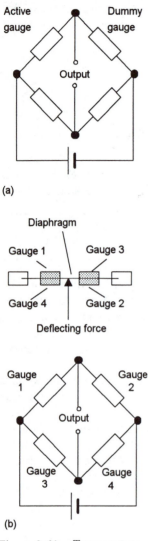

(a)

(b)

Figure 2.61 *Temperature compensation, (a) dummy gauge, (b) with four active strain gauges*

change in resistance = $R_t - R_0 = R_0 \alpha t$

Thus, for a one degree change in temperature:

change in resistance = $100 \times 0.0039 \times 1 = 0.39\ \Omega$

Since this resistance change is small compared to the 100 Ω, the approximate equation for the output voltage can be used. Hence, the change in output per degree change in temperature is:

$$\delta V_o \approx V_s\left(\frac{\delta R_1}{R_1 + R_2}\right) = \frac{6.0 \times 0.39}{100 + 100} = 0.012\ \text{V}$$

2.9.2 Temperature compensation

The electrical resistance strain gauge is a resistance element which changes resistance when subject to strain. However, it will also change resistance when subject to a temperature change. Thus, in order to use it to determine strain, compensation has to be made for temperature effects. One way of eliminating the temperature effect is to use a *dummy strain gauge*. This is a strain gauge identical to the one under strain, the active gauge, which is mounted on the same material as the active gauge but not subject to the strain. It is positioned close to the active gauge so that it suffers the same temperature changes. As a result, a temperature change will cause both gauges to change resistance by the same amount. The active gauge is mounted in one arm of a Wheatstone bridge (Figure 2.61(a)) and the dummy gauge in an opposite arm so that the effects of temperature-induced resistance changes cancel out.

Strain gauges are often used with other sensors such as diaphragm pressure gauges or load cells. Temperature compensation is still required. While dummy gauges could be used, a better solution is to·use four strain gauges with two of them attached so that the applied forces put them in tension and the other two in compression. The gauges, e.g. gauges 1 and 3, that are in tension will increase in resistance while those in compression, gauges 2 and 4, will decrease in resistance. The gauges are connected as the four arms of a Wheatstone bridge (Figure 2.61(b)). As all the gauges and so all the arms of the bridge will be equally affected by any temperature changes the arrangement is temperature compensated. The arrangement also gives a much greater output voltage than would occur with just a single active gauge.

2.9.3 Thermocouple compensation

With a thermocouple, one junction should be kept at 0°C; the temperature can then be obtained by looking up in tables the e.m.f. produced by the thermocouple. Figure 2.62 illustrates what is required. This keeping of a junction at 0°C, i.e. in a mixture of ice and water, is not always feasible or very convenient and the cold junction is often allowed to be at the ambient temperature. To take account of this a compensation voltage has to be added to the thermocouple. This voltage is the same as the e.m.f. that would be generated by the thermocouple

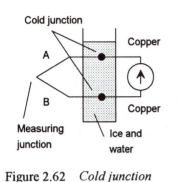

Figure 2.62 *Cold junction at 0°C*

with one junction at 0°C and the other at the ambient temperature. We thus need a voltage which will depend on the ambient temperature. Such a voltage can be produced by using a resistance temperature sensor in a Wheatstone bridge. Figure 2.63 illustrates this. The bridge is balanced at 0°C and the output voltage from the bridge provides the correction potential difference at other temperatures. By a suitable choice of resistance temperature sensor, the appropriate voltage can be obtained.

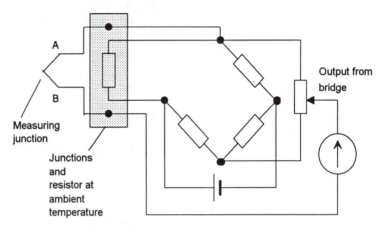

Figure 2.63 *A Wheatstone bridge compensation circuit*

The resistance of a metal resistance temperature sensor is given by:

$$R_t = R_0(1 + at)$$

where R_t is the resistance at t°C, R_0 the resistance at 0°C and a the temperature coefficient of resistance. When there is a change in temperature:

$$\text{change in resistance} = R_t - R_0 = R_0 at$$

The output voltage for the bridge, taking R_1 to be the resistance temperature sensor, is given by:

$$\delta V_o \approx V_s\left(\frac{\delta R_1}{R_1 + R_2}\right) = \frac{V_s R_0 at}{R_0 + R_2}$$

This voltage must be the same as that given by the thermocouple with one junction at 0°C and the other at the ambient temperature. The thermocouple e.m.f. e is likely to vary with temperature t in a reasonably linear manner over the small temperature range being considered, i.e. from 0°C to the ambient temperature. Thus we can write $e = at$, where a is a constant, i.e. the e.m.f. produced per degree change in temperature. Hence for compensation we must have:

$$at = \frac{V_s R_0 at}{R_0 + R_2}$$

and so the condition:

$$aR_2 = R_0(V_s a - a)$$

Example

Determine the value of the resistance R_2 in Figure 2.63 if compensation is to be provided for an iron–constantan thermocouple giving 51 μV/°C. The compensation is to be provided by a nickel resistance element with a resistance of 10 Ω at 0°C and a temperature coefficient of resistance of 0.0067 K^{-1}. Take the supply voltage for the bridge to be 2.0 V.

Using the equation developed above, $aR_2 = R_0(V_s a - a)$, then:

$$51 \times 10^{-6} \times R_2 = 10(2 \times 0.0067 - 51 \times 10^{-6})$$

Hence R_2 is 2617 Ω.

2.9.4 Protection

An important element that is often present with signal processing is *protection against high currents or high voltages*. For example, sensors when connected to a microprocessor can damage it if high currents or high voltages are transmitted to the microprocessor. A high current can be protected against by the incorporation in the input line of a series resistor to limit the current to an acceptable level and a fuse to break if the current does exceed a safe level (Figure 2.64). Protection against high voltages and wrong polarity voltages may be obtained by the use of a *Zener diode circuit* (Figure 2.65). The Zener diode with a reverse voltage connected across it has a high resistance up to some particular voltage at which it suddenly breaks down and becomes conducting (Figure 2.65(a)). Zener diodes are given voltage ratings, the rating indicating at which voltage they become conducting. For example, to allow a maximum voltage of 5 V but stop voltages above 5.1 V being applied to the following circuit, a Zener diode with a voltage rating of 5.1 V might be chosen. For voltages below 5.1 V the Zener diode, in reverse voltage connection, has a high resistance. When the voltage rises to 5.1 V the Zener diode breaks down and its resistance drops to a very low value. Thus, with the circuit shown in Figure 2.65(b), with the applied voltage below 5.1 V, the Zener diode, in reverse voltage connection, has a much higher resistance than the other resistor and so virtually all the applied voltage is across the Zener diode. When the applied voltage rises to 5.1 V, the Zener diode breaks down and has a low resistance. As a consequence, most of the voltage is then dropped across the resistor, the voltage across the diode drops and so the output voltage drops. Because the Zener diode is a diode with a low resistance for current in one direction through it and a high resistance for the opposite direction, it also provides protection against wrong polarity.

To ensure protection, it is often necessary to completely isolate circuits so that there are no electrical connections between them. This can be

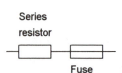

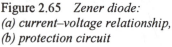

Figure 2.64 *Protection against high currents*

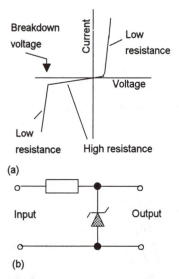

(a)

(b)

Figure 2.65 *Zener diode: (a) current–voltage relationship, (b) protection circuit*

done using an *optoisolator*. Such a device converts an electrical signal into an optical signal, transmits it to a detector which then converts it back into an electrical signal (Figure 2.66). The input signal passes through an infrared light-emitting diode (LED) and so produces a beam of infrared radiation. This infrared signal is then detected by a phototransistor. To prevent the LED having the wrong polarity or too high an applied voltage, it is likely to be protected by the Zener diode circuit (of the type shown above in Figure 2.65). Also, if there is likely to be an alternating signal in the input a diode would be put in the input line to rectify it.

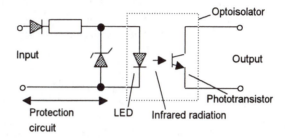

Figure 2.66 *An optoisolator*

2.9.5 Analogue-to-digital conversions

The electrical output from sensors such as thermocouples, resistance elements used for temperature measurement, strain gauges, diaphragm pressure gauges, LVDTs, etc. is in analogue form. Microprocessors require digital inputs. Thus, where a microprocessor is used it has to be converted into a digital form before it can be used as an input to the microprocessor. The output from a microprocessor is digital. Most control elements require an analogue input and so the digital output from a microprocessor has to be converted into an analogue form before it can be used by them. Thus there is a need for analogue-to-digital converters (ADC) and digital-to-analogue converters (DAC). Microcontrollers are microprocessors with input and output signal processing incorporated on the same chip and often incorporate analogue-to-digital and digital-to-analogue converters.

An *analogue signal* (Figure 2.67(a)) is one that is continuously variable, changing smoothly over a range of values. The signal is an analogue, i.e. a scaled version, of the quantity it represents. A *digital signal* increases in jumps, being a sequence of pulses, often just on-off signals (Figure 2.67(b)). The value of the quantity instead of being represented by the height of the signal, as with analogue, is represented by the sequence of on-off signals.

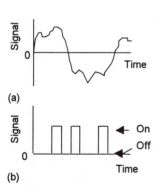

Figure 2.67 *Signals:*
(a) analogue, (b) digital

Analogue-to-digital conversion involves a number of stages. The first stage is to take samples of the analogue signal (Figure 2.68(a)). A clock supplies regular time signal pulses (Figure 2.68(b)) to the analogue-to-

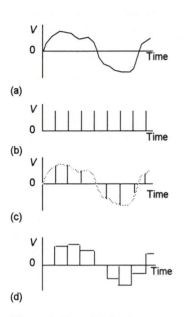

Figure 2.68   *(a) Analogue signal, (b) time signal, (c) sampled signal, (d) sampled and held signal*

digital converter and every time it receives a pulse it samples the analogue signal. The result is a series of narrow pulses with heights which vary in accord with the variation of the analogue signal (Figure 2.68(c)). This sequence of pulses is changed into the signal form shown in Figure 2.68(d) by each sampled value being held until the next pulse occurs. It is necessary to hold a sample of the analogue signal so that conversion can take place to a digital signal at an analogue-to-digital converter. This converts each sample into a sequence of pulses representing the value. For example, the first sampled value might be represented by 101, the next sample by 011, etc. The 1 represents an 'on' or 'high' signal, the 0 an 'off' or 'low' signal. Analogue-to-digital conversion thus involves a sample and hold unit followed by an analogue-to-digital converter (Figure 2.69).

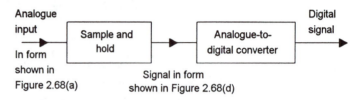

Figure 2.69 *Analogue-to-digital conversion*

To illustrate the action of the analogue-to-digital converter, consider one that gives an output restricted to three bits. The binary digits of 0 and 1, i.e. the 'low' and 'high' signals, are referred to as *bits*. A group of bits is called a *word*. Thus the three bits give the *word length* for this particular analogue-to-digital converter. The word is what represents the digital version of the analogue voltage. The position of bits in a word has the significance that the least significant bit is on the right end of the word and the most significant bit on the left. This is just like counting in tens, 435 has the 5 as the least significant number and the 4 as the most significant number, the least significant number contributing least to the overall value of the 435 number. The position of the digit in a decimal number is significant; the least significant digit having its value multiplied by 10^0, the next by 10^1, the next by 10^2, and so on. Likewise, the position of bits in a binary word is significant; the least significant bit having its value multiplied by 2^0, the next by 2^1, the next by 2^2, and so on. For a binary word of n bits :

$$2^{n-1}, ..., 2^3, 2^2, 2^1, 2^0$$
$$\uparrow \qquad\qquad\qquad \uparrow$$

Most significant   Least significant
bit (MSB)   bit (LSB)

With binary numbers we have the basic rules: $0 + 0 = 0$, $0 + 1 = 1$, $1 + 1 = 10$. Thus if we start with 000 and add 1 we obtain 001. If we add a further 1 we have 010. Adding another 1 gives 011. With three bits in a word we thus have the possible words of:

000 001 010 011 100 101 110 111

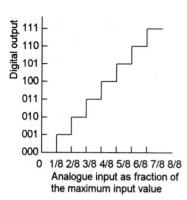

Figure 2.70 *Digital output from an ADC*

There are eight possible words which can be used to represent the analogue input; the number of possible words with a word length of n bits is 2^n. Thus we divide the maximum analogue voltage into eight parts and one of the digital words corresponds to each. Each rise in the analogue voltage of (1/8) of the maximum analogue input then results in a further bit being generated. Thus for word 000 we have 0 V input. To generate the next digital word of 001 the input has to rise to 1/8 of the maximum voltage. To generate the next word of 010 the input has to rise to 2/8 of the maximum voltage. Figure 2.70 illustrates this conversion of the sampled and held input voltage to a digital output.

Thus if we had a sampled analogue input of 8 V, the digital output would be 000 for a 0 V input and would remain at that output until the analogue voltage had risen to 1 V, i.e. 1/8 of the maximum analogue input. It would then remain at 001 until the analogue input had risen to 2 V. This value of 001 would continue until the analogue input had risen to 3 V. The smallest change in the analogue voltage that would result in a change in the digital output is thus 1 V. This is termed the *resolution* of the converter.

The word length possible with an analogue-to-digital converter determines its *resolution*. With a word length of n bits the maximum, or full scale, analogue input V_{FS} is divided into 2^n pieces. The minimum change in input that can be detected, i.e. the *resolution*, is thus $V_{FS}/2^n$. With an analogue-to-digital converter having a word length of 10 bits and the maximum analogue signal input range 10 V, then the maximum analogue voltage is divided into $2^{10} = 1024$ pieces and the resolution is $10/1024 = 9.8$ mV.

There are a number of forms of analogue-to-digital converter; the most commonly used being successive approximations, ramp and flash. The *flash form* is much faster than either the successive approximations form or the ramp form. The term *conversion time* is used to specify the time it takes a converter to generate a complete digital word when supplied with the analogue input.

Application

Analogue-to-digital converters are generally purchased as integrated circuits. Figure 2.71 shows an example of the pin connections for the ZN439, a successive approximation form of ADC. Its specification includes:

Resolution 8 bits
Conversion time 5 ms
Linearity error ± ½ LSB
Power dissipation 150 mW

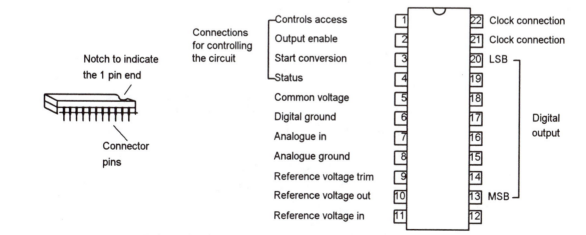

Figure 2.71 *The GEC Plessey ZN439E 8-bit analogue-to-digital converter*

Example

A thermocouple gives an output of 0.4 mV for each degree change in temperature. What will be the word length required when its output passes through an analogue-to-digital converter if temperatures from 0 to 200°C are to be measured with a resolution of 0.5°C?

The full scale output from the sensor is $200 \times 0.4 = 80$ mV. With a word length n there are 2^n digital numbers. Thus this voltage will be divided into 2^n levels and so the minimum voltage change that can be detected is $80/2^n$ mV. For a resolution of 0.5°C we must be able to detect a signal from the sensor of $0.5 \times 0.4 = 0.20$ mV. Hence:

$$0.20 = \frac{80}{2^n}$$

and so $2^n = 400$ and $n = 8.6$. Thus a 9-bit word length is required.

2.9.6 Digital-to-analogue conversions

The input to a digital-to-analogue converter is a binary word and the output its equivalent analogue value. For example, if we have a full scale output of 7 V then a digital input of 000 will give 0 V, 001 give 1 V, ... and 111 the full scale value of 7 V. Figure 2.72 illustrates this.

The basic form of a *digital-to-analogue converter* involves the digital input being used to activate electronic switches such that a 1 activates a switch and a 0 does not, the position of the 1 in the word determining which switch is activated. Thus when, with say a 3-bit converter, 001 is received we have a voltage of, say, 1 V switched to the output, when 010 is received we have 2 V, switched to the output, and when 100 is received we have 4 V switched to the output. Hence if we have the digital word 011 we have the least significant bit 001 switching 1 V to the output and the 010 bit 2 V to the output to give a summed output of 3 V (Figure 2.73).

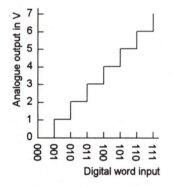

Figure 2.72 *Digital-to-analogue conversion*

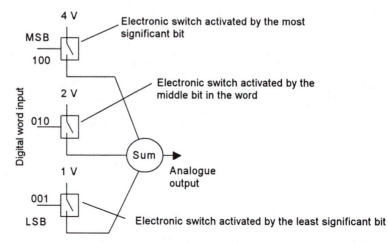

Figure 2.73 *The principle of a 3-bit digital-to-analogue converter*

Application

Figure 2.74 shows details of the GEC Plessey ZN558D 8-bit latched input digital-to-analogue converter. After the conversion is complete, the 8-bit result is placed in an internal latch until the next conversion is complete. A latch is just a device to retain the output until a new one replaces it. The settling time is the time taken for the analogue output voltage to settle within a specified band, usually ± LSB/2, about its final value when the digital word suddenly changes.

Resolution 8 bits
Settling time 800 ns
Non-linearity 0.5% of full scale
Power dissipation 100 mW

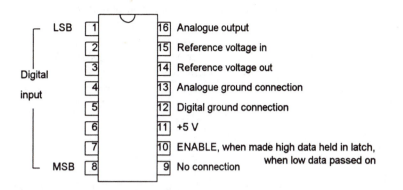

Figure 2.74 *The GEC Plessey ZN558D 8-bit latched digital-to-analogue converter*

Example

A microprocessor gives an output of an 8-bit word. This is fed through an 8-bit digital-to-analogue converter to a control valve which requires 6.0 V to be fully open. If the fully open state is to be indicated by the output of the digital word 11111111 what will be the change in output to the valve when there is a change of 1 bit?

The output voltage will be divided into 2^8 intervals. Since there is to be an output of 6.0 V when the output is 2^8 of these intervals, a change of 1 bit is a change in the output voltage of $6.0/2^8 = 0.023$ V.

2.9.7 Op-amps

The *operational amplifier* (op-amp) is a very high gain d.c. amplifier, the gain typically being of the order of 100 000 or more, which is supplied as an integrated circuit on a silicon chip. It has two inputs, known as the inverting input (−) and the non-inverting input (+). In addition there are inputs for a negative voltage supply, a positive voltage supply and two inputs termed offset null, these being to enable corrections to be made for the non-ideal behaviour of the amplifier. Figure 2.75 shows the pin connections for a 741 type operational amplifier with the symbol for the operational amplifier shown superimposed. On the symbol the + sign indicates the non-inverting input and the − sign the inverting input.

The operational amplifier is a very widely used element in signal conditioning and processing circuits and the following indicates common examples of such circuits.

Figure 2.76 shows the connections made to the amplifier when it is used as an *inverting amplifier*, such a form of amplifier giving an output which is an inverted form of the input, i.e. it is out of phase by 180°. The input is taken to the inverting input through a resistor R_1 with the non-inverting input being connected to ground. A feedback path is provided from the output, via the resistor R_2 to the inverting input. The operational amplifier has a very high voltage gain of about 100 000 and

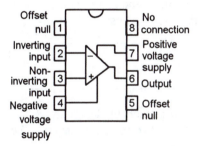

Figure 2.75 *Pin connections for a 741 op-amp*

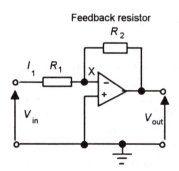

Figure 2.76 *Inverting amplifier*

the change in output voltage is limited to about ±10 V. Thus the input voltage at point X must be between +0.0001 V and −0.0001 V. This is virtually zero and so point X is at virtually earth potential and hence is termed a *virtual earth*. With an ideal operational amplifier with an infinite gain, point X is at zero potential. The potential difference across R_1 is $(V_{in} - V_X)$. Hence, for an ideal operational amplifier with an infinite gain, and hence $V_X = 0$, the input potential V_{in} can be considered to be across R_1. Thus:

$$V_{in} = I_1 R_1$$

The operational amplifier has a very high impedance between its input terminals, for a 741 this is about 2 MΩ. Thus virtually no current flows through X into it. For an ideal operational amplifier the input impedance is taken to be infinite and so there is no current flow through X into the amplifier input. Hence, since the current entering the junction at X must equal the current leaving it, the current I_1 through R_1 must be the current through R_2. The potential difference across R_2 is $(V_X - V_{out})$ and thus, since V_X is zero for the ideal amplifier, the potential difference across R_2 is $-V_{out}$. Thus:

$$-V_{out} = I_1 R_2$$

Dividing these two equations gives the ratio of the output voltage to the input voltage, i.e. the voltage gain of the circuit. Thus:

$$\text{voltage gain of circuit} = \frac{V_{out}}{V_{in}} = -\frac{R_2}{R_1}$$

The voltage gain of the circuit is determined solely by the relative values of R_2 and R_1. The negative sign indicates that the output is inverted, i.e. 180° out of phase, with respect to the input.

To illustrate the above, consider an inverting operational amplifier circuit which has a resistance of 10 kΩ in the inverting input line and a feedback resistance of 100 kΩ. The voltage gain of the circuit is:

$$\text{voltage gain of circuit} = \frac{V_{out}}{V_{in}} = -\frac{R_2}{R_1} = -\frac{100}{10} = -10$$

Figure 2.77 shows the operational amplifier connected as a *non-inverting amplifier*. Since the operational amplifier has a very high input impedance, there is virtually no current flowing into the inverting input. The inverting voltage input voltage is V_{in}. Since there is virtually no current through the operational amplifier between the two inputs there can be virtually no potential difference between them. Thus, with the ideal operational amplifier, we must have $V_X = V_{in}$. The output voltage is generated by the current I which flows from earth through R_1 and R_2. Thus:

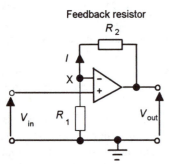

Feedback resistor

R_2

Figure 2.77 *Non-inverting amplifier*

$$I = -\frac{V_{out}}{R_1 + R_2}$$

But X is at the potential V_{in}, thus the potential difference across the feedback resistor R_2 is $(V_{in} - V_{out})$ and we must have:

$$I = \frac{V_{in} - V_{out}}{R_2}$$

Thus:

$$-\frac{V_{out}}{R_1 + R_2} = \frac{V_{in} - V_{out}}{R_2}$$

Hence:

$$\text{voltage gain of circuit} = \frac{V_{out}}{V_{in}} = \frac{R_1 + R_2}{R_1} = 1 + \frac{R_2}{R_1}$$

A particular form of this amplifier which is often used has the feedback loop as a short circuit, i.e. $R_2 = 0$. The voltage gain is then just 1. However, the input voltage to the circuit is across a large resistance, the input resistance of an operational amplifier such as the 741 typically being 2 MΩ. The resistance between the output terminal and the ground line is, however, much smaller, typically 75 Ω. Thus the resistance in the circuit that follows is a relatively small one compared with the resistance in the input circuit and so affects that circuit less. Such a form of amplifier circuit is referred to as a *voltage follower* and is typically used for sensors which require high impedance inputs such as piezoelectric sensors.

Figure 2.78 shows how the standard inverting amplifier can be used as a *current-to-voltage converter*. Point X is the virtual earth. Thus any input current has to flow through the feedback resistor R_2. The voltage drop across R_2 must therefore produce the output voltage and so $V_{out} = -IR_2$. So the output voltage is just the input current multiplied by the scaling factor R_2. The advantage of this method of converting a current to a voltage, compared with just passing a current through a resistor and taking the potential difference across it, is that there is a high impedance across the input and so there is less likelihood of loading problems.

Situations often arise where the output needs to be a current in order to drive perhaps an electromechanical device such as a relay or possibly give a display on a moving coil meter. A *voltage-to-current converter* is provided by the basic inverting amplifier circuit with the device through which the current is required being the feedback resistor (Figure 2.79). Since X is a virtual earth, the potential difference across R_1 is V_{in} and the current through it I_1. Hence $I_1 = V_{in}/R_1$. The current through R_2 is I_1. Thus the input voltage has been converted to the current I_1 through the feedback resistor, with the current being V_{in}/R_1.

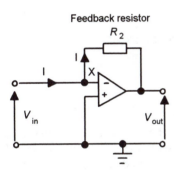

Figure 2.78 *Current-to-voltage converter*

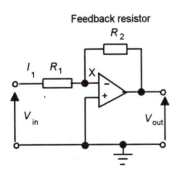

Figure 2.79 *Voltage-to-current converter*

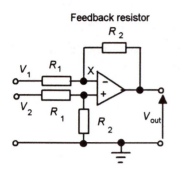

Figure 2.80 *Differential amplifier*

Figure 2.80 shows how an operational amplifier can be used as a *differential amplifier*, amplifying the difference between two input signals. Since the operational amplifier has high impedance between its two inputs, there will be virtually no current through the operational amplifier between the two input terminals. There is thus no potential difference between the two inputs and therefore both will be at the same potential, that at X. The voltage V_2 is across resistors R_1 and R_2 in series. We thus have a potential divider circuit with the potential at the non-inverting input, which must be the same as that at X of V_X, as:

$$\frac{V_X}{V_2} = \frac{R_2}{R_1 + R_2}$$

The current through the feedback resistance must be equal to that from V_1 through R_1. Hence:

$$\frac{V_1 - V_X}{R_1} = \frac{V_X - V_{out}}{R_2}$$

This can be rearranged to give:

$$\frac{V_{out}}{R_2} = V_X\left(\frac{1}{R_2} + \frac{1}{R_1}\right) - \frac{V_1}{R_1}$$

Hence, substituting for V_X using the earlier equation:

$$V_{out} = \frac{R_2}{R_1}(V_2 - V_1)$$

The output is thus proportional to the difference between the two input voltages. Such a circuit might be used with a thermocouple to amplify the difference in e.m.f.s between the hot and cold junctions. Suppose we require there to be an output of 1 mV/°C. With an iron–constantan thermocouple with the cold junction at 0°C, the e.m.f. produced between the hot and cold junctions is about 53 μV/°C. Thus, for a 1°C temperature difference between the junctions, the above equation gives:

$$1 \times 10^{-3} = \frac{R_2}{R_1} \times 53 \times 10^{-6}$$

Hence we must have $R_2/R_1 = 18.9$. Thus if we take for R_1 a resistance of 10 kΩ then R_2 must be 189 kΩ.

The differential amplifier is the simplest form of what is often termed an *instrumentation amplifier*. A more usual form involves three operational amplifiers (Figure 2.81). Such a circuit is available as a single integrated circuit. The first stage involves the amplifiers A_1 and A_2. These amplify the two input signals without any increase in the common mode voltage before amplifier A_3 is used to amplify the differential signal. The differential amplification produced by A_1 and A_2 is $(R_1 + R_2 + R_3)/R_1$ and that produced by A_3 is R_5/R_4 and so the overall amplification is the product of these two amplifications. The overall gain is usually set by varying the value of R_1. Normally the circuit has $R_2 = R_3$, $R_4 = R_6$ and $R_5 = R_7$.

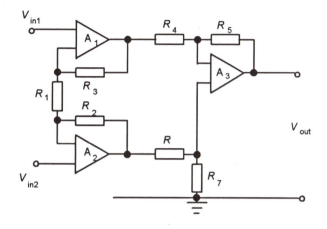

Figure 2.81 *Three op-amp form of an instrumentation amplifier*

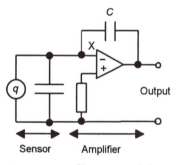

Figure 2.82 *Charge amplifier*

A *charge amplifier* provides an output voltage which is proportional to the charge stored on a device connected to its input terminals and is widely used with sensors employing piezoelectric crystals. Basically, a charge amplifier can be considered to be an op-amp with a capacitor in the feedback path (Figure 2.82). The potential difference across the capacitor is $(v_X - v_{out})$ and since v_X is effectively zero, being the virtual earth, it is $-v_{out}$. The charge on this capacitor is Cv_{out} and thus the current through it is the rate of movement of charge and so $-Cdv_{out}/dt$. But this is the same current as supplied by the sensor and this is dq/dt, where q is the charge on sensor. Thus, $-Cdv_{out}/dt = dq/dt$ and the output voltage is $-(1/C)$ times the charge on the sensor.

2.9.8 Pressure-to-current converter

Control systems generally are electrical and so there is often a need to convert a pressure into an electrical current. Figure 2.83 shows the basic principle of such a converter. The input pressure causes the bellows to extend and so apply a force to displace the end of the pivoted beam. The movement of the beam results in the core being moved in a linear variable differential transformer (LVDT). This gives an electrical output which is amplified. The resulting current is then passed through a solenoid. The current through the solenoid produces a magnetic field which is used to attract the end of the pivoted beam to bring the beam back to its initial horizontal position. When the beam is back in the position, the solenoid current maintaining it in this position is taken as a measure of the pressure input.

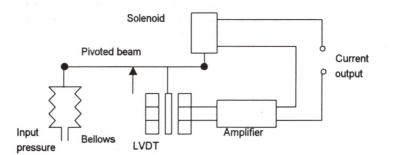

Figure 2.83 *Pressure-to-current converter*

2.10 Signal transmission

Measurement signals often have to be transmitted over quite large distances from the place of measurement to a display unit and/or a process control unit. Methods used for such transmission are:

1 *Analogue voltage transmission*

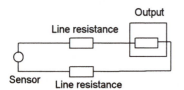

Figure 2.84 *Analogue voltage transmission*

Analogue voltage signals can suffer corruption due to induced noise and the resistance of the connecting cables can result in attenuation of the voltage, the voltage drop across the output being reduced by that across the line resistance (Figure 2.84). Such effects can be reduced by the use of signal amplification and shielding with the connecting cables. However, because of these problems, such signals are not generally used for large distance transmission.

2 *Current loop transmission*

The attenuation which occurs with voltage transmission can be minimised if signals are transmitted as varying current signals. This form of transmission is known as current loop transmission and uses currents in the range 4 mA to 20 mA to represent the levels of the analogue signal. The level of 4 mA, rather than 0 mA, is used to indicate the zero signal level since, otherwise, it would not be possible to distinguish between a zero value signal and a break in the transmission line. Figure 2.85 shows the basic arrangement with the signal from the sensor being converted to a current signal by a voltage-to-current converter, e.g. that shown in Figure 2.79, transmitted and then converted to a voltage signal at the display.

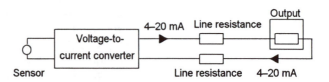

Figure 2.85 *Current loop transmission*

3 *Digital voltage signals*

Digital signals can be transmitted over transmission lines using either serial or parallel communication. With serial communication, the sequence of bits used to describe a value is sent in sequence

Table 2.2 *ASCII code*

ASCII	ASCII	ASCII
A 100 0001	N 100 1110	0 011 0000
B 100 0010	O 100 1111	1 011 0001
C 100 0011	P 101 0000	2 011 0010
D 100 0100	Q 101 0001	3 011 0011
E 100 0101	R 101 0010	4 011 0100
F 100 0110	S 101 0011	5 011 0101
G 100 0111	T 101 0100	6 011 0110
H 100 1000	U 101 0101	7 011 0111
I 100 1001	V 101 0110	8 011 1000
J 100 1010	W 101 0111	9 011 1001
K 100 1011	X 101 1000	
L 100 1100	Y 101 1001	
M 100 1101	Z 101 1010	

Application
The RS-232 is a widely used serial data interface standard in which the electrical signal characteristics, such as voltage levels, the forms of plug and sockets for interconnections, and the interchange circuits are all specified. Because RS-232 is limited to distances less than about 15 m, another standard such as RS-485 tends to be used in many control systems with distances up to about 1200 m being possible.

Light source   Fibre

Figure 2.86 *Optical fibres*

along a single transmission line. With parallel transmission, each of the bits is sent along a separate parallel transmission line. For long-distance communication, serial communication is used.

In order to transfer data, both the sender and receiver have to agree on the meaning of the transmitted binary digital patterns. The most commonly used character set is the American Standard Code for Information Interchange (ASCII), thus using 7 bits to represent each character (Table 2.2). The format used for sending such data has to be standardised. For example, with the RS-232 form of serial transmission, a sequence of 10 bits is used, the first bit being a start of message signal, then seven bits for the data, then a pairy bit to identify whether errors have occurred in the transmission, and finally a stop bit to indicate the end of the message.

Digital signal transmission has a great advantage when compared with analogue transmission in that signal corruption effects can be considerably reduced. With digital transmission, error coding is used to detect whether corruption has occurred. These are bits added to the sequence of bits used to represent the value and are check values which are not likely to tally with the received bits if corruption has occurred, the receiver can then request the message be sent again. For example, the sequence 1010 might be transmitted and corruption result in 1110 being received. In order to detect such errors one form of check uses a parity bit which is added at transmission. With even-parity, the bit is chosen so that the total number of 1s in the transmission, including the parity bit, is an even number. This 1010 would be transmitted as 10100. If it is corrupted and received as 11100 then the parity bit shows there is an error.

3   *Pneumatic transmission*
Pneumatic transmission involves converting the sensor output to a pneumatic pressure in that range 20 to 100 kPa or 20 to 180 kPa. The lower limit gives the zero sensor signal and enables the zero value to be distinguished from a break in the circuit. Such pressure signals can then be transmitted though plastic or metal piping, the distances being limited to about 300 m because of the limitations of speed of response at larger distances.

4   *Fibre-optic transmission*
An optical fibre is a light conductor in the form of a long fibre along which light can be transmitted by internally being reflected of the sides of the fibre (Figure 2.86). The light sources used are LEDs or semiconductor laser diodes. Digital electrical signals are converted into light pulses which travel down the fibre before being detected by a photodiode or phototransistor and converted back into an electrical signal. Fibre optics has the advantages that they are immune to electromagnetic interference, data can be transmitted with much lower losses than with electrical cables, the fibres are smaller and less heavy than copper cables and are more inert in hazardous areas.

2.10.1 Noise

The term *noise* is used, in this context, for the unwanted signals that may be picked up by a measurement system and interfere with the signals being measured. There are two basic types of electrical noise:

1   *Interference*
    This is due to the interaction between external electrical and magnetic fields and the measurement system circuits, e.g. the circuit picking up interference from nearby mains power circuits.

2   *Random noise*
    This is due to the random motion of electrons and other charge carriers in components and is determined by the basic physical properties of components in the system.

The three main types of interference are:

1   *Inductive coupling*
    A changing current in a nearby circuit produces a changing magnetic field which can induce e.m.f.s, as a result of electromagnetic induction, in conductors in the measurement system.

2   *Capacitive coupling*
    Nearby power cables, the earth, and conductors in the measurement system are separated from each other by a dielectric, air. There can thus be capacitance between the power cable and conductors, and between the conductors and earth. These capacitors couple the measurement system conductors to the other systems and thus signals in the other systems affecting the charges on these capacitors can result in interference in the measurement system.

3   *Multiple earths*
    If the measurement system has more than one connection to earth, there may be problems since there may be some difference in potential between the earth points. If this occurs, an interference current may arise in the measurement system.

Methods of reducing interference are:

1   *Twisted pairs of wires*
    This involves the elements of the measurement system being connected by twisted wire pairs (Figure 2.87). A changing magnetic field will induce e.m.f.s in each loop, but because of the twisting the directions of the e.m.f.s will be in one direction for one loop and in the opposite direction for the next loop and so cancel out.

2   *Electrostatic screening*
    Capacitive coupling can be avoided by completely enclosing the system in an earthed metal screen. Problems may occur if there are multiple earths. Coaxial cable gives screening of connections

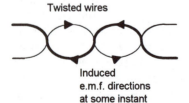

Twisted wires

Induced
e.m.f. directions
at some instant

Figure 2.87   *Twisted pairs*

between elements, however, the cable should only be earthed at one end if multiple earths are to be avoided.

3 *Single earth*
Multiple earthing problems can be avoided if there is only a single earthing point.

4 *Differential amplifiers*
A differential amplifier can be used to amplify the difference between two signals. Thus if both signals contain the same interference, then the output from the amplifier will not have amplified any interference signals.

5 *Filters*
A filter can be selected which transmits the measurement signal but rejects interference signals.

2.11 Smart systems

It is possible to have a measurement system where the sensor and signal processing such as amplification and analogue-to-digital conversion are carried out with separate components. However, these are often available combined in a single integrated sensor circuit. However, often the output from such a system needs further data processing and the resulting combination of sensor, signal processing and a microprocessor to give 'intelligent' processing of sensor inputs results in what is termed a *smart* or *intelligent sensor*. Such a microprocessor-equipped sensor can have the functions to give such functions as compensation for random errors, automatic calculation of measurement accuracy, automatic self-calibration, adjustment for non-linearities to give a linear output and self-diagnosis of faults.

Smart sensors have the ability to 'talk', to 'listen', and to interact with data.

This 'intelligent' processing is most likely to be accomplished by the use of a microprocessor.

In a process plant there are likely to be a large number of smart sensors, each providing information which has to be fed back to a control panel. To avoid using separate cables for each sensor to transmit their data, a *bus* system can be used. A *bus* is a common highway for signals which is used to link components. Thus each sensor would put its information onto the common highway for transmission to the control panel. The *Hart communication protocol* is widely used for such transmissions. This involves the digital signal from a smart sensor being superimposed on an analogue 4–20 mA current loop signal. With this protocol, a 0 is represented by a 2200 Hz frequency and a 1 by a 1200 Hz frequency and these are superimposed on the d.c. signal to give simultaneous digital and analogue transmission. The digital data transfer rate is 1200 bits/s. The arrangement is that a master, such as a display terminal, sends a message with a request for data to a device, the device interprets the request and replies with the data.

2.12 Data presentation elements

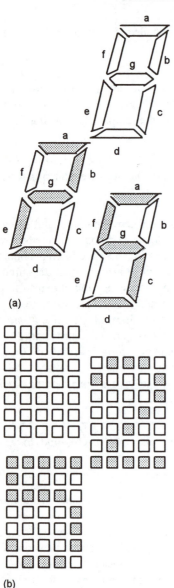

(a)

(b)

Figure 2.88 *(a) 7-segment format with examples of the numbers 2 and 5, (b) 7 by 5 dot-matrix format with examples of the numbers 2 and 5*

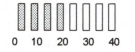

0   10   20   30   40

Figure 2.89 *Bar type of display*

The elements that can be used for the presentation of data can be classified into three groups: indicators, illuminative displays and recorders. *Indicators* and *illuminative displays* give an instant visual indication of the sensed variable while *recorders* record the output signal over a period of time and give automatically a permanent record. A recorder will be the most appropriate choice if the event is high speed or transient and cannot be followed by an observer, or there are large amounts of data, or it is essential to have a record of the data. The following are some brief notes about some of the characteristics of commonly used data presentation methods.

2.12.1 Indicator

The *moving coil meter* is an analogue data presentation element involving a pointer moving across a fixed scale. The basic instrument movement is a d.c. microammeter with shunts, multipliers and rectifiers being used to convert it to other ranges of direct current and alternating current, direct voltage and alternating voltage. With alternating current and voltages, the instrument is restricted to frequencies between about 50 Hz and 10 kHz. The overall accuracy is generally of the order of ±0.1 to ±5%. The time taken for a moving coil meter to reach a steady deflection is typically in the region of a few seconds. The low resistance of the meter can present loading problems.

2.12.2 Illuminative displays

Commonly used illuminative display systems generally use *light-emitting diodes* (*LEDs*) or *liquid crystal displays*. Light-emitting diodes require low voltages and low currents in order to emit light and are cheap. The most commonly used LEDs can give red, yellow or green colours. The term *alphanumeric display* is used for one that can display the letters of the alphabet and numbers. Two basic types of array are used to generate alphanumeric displays, segmented and dot matrix. The 7-segment display (Figure 2.88(a)) is a common form. By illuminating different segments of the display the full range of numbers and a small range of alphabetical characters can be formed. For example, to form a 2 the segments a, b, d, e and g are illuminated. The 5 × 7 dot-matrix (Figure 2.88(b)) display enables a full range of numbers and alphabetical characters to be produced by illuminating different segments in a rectangular array.

LEDs can also be arranged in other formats. For example, they can be arranged in the form of bars, the length of the illuminated bar then being a measure of some quantity (Figure 2.89). A speedometer might use this form of display.

Liquid crystal displays do not produce any light of their own but use reflected light and can be arranged in segments like the LEDs shown above. The crystal segments are on a reflecting plate (Figure 2.90). When an electric field is applied to a crystal, light is no longer passed through it and so there is no reflected light. That segment then appears dark.

Application

An example of an instrument using LED or crystal forms of display is the *digital voltmeter*. This gives its reading in the form of a sequence of digits and is essentially just a sample and hold unit feeding an analogue-to-digital converter with its digital output counted and the count displayed (Figure 2.91). It has a high resistance, of the order of 10 MΩ, and so loading effects are less likely than with the moving coil meter with its lower resistance. The sample and hold unit takes samples and thus the specification of the sample rate with such an instrument gives the time taken for the instrument to process the signal and give a reading. Thus, if the input voltage is changing at a rate which results in significant changes during the sampling time the voltmeter reading can be in error.

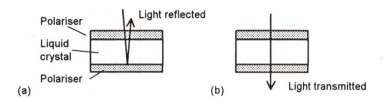

Figure 2.90 *Electric field: (a) applied, (b) not applied*

Figure 2.91 *Principle of the digital voltmeter*

Large screen displays, termed *visual display units* (VDUs), are basically just a form of cathode ray tube which is used to display alphanumeric, graphic and pictorial data. The *cathode ray tube* (Figure 2.92) consists of an electron gun which produces a focused beam of electrons and a deflection system. The beam of electrons in the cathode ray tube is deflected in the Y direction by a potential difference applied between the Y-deflection plates and in the X direction by a potential difference between the X-deflection plates.

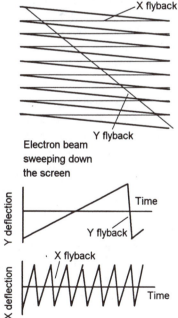

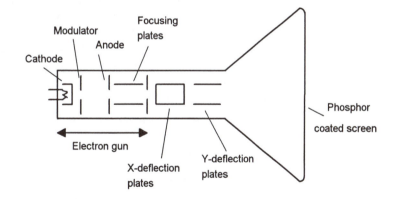

Figure 2.92 *The basic form of the cathode ray tube*

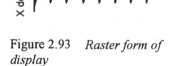

Figure 2.93   *Raster form of display*

With the *raster form* of VDU, saw-tooth signals are applied to both the X- and the Y-deflection plates. Figure 2.93 illustrates the principle. The Y signal causes the beam to move at a constant rate from top to bottom of the screen before flying back to the top again. The X signal causes the beam to move at a constant rate from left to right of the screen before flying back to the left again. The consequence of both these signals is that the beam pursues a zigzag path down the screen before flying back to the top left corner and then resuming its zigzag path down the screen. During its travel down the screen, the electron beam is switched on or off, with the result that a picture or character can be

Figure 2.94 *An interlaced display*

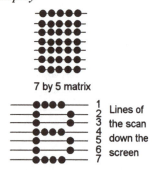

7 by 5 matrix

Figure 2.95 *Character build-up by pixels*

'painted' on the screen. The standard monochrome VDU has a 312-line raster display.

The raster form of display illustrated in Figure 2.93 is said to be *non-interlaced*, the electron beam just following a single zigzag path down the screen. An *interlaced* display has two beams following a zigzag pattern down the screen (Figure 2.94).

The screen of the visual display unit is coated with a large number of phosphor dots, these dots forming the *pixels*. The term pixel is used for the smallest addressable dot on a display device. Character generation is by the selective illuminations of these pixels. Thus for a 7 by 5 matrix, Figure 2.95 shows how characters are built up by the electron beam moving in its zigzag path down the screen.

The input data to the VDU is usually in digital *ASCII (American Standard Code for Information Interchange)* format so that as the electron beam sweeps across the screen it is subject to on-off signals which end up 'painting' the characters on the screen. The ASCII code is a 7-bit code and so can be used to represent $2^7 = 128$ characters. This enables all the standard keyboard characters to be covered, as well as some control functions such as RETURN which is used to indicate the return from the end of a line to the start of the next line (see Table 2.2 for an abridged list of the code).

The *cathode-ray oscilloscope* is a voltage-measuring instrument using a cathode ray tube. The deflection of the electron beam is a measure of the input voltage. A general-purpose instrument can respond to signals up to about 10 MHz while more specialist instruments can respond up to about 1 GHz. Double-beam oscilloscopes enable two separate traces to be observed simultaneously on the screen while storage oscilloscopes enable the trace to remain on the screen after the input signal has ceased, only being removed by a deliberate action of erasure.

2.12.3 Recorders

The *galvanometric type* of chart recorder (Figure 2.96) works on the same principle as the moving coil meter movement. A coil is suspended between two fixed points by a suspension wire and in the magnetic field produced by a permanent magnet. When a current passes through the coil a torque acts on it, causing it to rotate and twist the suspension. The coil rotates to an angle at which the torque is balanced by the opposing torque resulting from the twisting of the suspension. The rotation of the coil results in a pen being moved across a chart.

The *ultraviolet galvanometric chart recorder* works on the same principle but instead of using a pointer moving a pen across the chart, a small mirror is attached to the suspension and reflects a beam of ultraviolet light onto sensitive paper. When the coil rotates, the suspension twists and the mirror rotates and so moves the beam across the chart.

While analogue chart recorders give records in the form of a continuous trace, digital printers give records in the form of numbers, letters or special characters. Such printers are known as *alphanumeric printers*. These can be dot-matrix, ink-jet or laser printers.

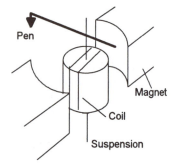

Figure 2.96 *The principle of the galvanometric chart recorder*

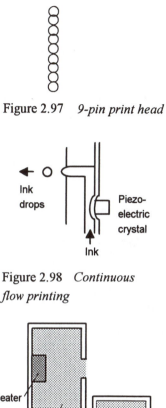

Figure 2.97   *9-pin print head*

Figure 2.98   *Continuous flow printing*

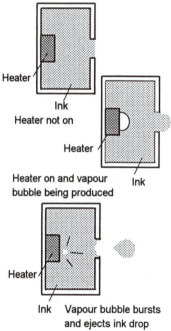

Figure 2.99   *Principle of the on-demand system*

The *dot-matrix printer* has a print head of either 9 or 24 pins in a vertical line (Figure 2.97). Each pin is controlled by an electromagnet which when turned on presses the pin onto the inking ribbon and so gives a small blob of ink on the paper behind the ribbon. The alphanumeric characters are formed by moving the print head across the paper and firing the appropriate pins.

The *ink jet printer* uses conductive ink being forced through a small nozzle to produce a jet of drops of ink. Very small, constant diameter, drops of ink are produced at a constant frequency and so regularly spaced in the jet. With one form a constant stream of ink passes along a tube and is pulsed to form fine drops by a piezoelectric crystal which vibrates at a frequency of about 100 kHz (Figure 2.98). Another form uses a small heater in the print head with vaporised ink in a capillary tube, so producing gas bubbles which push out drops of ink (Figure 2.99). In one form each drop of ink is given a charge as a result of passing through a charging electrode. The charged drops then pass between deflection plates which can deflect the stream of drops in a vertical direction, the amount of deflection depending on the charge on the drops. In another form, a vertical stack of nozzles is used and the jets of each just switched on or off on demand.

The *laser printer* has a photosensitive drum which is coated with a selenium-based material that is light sensitive (Figure 2.100). In the dark the selenium has a high resistance and becomes charged as it passes close to the charging wire. This is a wire which is held at a high voltage and charge leaks off it. A light beam is made to scan along the length of the drum by a small eight-sided mirror which rotates and so reflects the light that it scans across the drum. When light strikes the selenium its resistance drops and it can no longer remain charged. By controlling the brightness of the beam of light, so points on the drum can be discharged or left charged. As the drum passes the toner reservoir, the charged areas attract particles of toner which then stick to its surface to give a pattern of toner on the drum with toner on the areas that have not been exposed to light and no toner on the areas exposed to light. The paper is given a charge as it passes another charging wire, the so-called corona wire, with the result that as the paper passes close to the drum it attracts the toner off the drum. A hot fusing roller is then used to melt the toner particles so that, after passing between rollers, they firmly adhere to the paper.

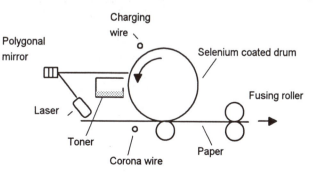

Figure 2.100 *Basic elements of a laser printer*

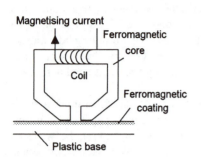

Figure 2.101 *A magnetic recording write/read head*

Magnetic recording involves data being stored in a thin layer of magnetic material as a sequence of regions of different magnetism. The material may be in the form of magnetic tape or disks, the disks being referred to as hard disks or floppy disks. Figure 2.101 shows the basic principle of magnetic recording. The recording current is passed through a coil wrapped round a ferromagnetic core. This core has a small non-magnetic gap. The proximity of the magnetic tape or disk to the gap means that the magnetic flux in the core is readily diverted through it. The magnetic tape of disk consists of a plastic base coated with a ferro-magnetic powder. When magnetic flux passes through this material it becomes permanently magnetised. Thus a magnetic record can be produced of the current through the coil. The patterns of magnetism on a tape or disk can be read by passing it under a similar head to the recording head. The movement of the magnetised material under the head results in magnetic flux passing through the core of the head and electromagnetic induction producing a current through the coil wrapped round the head.

Problems

Questions 1 to 19 have four answer options: A, B, C and D. Choose the correct answer from the answer options.

Questions 1 to 6 relate to the following information. The outputs from sensors can take a variety of forms. These include changes in:

A Displacement
B Resistance
C Voltage
D Capacitance

Select the form of output from above which is concerned with the following sensors:

1 A thermocouple which has an input of a temperature change.
2 A thermistor which has an input of a temperature change.
3 A diaphragm pressure cell which has an input of a change in the pressure difference between its two sides.
4 A LVDT which has an input of a change in displacement.
5 A strain gauge which has an input of a change in length.
6 A Bourdon gauge which has an input of a pressure change.
7 Decide whether each of these statements is True (T) or False (F).

In selecting a temperature sensor for monitoring a rapidly changing temperature, it is vital that the sensor has:
(i) A small thermal capacity.
(ii) High linearity.

Which option BEST describes the two statements?

A (i) T (ii) T
B (i) T (ii) F
C (i) F (ii) T
D (i) F (ii) F

8 A copper–constantan thermocouple is to be used to measure temperatures between 0 and 200°C. The e.m.f. at 0°C is 0 mV, at 100°C it is 4.277 mV and at 200°C it is 9.286 mV. If a linear relationship is assumed between e.m.f. and temperature over the full range, the non-linearity error at 100°C is:

A −3.9°C
B −7.9°C
C +3.9°C
D +7.9°C

9 The change in resistance of an electrical resistance strain gauge with a gauge factor of 2.0 and resistance 50 Ω when subject to a strain of 0.001 is:

A 0.0001 Ω
B 0.001 Ω
C 0.01 Ω
D 0.1 Ω

10 An incremental shaft encoder gives an output which is a direct measure of:

A The absolute angular position of a shaft.
B The change in angular rotation of a shaft.
C The diameter of the shaft.
D The change in diameter of the shaft.

11 A pressure sensor consisting of a diaphragm with strain gauges bonded to its surface has the following information in its specification:

Range: 0 to 1000 kPa
Non-linearity error: ±0.15% of full range
Hysteresis error: ±0.05% of full range

The total error due to non-linearity and hysteresis for a reading of 200 kPa is:

A ±0.2 kPa
B ±0.4 kPa
C ±2 kPa
D ±4 kPa

12 The water level in an open vessel is to be monitored by a diaphragm pressure cell responding to the difference in pressure between that at the base of the vessel and the atmosphere. The range of pressure differences across the diaphragm that the cell will have to respond to if the water level can vary between zero height above the cell measurement point and 1 m above it is (take the acceleration due to gravity to be 9.8 m/s² and the density of the water as 1000 kg/m³):

A 102 Pa
B 102 kPa
C 9800 Pa
D 9800 kPa

13 Decide whether each of these statements is True (T) or False (F).

A float sensor for the determination of the level of water in a container is cylindrical with a mass 1.0 kg, cross-sectional area 20 cm^2 and a length of 0.5 m. It floats vertically in the water and presses upwards against a beam attached to its upward end.
(i)   The maximum force that can act on the beam is 9.8 N.
(ii) The minimum force that can act on the beam is 8.8 N.

Which option BEST describes the two statements?

A (i) T (ii) T
B (i) T (ii) F
C (i) F (ii) T
D (i) F (ii) F

14 A Wheatstone bridge when used as a signal processing element can have an input of a change in resistance and an output of:

A A bigger resistance change.
B A digital signal.
C A voltage.
D A current.

15 The resolution of an analogue-to-digital converter with a word length of 8 bits and an analogue signal input range of 10 V is:

A 39 mV
B 625 mV
C 1.25 V
D 5 V

16 A sensor gives a maximum analogue output of 5 V. The word length is required for an analogue-to-digital converter if there is to be a resolution of 10 mV is:

A 500 bits
B 250 bits
C 9 bits
D 6 bits

17 Decide whether each of these statements is True (T) or False (F).

A cold junction compensator circuit is used with a thermocouple if it has:
(i)   No cold junction.
(ii) A cold junction at the ambient temperature.

Which option BEST describes the two statements?

A (i) T (ii) T
B (i) T (ii) F
C (i) F (ii) T
D (i) F (ii) F

18 Decide whether each of these statements is True (T) or False (F).

A data presentation element which has an input which results in a pointer moving across a scale is an example of:
(i)   An analogue form of display.
(ii) An indicator form of display.

Which option BEST describes the two statements?

A (i) T (ii) T
B (i) T (ii) F
C (i) F (ii) T
D (i) F (ii) F

19 Suggest sensors which could be used in the following situations:
(a) To monitor the rate at which water flows along a pipe and given an electrical signal related to the flow rate.
(b) To monitor the pressure in a pressurised air pipe, giving a visual display of the pressure.
(c) To monitor the displacement of a rod and give a voltage output.
(d) To monitor a rapidly changing temperature.

20 Suggest the type of signal processing element that might be used to:
(a) Transform an input of a resistance change into a voltage.
(b) Transform an input of an analogue voltage into a digital signal.

21 A potentiometer with a uniform resistance per unit length of track is to have a track length of 100 mm and used with the output being measured with an instrument of resistance 10 kΩ. Determine the resistance required of the potentiometer if the maximum error is not to exceed 1% of the full-scale reading.

22 A platinum resistance coil has a resistance at 0°C of 100 Ω. Determine the change in resistance that will occur when the temperature rises to 30°C if the temperature coefficient of resistance is 0.0039 K^{-1}.

23 A platinum resistance thermometer has a resistance of 100.00 Ω at 0°C, 138.50 Ω at 100°C and 175.83 Ω at 200°C. What will be the non-linearity error at 100°C if a linear relationship is assumed between 0°C and 200°C?

24 An electrical resistance strain gauge has a resistance of 120 Ω and a gauge factor of 2.1. What will be the change in resistance of the gauge when it experiences a uniaxial strain of 0.0005 along its length?

25 A capacitive sensor consists of two parallel plates in air, the plates each having an area of 1000 mm^2 and separated by a distance of 0.3

mm in air. Determine the displacement sensitivity of the arrangement if the dielectric constant for air is 1.0006.

26 A capacitive sensor consists of two parallel plates in air, the plates being 50 mm square and separated by a distance of 1 mm. A sheet of dielectric material of thickness 1 mm and 50 mm square can slide between the plates. The dielectric constant of the material is 4 and that for air may be assumed to be 1. Determine the capacitance of the sensor when the sheet has been displaced so that only half of it is between the capacitor plates.

27 A chromel–constantan thermocouple has a cold junction at 20°C. What will be the thermoelectric e.m.f. when the hot junction is at 200°C? Tables give for this thermocouple: 0°C, e.m.f. 0.000 mV; 20°C, e.m.f. 1.192 mV; 200°C, e.m.f. 13.419 mV.

28 An iron–constantan thermocouple has a cold junction at 0°C and is to be used for the measurement of temperatures between 0°C and 400°C. What will be the non-linearity error at 100°C, as a percentage of the full-scale reading, if a linear relationship is assumed over the full range? Tables give for this thermocouple: 0°C, e.m.f. 0.000 mV; 100°C, e.m.f. 5.268 mV; 400°C, e.m.f. 21.846 mV.

29 Show that the output voltage for a Wheatstone bridge with a single strain gauge in one arm of the bridge and the other arms all having the same resistance as that of the unstrained strain gauge is $\frac{1}{4}V_sG\varepsilon$, where V_s is the supply voltage to the bridge, G the gauge factor of the strain gauge and ε the strain acting on the gauge.

30 A Wheatstone bridge has a platinum resistance temperature sensor with a resistance of 120 Ω at 0°C in one arm of the bridge. At this temperature the bridge is balanced with each of the other arms being 120 Ω. What will be the output voltage from the bridge for a change in temperature of 20°C? The supply voltage to the bridge is 6.0 V and the temperature coefficient of resistance of the platinum' is 0.0039 K^{-1}.

31 A diaphragm pressure gauge employs four strain gauges to monitor the displacement of the diaphragm. A differential pressure applied to the diaphragm results in two of the gauges on one side of the diaphragm being subject to a tensile strain of 1.0×10^{-5} and the two on the other side a compressive strain of 1.0×10^{-5}. The gauges have a gauge factor of 2.1 and resistance 120 Ω and are connected in the bridge with the gauges giving subject to the tensile strains in arms 1 and 3 and those subject to compressive strain in arms 2 and 4 (Figure 2.61). If the supply voltage for the bridge is 10 V, what will be the voltage output from the bridge?

32 A thermocouple gives an e.m.f. of 820 μV when the hot junction is at 20°C and the cold junction at 0°C. Explain how a Wheatstone bridge incorporating a metal resistance element can be used to compensate for when the cold junction is at the ambient temperature rather than 0°C and determine the parameters for the bridge if a nickel resistance element is used with a resistance of 10 Ω at 0°C and a temperature coefficient of resistance of 0.0067 K^{-1} and the bridge voltage supply is 2 V.

33 An operational amplifier circuit is required to produce an output that ranges from 0 to −5 V when the input goes from 0 to 100 mV. By what factor is the resistance in the feedback arm greater than that in the input?

34 What will be the feedback resistance required for an inverting amplifier which is to have a voltage gain of 50 and an input resistance of 10 kΩ?

35 What will be the feedback resistance required for a non-inverting amplifier which is to have a voltage gain of 50 and an input resistance of 10 kΩ?

36 A differential amplifier is to have a voltage gain of 100 and input resistances of 1 kΩ. What will be the feedback resistance required?

37 A differential amplifier is to be used to amplify the voltage produced between the two junctions of a thermocouple. The input resistances are to be 1 kΩ. What value of feedback resistance is required if there is to be an output of 10 mV for a temperature difference between the thermocouple junctions of 100°C with a copper–constantan thermocouple. The thermocouple can be assumed to give an output of 43 µV/°C?

38 What is the resolution of an analogue-to-digital converter with a word length of 12 bits?

39 A sensor gives a maximum analogue output of 5 V. What word length is required for an analogue-to-digital converter if there is to be a resolution of 10 mV?

40 What is the voltage resolution of an 8-bit DAC when it has a full-scale input of 5 V?

3 Instrumentation case studies

3.1 Introduction

In designing measurement systems, there are a number of steps that need to be considered:

1   Identification of the *nature of the measurement* required.
    For example, what is the variable to be measured, its nominal value, the range of values that might have to be measured, the accuracy required, the required speed of measurement, the reliability required, the environmental conditions under which the measurement is to be made, etc.

2   Identification of *possible sensors*.
    This means taking into account such factors as their range, accuracy, linearity, speed of response, reliability, maintainability, life, power supply requirements, ruggedness, availability, cost. The sensor needs to fit the requirements arrived at in 1 and also be capable, with suitable signal processing, to give the required output for use in a control system and/or display.

3   Selection of appropriate *signal processing*.
    This element needs to take the output signal from the sensor and modify it in such a way as to enable it to drive the required display or be suitable for control of some device. For example, control applications might require a 4 to 20 mA current to drive an actuator.

4   Identification of the *required display*.
    This means considering the form of display that is required. Is it to be an indicator or recorder? What is the purpose of the display?

This chapter is a consideration of some examples of instrumentation systems and the selection of the elements in such systems.

3.2 Case studies

The following gives case studies of instrumentation systems.

3.2.1 A temperature measurement

Requirement: Determination of temperature of a liquid in the range 0°C to 100°C where only rough accuracy is required. The situation might be the determination of the temperature of the cooling water for a car engine and its display as a pointer moving across a scale marked to indicate safe and unsafe operating temperatures.

Sensor: A solution might be to use a *thermistor* as a sensor. It is cheap and robust. This is the commonly used solution with car engine coolant.

Signal processing: The resistance change of the thermistor has to be converted into a voltage which can then be applied across a meter and so converted to a current through it and hence a reading on the meter related to the temperature. Figure 3.1 shows a possible solution involving a *potential divider circuit* to convert the resistance change into a voltage change. Suppose we use a 4.7 kΩ bead thermistor. This has a resistance of 4.7 kΩ at 25°C, 15.28 kΩ at 0°C and 0.33 kΩ at 100°C. The variable resistor might be 0 to 10 kΩ. It enables the sensitivity of the arrangement to be altered. However, if the variable resistor was set to zero resistance then, without a protective resistor, we could possibly have a large current passed through the thermistor. The *protective resistor* is there to prevent this occurring. The maximum power that the thermistor can withstand is specified as 250 mW. Thus, with a 6 V supply, the variable resistor set to zero resistance, the protective resistance of R, and the thermistor at 100°C, the current I through the thermistor is given by $V = IR$ as $6 = I(0 + R + 330)$, and so:

$$I = \frac{6}{R + 330}$$

The power dissipated by the thermistor is $I^2 \times 330$ and so if we want this to be significantly below the maximum possible, say 100 mW, then we have:

$$0.100 = \left(\frac{6}{R + 330}\right)^2 \times 330$$

Hence R needs to be about 15 Ω.

Display: When the temperature of the thermistor is 0°C its resistance is 15.28 kΩ. If we set the variable resistor as, say, 5 kΩ and the protective resistor as 15 Ω then the voltage output when the supply is 6 V is:

$$\text{output voltage} = \frac{5.015}{15.28 + 5.015} \times 6 = 1.48 \text{ V}$$

When the temperature rises to 100°C the output voltage becomes:

$$\text{output voltage} = \frac{5.015}{0.33 + 5.015} \times 6 = 5.63 \text{ V}$$

Thus, over the required temperature range, the voltage output varies from 1.48 V to 5.63 V. A voltmeter to cover this range could be used to display the output.

In general, calibration of thermometers is by determining their response at temperatures which are specified as the standard values

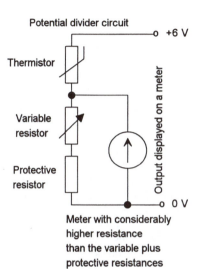

Potential divider circuit

Thermistor

Variable resistor

Protective resistor

Output displayed on a meter

+6 V

0 V

Meter with considerably higher resistance than the variable plus protective resistances

Figure 3.1 *Temperature measurement*

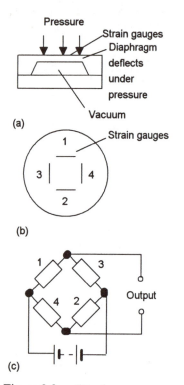

(a)

(b)

(c)

Figure 3.2 *Diaphragm pressure gauge: (a) basic form of sensor, (b) possible arrangement of strain gauges on diaphragm, (c) the Wheatstone bridge signal processor*

for freezing points and boiling points for pure materials. Alternatively, calibration at these and other temperatures within the range can be obtained by comparison with the readings given by a standard thermometer.

3.2.2 An absolute pressure measurement

Requirement: Measurement of the manifold absolute pressure in a car engine as part of the electronic control of engine power.

Sensor: A sensor that is used for such a purpose is a *diaphragm pressure gauge*. Figure 3.2(a) shows the basic form of a diaphragm pressure gauge which is often used in such circumstances. The diaphragm is made of silicon with the strain gauges diffused directly into its surface. Four strain gauges are used and so arranged that when two are in tension the other two are in compression (Figure 3.2(b)).

Signal processing: The four gauges are so connected as to form the arms of a Wheatstone bridge (Figure 3.2(c)). This gives temperature compensation since a change in temperature affects all the gauges equally. Thus the output from the sensor with its signal processing is a voltage which is a measure of the pressure.

Display: If required, the output voltage could be displayed on a meter, possibly following some amplification. Calibration of pressure gauges is usually with a *dead-weight pressure system*. Figure 3.3 shows the basic form of such a system. The calibration pressures are generated by adding standard weights to the piston tray, the pressure then being W/A, where W is the total weight of the piston, tray and standard weights and A the cross-sectional area of the piston. After the weights have been placed on the tray, the screw-driven plunger is screwed up to force the oil to lift the piston–weight assembly. Then the oil is under the pressure given by the piston–weight assembly since it is able to support that weight. By adding weights to the piston tray a gauge can be calibrated over its range.

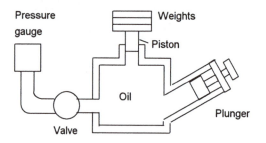

Figure 3.3 *Dead-weight pressure gauge calibration*

3.2.3 Detection of the angular position of a shaft

Requirement: Detection of the angular position of the throttle shaft of a car to give an indication of the throttle opening, and hence the

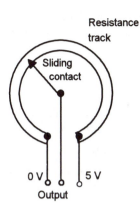

Figure 3.4 *Rotary potentiometer*

driver's power demand on the engine, as part of a car engine management system.

Sensor: A rotary potentiometer (Figure 3.4) is generally used with the potentiometer wiper being rotated over the potentiometer track.

Signal processing: For a 5 V d.c. voltage connected across the potentiometer, with the throttle closed and the engine idling the wiper can be at a position close to the 0 V terminal and so give a small voltage output, typically about 0.5 V. As the throttle is opened, the shaft rotates and the wiper moves over the track so that at wide-open throttle the wiper is nearly at the end of its track and the output voltage has risen to about 4.3 V. The engine management system uses an operational amplifier to compare the output from the potentiometer with a fixed voltage of 0.5 V so that the op-amp gives a high output when the potentiometer output is 0.5 V or lower and a low output when higher. This high–low signal, together with signals from other sensors, is fed to a microprocessor which then can give an output to control the engine idle speed.

3.2.4 Air flow rate determination

Requirement: Measurement of the flow rate for the inflow of air in a car manifold in an electronic controlled engine. A simple, and cheap, measurement of the mass rate of flow of air is required with the output being an electrical signal which can be used for control purposes.

Sensor: One method that is used with cars is the *hot-wire anemometer*. This sensor consists of a platinum wire which is heated by an electrical current passing through it to about 100 to 200°C. The temperature of the wire will depend on the cooling generated by the flow of air over the wire. Thus, since the electrical resistance of the wire will depend on its temperature, the resistance is a measure of the rate of flow of air over the heated wire. Figure 3.5(a) shows the basic form of such a sensor.

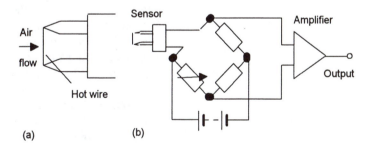

Figure 3.5 *Hot-wire anemometer and its signal processing*

Signal processing: The resistance change is transformed into a voltage change by incorporating the sensor as one of the arms of a Wheatstone bridge (Figure 3.5(b)). The bridge is balanced at zero rate of flow and then the out-of-balance voltage is a measure of the rate of flow. This voltage is fairly small and so has generally to be amplified.

An alternative arrangement which is used is:

Sensor: The vortex flow sensor (see Section 2.5.4) with the vortex frequency measured by means of a semiconductor pressure sensor.

Signal processing: The output of the pressure sensor is typically a frequency of about 100 Hz when the engine is idling and rising to about 2000 Hz at high engine speed. Signal conditioning is used to transform this output into a square-wave signal which varies between 0.6 V and 4.8 V and can then be processed by the engine control unit.

3.2.5 Fluid level monitoring

Requirement: Monitoring the level of a liquid to indicate when the level falls below some critical value.

Sensor: One method would be to use a magnetic float (Figure 3.6) which rises with the liquid level and opens a reed switch (see Section 2.2.8) when the level falls too low.

Signal conditioning: The reed switch is in series with a 39 Ω resistor so that this is switched in parallel with a 1 kΩ resistor by the action of the reed switch. Opening the reed switch thus increases the resistance from about 37 Ω to 1 kΩ. Such a resistance change can be further transformed by signal conditioning to give suitable light on–off signals.

3.2.6 Measurement of relative humidity

Requirement: Direct measurement of relative humidity without the need for using the operator to use tables to convert temperature values to relative humidity. The traditional method of measuring relative humidity involves two thermometers, one with its bulb directly exposed to the air and giving the 'dry temperature' and the other with its bulb covered with muslin which dips into water. The rate of evaporation from the wet muslin depends on the amount of water vapour present in the air; when the air is far from being saturated then the water evaporates quickly, when saturated there is not net evaporation. This rate of evaporation affects the temperature indicated by the thermometer, so giving the 'wet temperature'. Tables are then used to convert these readings into the humidity.

Sensor: Rather than use a 'wet' thermometer element, a capacitive humidity sensor can be used. The sensor (Figure 3.7(a)) consists of an aluminium substrate with its top surface oxidised to form a porous layer of aluminium oxide. On top of the oxide a very thin gold layer is deposited, this being permeable to water vapour. Electrical connections are made to the gold layer and the aluminium

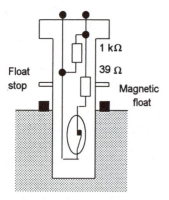

Figure 3.6 *Liquid level monitoring*

substrate, the arrangement being a capacitor with an aluminium oxide dielectric. Water vapour enters the pores of the aluminium oxide and changes its dielectric constant and hence the capacitance of the capacitor. The capacitance thus gives a measure of the amount of water vapour present in the air.

Signal processing: Figure 3.7(b) shows the type of system that might be used with such a sensor. For the capacitive sensor, signal conditioning is used to transform the change in capacitance to a suitable size voltage signal. A temperature sensor is also required since the maximum amount of water vapour that air can hold depends on the temperature and thus to compute the humidity the microprocessor needs to know the temperature, this also requiring signal conditioning to get a signal of the right size. An ADC is then used to convert the signals to digital for processing by a microprocessor system; a microcontroller is likely to be used with an integrated ADC, microprocessor and memory on a single chip, there then being a number of input connections for analogue signals to the system. The microprocessor takes the values for the two inputs and can use a 'look-up' table in its memory to determine the value of the relative humidity. This is then outputted to a digital meter.

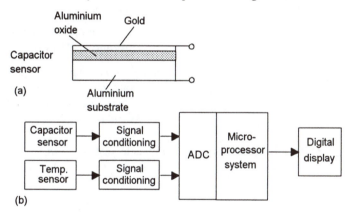

Figure 3.7   *Relative humidity measurement*

3.2.7 Dimension checking

Requirement: A method by which the dimensions of components can be checked.

Sensor: LVDT displacement sensors can be used (see Section 2.2.4).

Signal processing: The e.m.f. induced in a secondary coil by a changing current I in the primary coil is given by $e = M \, di/dt$, where M is the mutual inductance, its value depending on the number of turns on the coils and the magnetic linkage between the coils, thus the material in the core of the coils. Thus, for a sinusoidal input current to the primary coil of an LVDT, alternating e.m.f.s are induced in the two secondary coils A and B. The two outputs are in series so that their difference is the output. Figure 3.8 shows how the size and

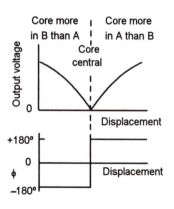

Figure 3.8   *LVDT output*

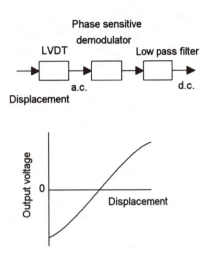

Figure 3.9   *LVDT d.c. output*

phase of the alternating output changes with the displacement of the core. The same amplitude output voltage is produced for two different displacements. To give an output voltage which distinguishes between these two situations, a phase sensitive demodulator, with a low pass filter, is used to convert the output into a d.c. voltage which gives a unique value for each displacement (Figure 3.9).

The coils with simple LVDT sensors and parallel-sided coils exhibit non-linearities as the ferrite core approaches the ends of the coils. This can be corrected for by using stepped windings or, more cheaply, by using a microprocessor system and programming it to compensate for such non-linearities. Such a microprocessor system could also be used to receive the inputs from a number of such LVDT sensors and compare the outputs with the required dimensions of the components and give outputs indicating divergence.

3.2.8 Temperature of a furnace

Requirement: To monitor the temperature of a furnace operating from room temperature up to 500°C with an accuracy of plus or minus a few degrees.

Sensor: A Chromel–aluminium thermocouple (see Table 2.1), with a sensitivity of 41 mV/°C and range −1890 to 1260°C, can be used.

Signal processing: Cold junction compensation and amplification are required (see Section 2.7.5) for the output to be displayed on a meter. Calibration can be against the fixed temperature points or a secondary standard thermometer.

3.3 Data acquisition systems

Frequently, modern instrumentation systems involve the use of PCs and the term data acquisition is used.

Data acquisition is the process by which data from sensors is transformed into electrical signals that are converted into digital form for processing and analysis by a computer.

A *data acquisition system* (DAQ) will thus include:

1   *Sensors*

2   *Signal processing* to get the signal into the required form and size for the data acquisition hardware.

3   *Data acquisition hardware* to collect and convert analogue signals to digital format and digital signals for transfer to the computer. Allow for signals from the computer to control the process.

4   A *computer* which is loaded with *data acquisition software* to enable analysis and display of the data.

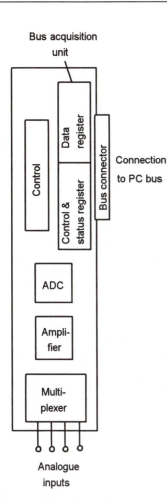

Bus acquisition unit

Connection to PC bus

Analogue inputs

Figure 3.10 *Basic elements of a digital acquisition card*

Personal computer-based systems frequently use *plug-in boards* for the interface between the computer and the sensors. These boards carry the components for the data acquisition so that the computer can store data, carry out data processing, control the data acquisition process and issue signals for use in control systems. The plug-in board is controlled by the computer and the digitised data transferred from the board to the memory of the computer. The processing of the data is then carried out within the computer according to how it is programmed.

Data acquisition boards are available for many computers and offer, on a plug-in board, various combinations of analogue, digital and timing/counting inputs and outputs for interfacing sensors with computers. The board is a printed circuit board that is inserted into an expansion slot in the computer, so connecting it into the computer bus system. Figure 3.10 shows the basic elements of a simple data acquisition board. Analogue inputs from the sensors are accessed through a multiplexer. A multiplexer is essentially an electronic switching device which enables a number of inputs to be sampled and each in turn fed to the rest of the system. An analogue-to-digital converter then converts the amplified sampled signal to a digital signal. The control element can be set up to control the multiplexer so that each of the inputs is sequentially sampled or perhaps samples are taken at regular time intervals or perhaps just a single sensor signal is used. The other main element is the bus interface element which contains two registers, the control and status register and the data register. The term register is used for memory locations within a microprocessor.

Consider now what happens when the computer is programmed to take a sample of the voltage input from a particular sensor. The computer first activates the board by writing a control word into the control and status register. This word indicates the type of operation that is to be carried out. The result of this is that the multiplexer is switched to the appropriate channel for the sensor concerned and the signal from that sensor passed, after amplification at an instrumentation amplifier, to the analogue-to-digital converter. The outcome from the ADC is then passed to the data register and the word in the control and status register changed to indicate that the conversion is complete. Following that signal, the computer sets the address bus to select the address of where the data is stored and then issues the signal on the control bus to read the data and so take it into storage in the computer for processing. In order to avoid the computer having to wait and do nothing while the board carries out the acquisition, an interrupt system can be used whereby the board signals the computer when the acquisition is complete and then the computer microprocessor can interrupt any program it was carrying out and jump from that program to a subroutine which stores its current position in the program, reads the data from the board and then jumps back to its original program at the point where it left it.

With the above system, the acquired data moves from the board to the microprocessor which has to interrupt what it is doing, set the memory address and then direct the data to that address in the memory. The microprocessor thus controls the movement. A faster system is to transfer the acquired data directly from the board to memory without involving the microprocessor. This is termed *direct memory access* (DMA). To carry this out, a DMA controller is connected to the bus. This controller supplies the memory address locations where the data is to be put and so enables the data to be routed direct to memory.

3.3.1 LabVIEW

The software for use with a data-acquisition system can be specially written for the data aquisition hardware concerned, developed using off-the-shelf software that is provided with the hardware used, or developed using a general package which provides a graphical interface for programming, e.g. LabVIEW. LabVIEW is a programming language and set of subroutines developed by National Instruments for data acquisition and scientific programming. It is a graphical programming language which has each subprogram and program structure represented by icons. All the programming is then done graphically by drawing lines between connection points on the various icons. The resulting picture shows the data flow. Figure 3.11 illustrates this for the simple program of getting input A, getting input B, adding A and B and then displaying the result.

Programs that use one or more LabVIEW functions are called *virtual instruments*. Each virtual instrument has a front panel and a diagram. The front panel may be thought of as representing the front panel of an instrument with the instrument controls and display. The diagram contains the actual program and is thus basically of the form shown in Figure 3.11.

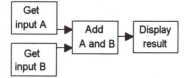

Figure 3.11   *Graphical programming*

3.3.2 Data loggers

Because sensors have often to be located some distance from the computer in which the data processing occurs, instead of running long wires from the sensors to the computer with degradation of the signal en-route, signal conditioning modules are often located close to the sensors and provide digital signal outputs which can then be fed to the computer over a common bus (Figure 3.12) or transferred to it on a portable memory card. Such modules are termed *data loggers*. Loggers are intelligent devices which, as well as data acquisition, are able to be programmed to make decisions based on system conditions.

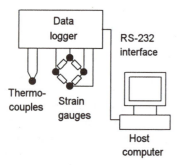

Figure 3.12   *Using a data logger*

3.3.3 Data transfer

Digital data can be communicated between devices by serial or parallel communication. With *serial communication* each word is sent in a sequence of bits along the same wire, with *parallel communication* a number of wires are used and each of the bits of a word is sent simultaneously along their own wires. For anything other than very short distances, parallel communication is too expensive and so serial communication is used. A communication link can be established using one of the serial interface standards such as RS-232 or RS-422. The standard interface most commonly used for parallel communication is the *general purpose instrument bus* (GPIB) IEEE-488. Such standards define the electrical and mechanical details of the communication link.

3.4 Testing In order to ensure that when a measurement system is installed it will correctly function, testing is required. Testing a measurement system on installation can be considered to fall into three stages:

1 *Pre-installation testing*
   This is the testing of each instrument and element for correct calibration and operation prior to it being installed as part of a measurement system.

2 *Cabling and piping testing*
   Cables and/or piping will be used to connect together the elements of measurement systems. The display might, for example, be in a control room. All the instrument cables should be checked for continuity and insulation resistance prior to the connection of any instruments or elements of the system. When the system involves pneumatic lines the testing involves blowing through with clear, dry, air prior to connection and pressure testing to ensure the lines are leak free.

3 *Precommissioning*
   This involves testing that the measurement system installation is complete, all the instrument and other components are in full operational order when interconnected and all control room panels or displays function.

3.4.1 Maintenance

When the system is in operation, maintenance will be required to ensure it continues to operate correctly.

If you own a car or motorcycle, then you will be involved with maintenance and testing, whether you carry out the procedures yourself or a garage does it for you. The function of maintenance is to keep the car/motorcycle in a serviceable condition to that it can continue to carry out its function of transporting you from one place to another. Maintenance is likely to take two forms. One form is *breakdown or corrective maintenance* in which repairs are only made when the car/motorcycle fails to work. Thus breakdown maintenance might be used with the exhaust with it only being replaced when it fails. The other form is *preventative maintenance* which involves anticipating failure and replacing or adjusting items before failure occurs. Preventative maintenance involves inspection and servicing. The inspection is intended to diagnose impending breakdown so that maintenance can prevent it. Servicing is an attempt to reduce the chance of breakdown occurring. Thus preventative maintenance might be used with the regular replacement of engine oil, whether it needs it or not at that time. Inspection of the brakes might be carried out to ascertain when new brake linings are going to be required so that they can be replaced before they wear out. In carrying out maintenance, testing will be involved.

Testing might involve checking the coolant level, brake fluid level, etc. and diagnosis tests in the case of faults to establish where the fault is.

In carrying out the maintenance of a measurement system, the most important aid is the *maintenance manual*. This includes such information as:

1   A description of the measurement system with an explanation of its use.

2   A specification of its performance.

3   Details of the system such as block diagrams illustrating how the elements are linked; photographs, drawings, exploded views, etc. giving the mechanical layout; circuit diagrams of individual elements; etc.

4   Preventative maintenance details, e.g. lubrication, replacement of parts, cleaning of parts and the frequency with which such tasks should be carried out.

5   Breakdown/corrective maintenance details, e.g. methods for dismantling, fault diagnosis procedures, test instruments, test instructions, safety precautions necessary to protect the service staff and precautions to be observed to protect sensitive components. With electrical systems the most commonly used test instruments are multirange meters, cathode ray oscilloscopes and signal generators to provide suitable test signals for injections into the system.

6   Spare parts list.

Maintenance can involve such activities as:

1   Inspection to determine where potential problems might occur or where problems have occurred. This might involve looking to see if wear has occurred or a liquid level is at the right level.

2   Adjustment, e.g. of contacts to prescribed separations or liquid levels to prescribed values.

3   Replacement, e.g. routine replacement of items as part of preventative maintenance and replacement of worn or defective parts.

4   Cleaning as part of preventative maintenance, e.g. of electrical contacts.

5   Calibration. For example, the calibration of an instrument might drift with time and so recalibration becomes necessary.

A record should be maintained of all maintenance activities, i.e. a maintenance activity log. With preventative maintenance this could take the form of a checklist with items being ticked as they are completed. Such a preventative maintenance record is necessary to ensure that such maintenance is carried out at the requisite times. The maintenance log should also include details of any adjustments made, recalibrations necessary or parts replaced. This can help in the diagnosis of future problems.

3.4.2 Common faults

The following are some of the commonly encountered tests and maintenance points that can occur with measurement systems:

1 *Sensors*
   A test is to substitute a sensor with a new one and see what effect this has on the results given by the measurement system. If the results change then it is likely that the original sensor was faulty. If the results do not change then the sensor was not at fault and the fault is elsewhere in the system. Where a sensor is giving incorrect results it might be because it is not correctly mounted or used under the conditions specified by the manufacturer's data sheet. In the case of electrical sensors their output can be directly measured and checked to see if the correct voltages/currents are given. They can also be checked to see whether there is electrical continuity in connecting wires.

2 *Switches and relays*
   A common source of incorrect functioning of mechanical switches and relay is dirt and particles of waste material between the switch contacts. A voltmeter used across a switch should indicate the applied voltage when the contacts are open and very nearly zero when they are closed. If visual inspection of a relay discloses evidence of arcing or contact welding then it might function incorrectly and so should be replaced. If a relay fails to operate checks can be made to see if the correct voltage is across the relay coil and, if the correct voltage is present, that there is electrical continuity within the coil with an ohmmeter.

3 *Hydraulic and pneumatic systems*
   A common cause of faults with hydraulic and pneumatic systems is dirt. Small particles of dirt can damage seals, block orifices, and cause moving parts to jam. Thus, as part of preventative maintenance, filters need to be regularly checked and cleaned. Also oil should be regularly checked and changed. Testing with hydraulic and pneumatic systems can involve the measurement of the pressure at a number of points in a system to check that the pressure is the right value. Leaks in hoses, pipes and fittings are common faults. Also, damage to seals can result in hydraulic and pneumatic cylinders leaking, beyond that which is normal, and result in a drop in system pressure.

Problems

Questions 1 to 6 have four answer options: A, B, C and D. Choose the correct answer from the answer options.

1 Decide whether each of these statements is True (T) or False (F).

As part of the electronic control system for a car engine, a thermistor is to be used to monitor the air temperature. The signal processing circuit that could be used with the thermistor in order to give an electrical voltage output is:
(i) A Wheatstone bridge.
(ii) A potential divider circuit.

A   (i) T   (ii) T
B   (i) T   (ii) F
C   (i) F   (ii) T
D   (i) F   (ii) F

2   Decide whether each of these statements is True (T) or False (F).

It is proposed to monitor the exhaust temperature of a diesel engine by using a thermocouple. In order to give an output which is a few volts in size and which is independent of the temperature of the surrounding air temperature, the output from the thermocouple:
(i) Requires amplification.
(ii) Requires cold junction compensation.

A   (i) T   (ii) T
B   (i) T   (ii) F
C   (i) F   (ii) T
D   (i) F   (ii) F

3   Decide whether each of these statements is True (T) or False (F).

It is proposed to monitor the transmission oil pressure in a car by using a diaphragm pressure gauge with the movement of the diaphragm monitored by means of a linear variable differential transformer (LVDT).
(i)   The output from the LVDT will be a resistance change which can be converted into a voltage change by a Wheatstone bridge.
(ii) The input to the LVDT is the displacement of the diaphragm.

A   (i) T   (ii) T
B   (i) T   (ii) F
C   (i) F   (ii) T
D   (i) F   (ii) F

4   Decide whether each of these statements is True (T) or False (F).

The signal processing needed for a system where the output from a thermocouple is to be fed into a microprocessor/computer includes:
(i) A digital-to-analogue converter.
(ii) Amplification.

A   (i) T   (ii) T
B   (i) T   (ii) F
C   (i) F   (ii) T
D   (i) F   (ii) F

5 Decide whether each of these statements is True (T) or False (F).

The signal processing needed for a system where the output from an optical encoder is to be fed into a microprocessor/computer includes:
(i) An analogue-to-digital converter.
(ii) A resistance-to-voltage converter.

A (i) T (ii) T
B (i) T (ii) F
C (i) F (ii) T
D (i) F (ii) F

6 Decide whether each of these statements is True (T) or False (F).

The term preventative maintenance is used when:
(i) Systems are inspected to diagnose possible points of failure before failure occurs.
(ii) Systems are regularly maintained with such things as items being lubricated and cleaned.

A (i) T (ii) T
B (i) T (ii) F
C (i) F (ii) T
D (i) F (ii) F

7 A driverless vehicle is being designed for operation in a factory where it has to move along prescribed routes transporting materials between machines. Suggest a system that could be used to direct the vehicle along a route.

8 Identify the requirements of the measurement system and hence possible functional elements that could be used to form such a system for the measurement of:
(a) The production of an electrical signal when a package on a conveyor belt has reached a particular position.
(b) The air temperature for an electrical meter intended to indicate when the temperature drops below freezing point.
(c) The production of an electrical signal which can be displayed on a meter and indicate the height of water in a large storage tank.

9 A data acquisition board has a 12-bit analogue-to-digital converter and is set for input signals in the range 0 to 10 V with the amplifier gain at 10. What is the resolution in volts?

10 A load cell has a sensitivity of 25 mV/kN and is connected to a digital acquisition board which has a 0 to 10 V, 12-bit analogue-to-digital converter. What amplifier gain should be used if the cell is to give an output for forces in the range 0.1 kN to 10 kN?

4 Control systems

4.1 Introduction

The term *automation* is used to describe the automatic operation or control of a process. In modern manufacturing there is an ever increasing use of automation, e.g. automatically operating machinery, perhaps in a production line with robots, which can be used to produce components with virtually no human intervention. Also, in appliances around the home and in the office there is an ever increasing use of automation. Automation involves carrying out operations in the required sequence and controlling outputs to required values.

The following are some of the key historical points in the development of automation, the first three being concerned with developments in the organisation of manufacturing which permitted the development of automated production:

1   Modern manufacturing began in England in the 18th century when the use of water wheels and steam engines meant that it became more efficient to organise work to take place in factories, rather than it occurring in the home of a multitude of small workshops. The impetus was thus provided for the development of machinery.

2   The development of powered machinery in the early 1900s meant improved accuracy in the production of components so that instead of making each individual component to fit a particular product, components were fabricated in identical batches with an accuracy which ensured that they could fit any one of a batch of a product. Think of the problem of a nut and bolt if each nut has to be individually made so that it fitted the bolt and the advantages that are gained by the accuracy of manufacturing nuts and bolts being high enough for any of a batch of nuts to fit a bolt.

3   The idea of production lines followed from this with Henry Ford, in 1909, developing them for the production of motor cars. In such a line, the production process is broken up into a sequence of set tasks with the potential for automating tasks and so developing an automated production line.

4   In the 1920s developments occurred in the theoretical principles of control systems and the use of feedback for exercising control. A particular task of concern was the development of control systems to steer ships and aircraft automatically.

5   In the 1940s, during the Second World War, developments occurred in the application of control systems to military tasks, e.g. radar tracking and gun control.

6   The development of the analysis and design of feedback amplifiers, e.g. the paper by Bode in 1945 on Network Analysis and Feedback

Amplifier design, was instrumental in further developing control system theory.

7   Numerical control was developed in 1952 whereby tool positioning was achieved by a sequence of instructions provided by a program of punched paper tape, these directing the motion of the motors driving the axes of the machine tool. There was no feedback of positional data in these early control systems to indicate whether the tool was in the correct position, the system being open-loop control.

8   The invention of the transistor in 1948 in the United States led to the development of integrated circuits, and, in the 1970s, microprocessors and computers which enabled control systems to be developed which were cheap and able to be used to control a wide range of processes. As a consequence, automation has spread to common everyday processes such as the domestic washing machine and the automatic focusing, automatic exposure, camera.

The automatic control of machines and processes is now a vital part of modern industry. The benefits of such control systems include greater consistency of product, reduced operating costs due to improved utilisation of plant and materials and a reduction in manpower, and greater safety for operating personnel.

This chapter is an introduction to the basic idea of a control system and the elements used.

4.2 Control systems   As an illustration of what control systems can do, consider the following:

Control a variable to obtain the required value

1   You set the required temperature for a room by setting to the required temperature the room thermostat of a central heating system. This is an example of a control system with the variable being controlled being the room temperature.

2   In a bottling plant the bottles are automatically filled to the required level. The variable being controlled is the liquid level in a bottle and control is exercised to ensure no difference between the required level and that which occurs.

3   A computer-numerical-control (CNC) machine tool is used to automatically machine a workpiece to the required shape, the control system ensuring that there is no difference between the required dimensions and that which occurs.

4   Packets of biscuits moving along a conveyor belt have their weights checked and those that are below the required minimum weight limit are automatically rejected. Control is being exercised over the weight.

Control the sequence of events

5   A belt is used to feed blanks to a pressing machine. As a blank reaches the machine, the belt is stopped, the blank positioned in the

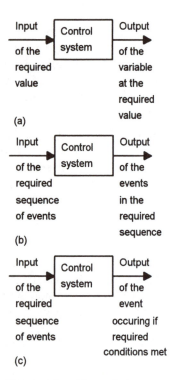

Figure 4.1 *Control systems (a) to control a variable, (b) to control a sequence of events, (c) control whether an event is to be allowed*

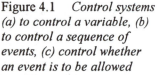

Figure 4.2 *Central heating system*

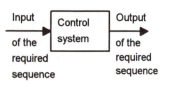

Figure 4.3 *Clothes washing machine system*

machine, the press activated to press the required shape, then the pressed item is ejected from the machine and the entire process repeated. A sequence of operations is being controlled with some operations controlled to occur only if certain conditions are met, e.g. activation of the press if there is a blank in place.

6 You set the dials on the automatic clothes washing machine to indicate that 'whites' are being washed and the machine then goes through the complete washing cycle appropriate to that type of clothing. This is an example of a control system where a controlled sequence of events occurs.

Control whether an event occurs or not

7 The automatic clothes washing machine has a safety lock on the door so that the machine will not operate if the power is off and the door open. The control is of the condition which allows the machine to operate.

A *control system* can be thought of as a system which for some particular input or inputs is used to control its output to some particular value (Figure 4.1(a)), give a particular sequence of events (Figure 4.1(b)) or give an event if certain conditions are met (Figure 4.1(c)).

As an example of the type of control system described by Figure 4.1(a), a central heating control system has as its input the temperature required in the house and as its output the house at that temperature (Figure 4.2). The required temperature is set on the thermostat and the control system adjusts the heating furnace to produce that temperature. The control system is used to control a variable to some set value.

As an example of the type of control system described by Figure 4.1(b), a clothes washing machine has as its input a set of instructions as to the sequence of events required to wash the clothes, e.g. fill the drum with cold water, heat the water to 40°C, tumble the clothes for a period of time, empty the drum of water, etc. The manufacturers of the machine have arranged a number of possible sequences which are selected by pressing a button or rotating a dial to select the appropriate sequence for the type of wash required. Thus the input is the information determining the required sequence and the output is the required sequence of events (Figure 4.3). The control system is used to control a sequence of events.

4.2.1 Open- and closed-loop control

Consider two alternative ways of heating a room to some required temperature. In the first instance there is an electric fire which has a selection switch which allows a 1 kW or a 2 kW heating element to be selected. The decision might be made that to obtain the required temperature it is only necessary to switch on the 1 kW element. The room will heat up and reach a temperature which is determined by the fact the 1 kW

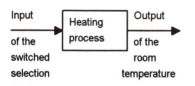

Figure 4.4 *Open-loop control*

element is switched on. The temperature of the room is thus controlled by an initial decision and no further adjustments are made. This is an example of *open-loop control*. Figure 4.4 illustrates this. If there are changes in the conditions, perhaps someone opening a window, no adjustments are made to the heat output from the fire to compensate for the change. There is no information *fed back* to the fire to adjust it and maintain a constant temperature.

Now consider the electric fire heating system with a difference. To obtain the required temperature, a person stands in the room with a thermometer and switches the 1 kW and 2 kW elements on or off, according to the difference between the actual room temperature and the required temperature in order to maintain the temperature of the room at the required temperature. There is a constant comparison of the actual and required temperatures. In this situation there is *feedback*, information being fed back from the output to modify the input to the system. Thus if a window is opened and there is a sudden cold blast of air, the feedback signal changes because the room temperature changes and so is fed back to modify the input to the system. This type of system is called *closed-loop*. The input to the heating process depends on the deviation of the actual temperature fed back from the output of the system from the required temperature initially set, the difference between them being determined by a comparison element. In this example, the person with the thermometer is the comparison element. Figure 4.5 illustrates this type of system.

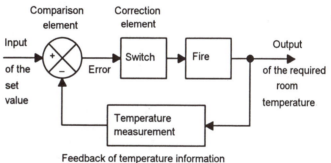

Figure 4.5 *The electric fire closed-loop system*

Note that the comparison element in the closed-loop control system is represented by a circular symbol with a + opposite the set value input and a − opposite the feedback signal. The circle represents a summing unit and what we have is the sum

+ set value − feedback value = error

This difference between the set value and feedback value, the so-called error, is the signal used to control the process. If there is a difference between the signals then the actual output is not the same as the desired output. When the actual output is the same as the required output then

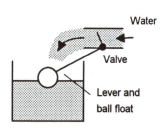

Figure 4.6 *Ball valve in a cistern*

there is zero error. Because the feedback signal is subtracted from the set value signal, the system is said to have *negative feedback*.

Consider an example of a ball valve in a cistern used to control the height of the water (Figure 4.6). The set value for the height of the water in the cistern is determined by the initial setting of the pivot point of the lever and ball float to cut the water off in the valve. When the water level is below that required, the ball moves to a lower level and so the lever opens the valve to allow water into the tank. When the level is at the required level the ball moves the lever to a position which operates the valve to cut off the flow of water into the cistern. Figure 4.7 shows the system when represented as a block diagram.

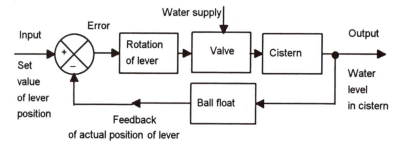

Figure 4.7 *Ball valve used to control water level in a cistern*

In an open-loop control system the output from the system has no effect on the input signal to the plant or process. The output is determined solely by the initial setting. In a closed-loop control system the output does have an effect on the input signal, modifying it to maintain an output signal at the required value.

Open-loop systems have the advantage of being relatively simple and consequently cheap with generally good reliability. However, they are often inaccurate since there is no correction for errors in the output which might result from extraneous disturbances. Closed-loop systems have the advantage of being relatively accurate in matching the actual to the required values. They are, however, more complex and so more costly with a greater chance of breakdown as a consequence of the greater number of components.

4.3 Basic elements

Figure 4.8 shows the basic elements of an open-loop control system. The system has three basic elements: control, correction and the process of which a variable is being controlled.

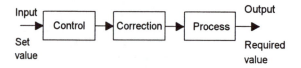

Figure 4.8 *Basic elements of an open-loop control system*

1　*Control element*
This determines the action to be taken as a result of the input of the required value signal to the system.

2　*Correction element*
This has an input from the controller and gives an output of some action designed to change the variable being controlled.

3　*Process*
This is the process of which a variable is being controlled.

There is no changing of the control action to account for any disturbances which change the output variable.

4.3.1 Basic elements of a closed-loop system

Figure 4.9 shows the general form of a basic closed-loop system.

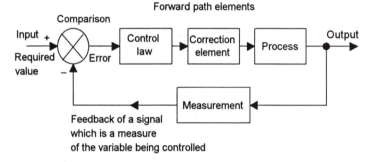

Figure 4.9 *Basic elements of a closed-loop control system*

The following are the functions of the constituent elements:

1　*Comparison element*
This element compares the required value of the variable being controlled with the measured value of what is being achieved and produces an error signal:

error = required value signal – measured actual value signal

Thus if the output is the required value then there is no error and so no signal is fed to initiate control. Only when there is a difference between the required value and the actual values of the variable will there be an error signal and so control action initiated.

2　*Control law implementation element*
The control law element determines what action to take when an error signal is received. The control law used by the element may be just to supply a signal which switches on or off when there is an error, as in a room thermostat, or perhaps a signal which is proportional to the size of the error so that if the error is small a small control signal is produced and if the error is large a large

proportional control signal is produced. Other control laws include *integral mode* where the control signal continues to increase as long as there is an error and *derivative mode* where the control signal is proportional to the rate at which the error is changing.

The term *control unit* or *controller* is often used for the combination of the comparison element, i.e. the error detector, and the control law implementation element. An example of such an element is a differential amplifier which has two inputs, one the set value and one the feedback signal, and any difference between the two is amplified to give the error signal. When there is no difference there is no resulting error signal.

3   *Correction element*
The correction element or, as it is often called, the *final control element*, produces a change in the process which aims to correct or change the controlled condition. The term *actuator* is used for the element of a correction unit that provides the power to carry out the control action.   Examples of correction elements are directional control valves which are used to switch the direction of flow of a fluid and so control the movement of an actuator such as the movement of a piston in a cylinder. Another example is an electric motor where a signal is used to control the speed of rotation of the motor shaft.

4   *Process*
The process is the system in which there is a variable that is being controlled, e.g. it might be a room in a house with the variable of its temperature being controlled.

5   *Measurement element*
The measurement element produces a signal related to the variable condition of the process that is being controlled. For example, it might be a temperature sensor with suitable signal processing.

The following are terms used to describe the various paths through the system taken by signals:

1   *Feedback path*
Feedback is a means whereby a signal related to the actual condition being achieved is fed back to modify the input signal to a process. The feedback is said to be *negative* when the signal which is fed back subtracts from the input value. It is negative feedback that is required to control a system. *Positive feedback* occurs when the signal fed back adds to the input signal.

2   *Forward path*
The term *forward path* is used for the path from the error signal to the output. In Figure 4.9 these forward path elements are the control law element, the correction element and the process element.

The term *process control* is often used to describe the control of variables, e.g. liquid level or the flow of fluids, associated with a process in order to maintain them at some value. Note also that the term *regulator* is sometimes used for a control system for maintaining a plant output constant in the presence of external disturbances. Hence the term *regulator* is sometimes applied to the correction unit.

4.4 Case studies

The following are examples of closed-loop control systems to illustrate how, despite the different forms of control being exercised, the systems all have the same basic structural elements.

4.4.1 Control of the speed of rotation of a motor shaft

Consider the motor system shown in Figure 4.10 for the control of the speed of rotation of the motor shaft and its block diagram representation in Figure 4.11.

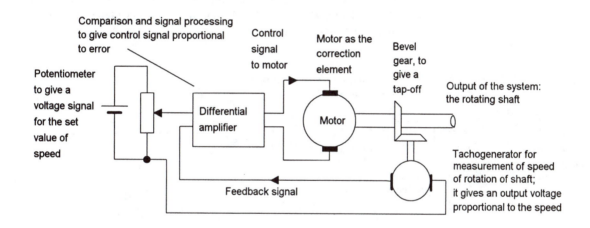

Figure 4.10 *Control of the speed of rotation of a shaft*

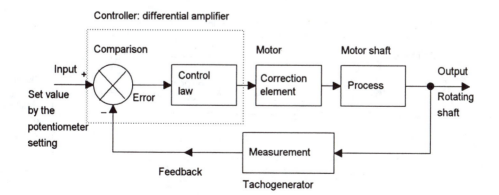

Figure 4.11 *Control of the speed of rotation of a shaft*

The input of the required speed value is by means of the setting of the position of the movable contact of the potentiometer. This determines what voltage is supplied to the comparison element, i.e. the differential amplifier, as indicative of the required speed of rotation. The differential amplifier produces an amplified output which is proportional to the difference between its two inputs. When there is no difference then the output is zero. The differential amplifier is thus used to both compare and implement the control law. The resulting control signal is then fed to a motor which adjusts the speed of the rotating shaft according to the size of the control signal. The speed of the rotating shaft is measured using a tachogenerator, this being connected to the rotating shaft by means of a pair of bevel gears. The signal from the tachogenerator gives the feedback signal which is then fed back to the differential amplifier.

4.4.2 Control of the position of a tool

Figure 4.12 shows a position control system using a belt driven by a stepper motor to control the position of a tool and Figure 4.13 its block diagram representation.

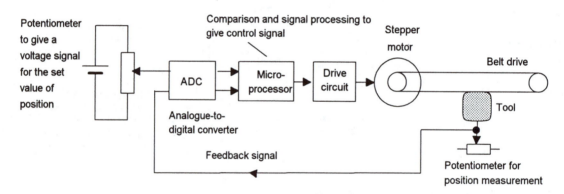

Figure 4.12 *Position control system*

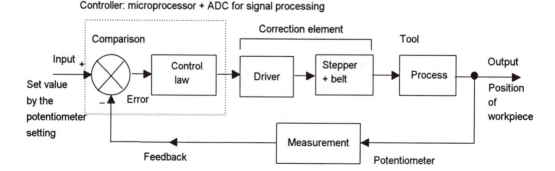

Figure 4.13 *Position control system*

The inputs to the controller are the required position voltage and a voltage giving a measure of the position of the workpiece, this being provided by a potentiometer being used as a position sensor. Because a microprocessor is used as the controller, these signals have to be processed to be digital. The output from the controller is an electrical signal which depends on the error between the required and actual positions and is used, via a drive unit, to operate a stepper motor. Input to the stepper motor causes it to rotate its shaft in steps, so rotating the belt and moving the tool.

4.4.3 Power steering

Control systems are used to not only maintain some variable constant at a required value but also to control a variable so that it follows the changes required by a variable input signal. An example of such a control system is the power steering system used with a car. This comes into operation whenever the resistance to turning the steering wheel exceeds a predetermined amount and enables the movement of the wheels to follow the dictates of the angular motion of the steering wheel. The input to the system is the angular position of the steering wheel. This mechanical signal is scaled down by gearing and has subtracted from it a feedback signal representing the actual position of the wheels. This feedback is via a mechanical linkage. Thus when the steering wheel is rotated and there is a difference between its position and the required position of the wheels, there is an error signal. The error signal is used to operate a hydraulic valve and so provide a hydraulic signal to operate a cylinder. The output from the cylinder is then used, via a linkage, to change the position of the wheels. Figure 4.14 shows a block diagram of the system.

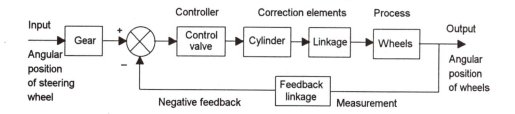

Figure 4.14 *Power assisted steering*

4.4.4 Control of fuel pressure

The modern car involves many control systems. For example, there is the *engine management system* aimed at controlling the amount of fuel injected into each cylinder and the time at which to fire the spark for ignition. Part of such a system is concerned with delivering a constant pressure of fuel to the ignition system. Figure 4.15(a) shows the elements involved in such a system. The fuel from the fuel tank is pumped through a filter to the injectors, the pressure in the fuel line being controlled to be 2.5 bar (2.5 × 0.1 MPa) above the manifold pressure by a regulator valve. Figure 4.15(b) shows the principles of such a valve. It consists of a diaphragm which presses a ball plug into the flow path of the fuel. The diaphragm has the fuel pressure acting on one side of it and on the other side is the manifold pressure and a spring. If the pressure is too high, the diaphragm moves and opens up the return path to the fuel tank for the excess fuel, so adjusting the fuel pressure to bring it back to the required value.

The pressure control system can be considered to be represented by the closed-loop system shown in Figure 4.16. The set value for the pressure is determined by the spring tension. The comparator and control law is given by the diaphragm and spring. The correction element is the ball in its seating and the measurement is given by the diaphragm.

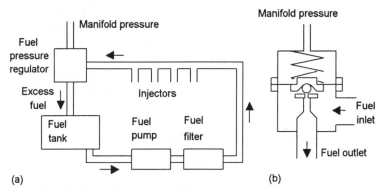

Figure 4.15 *(a) Fuel supply system, (b) fuel pressure regulator*

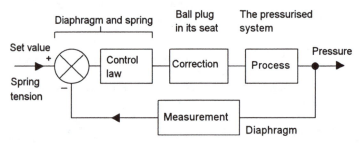

Figure 4.16 *Fuel supply control system*

4.4.5 Antilock brakes

Another example of a control system used with a car is the *antilock brake system (ABS)*. If one or more of the vehicle's wheels lock, i.e. begins to skid, during braking, then braking distance increases, steering control is lost and tyre wear increases. Antilock brakes are designed to eliminate such locking. The system is essentially a control system which adjusts the pressure applied to the brakes so that locking does not occur. This requires continuous monitoring of the wheels and adjustments to the pressure to ensure that, under the conditions prevailing, locking does not occur. Figure 4.17 shows the principles of such a system.

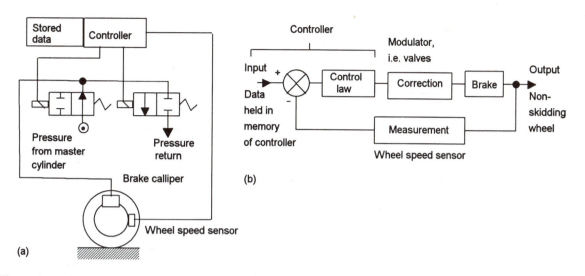

Figure 4.17 *Antilock brakes: (a) schematic diagram, (b) block form of the control system*

The two valves used to control the pressure are solenoid-operated directional-control valves, generally both valves being combined in a component termed the modulator. When the driver presses the brake pedal, a piston moves in a master cylinder and pressurises the hydraulic fluid. This pressure causes the brake calliper to operate and the brakes to be applied. The speed of the wheel is monitored by means of a sensor. When the wheel locks, its speed changes abruptly and so the feedback signal from the sensor changes. This feedback signal is fed into the controller where it is compared with what signal might be expected on the basis of data stored in the controller memory. The controller can then supply output signals which operate the valves and so adjust the pressure applied to the brake.

4.4.6 Thickness control

As an illustration of a *process control system*, Figure 4.18 shows the type of system that might be used to control the *thickness of sheet* produced by rollers, Figure 4.19 showing the block diagram description.

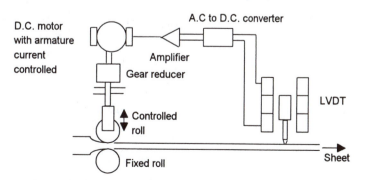

Figure 4.18 *Sheet thickness control system*

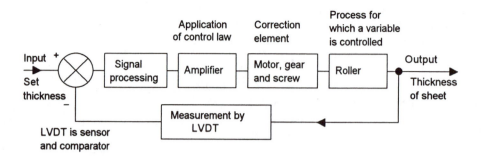

Figure 4.19 *Sheet thickness control system*

The thickness of the sheet is monitored by a sensor such as a linear variable differential transformer (LVDT). The position of the LVDT probe is set so that when the required thickness sheet is produced, there is no output from the LVDT. The LVDT produces an alternating current output, the amplitude of which is proportional to the error. This is then converted to a d.c. error signal which is fed to an amplifier. The amplified signal is then used to control the speed of a d.c. motor, generally being used to vary the armature current. The rotation of the shaft of the motor is likely to be geared down and then used to rotate a screw which alters the position of the upper roll, hence changing the thickness of the sheet produced.

4.4.7 Control of liquid level

Figure 4.20 shows a control system used to control the level of liquid in a tank using a float-operated pneumatic controller, Figure 4.21 showing a block diagram of the system.

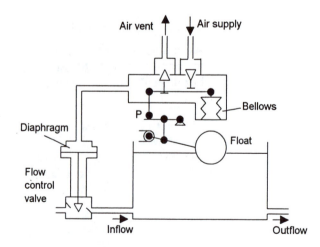

Figure 4.20 *Level control system*

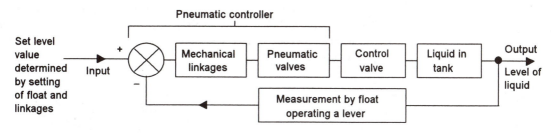

Figure 4.21 *Level control system*

When the level of the liquid in the tank is at the required level and the inflow and outflows are equal, then the controller valves are both closed. If there is a decrease in the outflow of liquid from the tank, the level rises and so the float rises. This causes point P to move upwards. When this happens, the valve connected to the air supply opens and the air pressure in the system increases. This causes a downward movement of the diaphragm in the flow control valve and hence a downward movement of the valve stem and the valve plug. This then results in the inflow of liquid into the tank being reduced. The increase in the air pressure in the controller chamber causes the bellows to become compressed and move that end of the linkage downwards. This eventually closes off the valve so that the flow control valve is held at the new pressure and hence the new flow rate.

If there is an increase in the outflow of liquid from the tank, the level falls and so the float falls. This causes point P to move downwards. When this happens, the valve connected to the vent opens and the air pressure in the system decreases. This causes an upward movement of the diaphragm in the flow control valve and hence an upward movement of the valve stem and the valve plug. This then results in the inflow of liquid into the tank being increased. The bellows react to this new air pressure by moving its end of the linkage, eventually closing off the exhaust and so holding the air pressure at the new value and the flow control valve at its new flow rate setting.

4.4.8 Robot gripper

The term *robot* is used for a machine which is a reprogrammable multi-function manipulator designed to move tools, parts, materials, etc. through variable programmed motions in order to carry out specified tasks. Here just one aspect will be considered, the gripper used by a robot at the end of its arm to grip objects. A common form of gripper is a device which has 'fingers' or 'jaws'. The gripping action then involves these clamping on the object. Figure 4.22 shows one form such a gripper can take if two gripper fingers are to close on a parallel sided object. When the input rod moves towards the fingers they pivot about their pivots and move closer together. When the rod moves outwards, the fingers move further apart. Such motion needs to be controlled so that the grip exerted by the fingers on an object is just sufficient to grip it, too little grip and the object will fall out of the grasp of the gripper and too

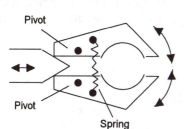

Figure 4.22 *An example of a gripper*

great might result in the object being crushed or otherwise deformed. Thus there needs to be feedback of the forces involved at contact between the gripper and the object. Figure 4.23 shows the type of closed-loop control system involved.

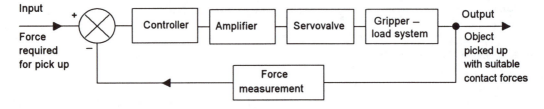

Figure 4.23 *Gripper control system*

The drive system used to operate the gripper can be electrical, pneumatic or hydraulic. Pneumatic drives are very widely used for grippers because they are cheap to install, the system is easily maintained and the air supply is easily linked to the gripper. Where larger loads are involved, hydraulic drives can be used. Sensors that might be used for measurement of the forces involved are piezoelectric sensors or strain gauges. Thus when strain gauges are stuck to the surface of the gripper and forces applied to a gripper, the strain gauges will be subject to strain and give a resistance change related to the forces experienced by the gripper when in contact with the object being picked up.

The robot arm with gripper is also likely to have further control loops to indicate when it is in the right position to grip an object. Thus the gripper might have a control loop to indicate when it is in contact with the object being picked up; the gripper can then be actuated and the force control system can come into operation to control the grasp. The sensor used for such a control loop might be a microswitch which is actuated by a lever, roller or probe coming into contact with the object.

4.4.9 Machine tool control

Machine tool control systems are used to control the position of a tool or workpiece and the operation of the tool during a machining operation. Figure 4.24 shows a block diagram of the basic elements of a closed-loop system involving the continuous monitoring of the movement and position of the work tables on which tools are mounted while the workpiece is being machined.

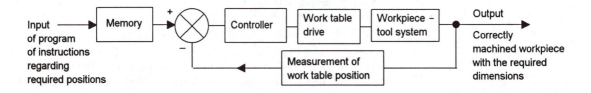

Figure 4.24 *Closed-loop machine tool control system*

The amount and direction of movement required in order to produce the required size and form of workpiece is the input to the system, this being a program of instructions fed into a memory which then supplies the information as required. The sequence of steps involved is then:

1   An input signal is fed from the memory store.

2   The error between this input and the actual movement and position of the work table is the error signal which is used to apply the correction. This may be an electric motor to control the movement of the work table. The work table then moves to reduce the error so that the actual position equals the required position.

3   The next input signal is fed from the memory store.

4   Step 2 is then repeated.

5   The next input signal is fed from the memory store and so on.

4.4.10 Fluid flow control

Figure 4.25 shows the elements of a control system used to control the rate of flow of liquid to some required value, regardless of any fluctuations in supply pressure or back pressure. Figure 4.26 shows a block diagram of the system.

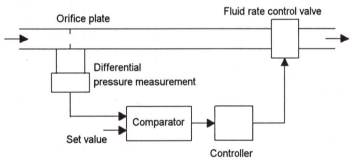

Figure 4.25 *Control system for controlling the rate of flow of a liquid*

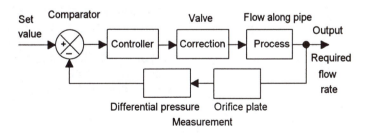

Figure 4.26 *Control system for controlling the rate of flow of a liquid*

4.5 Discrete-time control systems

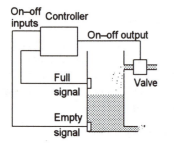

Figure 4.27 *Discrete-event control with the controller switching the valve open when empty signal received and closed when the full signal*

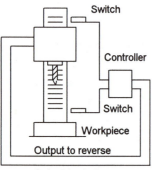

Figure 4.28 *An automatic drill*

Discrete-time control systems are control systems in which one or more inputs can change only at discrete instants of time, i.e. the inputs are effectively on–off signals and so in digital form rather than the analogue form which has been discussed earlier in this chapter. This form of control is often called *sequential control*. It describes control systems involving logic control functions, e.g. is there or is there not a signal from sensor A or perhaps an AND logic system where the issue is whether there is an input from sensor A and an input from sensor B, in order to determine whether to give an output and so switch some device on or off.

> *Discrete-time control systems* are control systems in which one or more inputs can change only at discrete instants of time and involve logic control functions.

As a simple illustration of sequential control, consider the automatic kettle. When the kettle is switched on, the water heats up and continues heating until a sensor indicates that boiling is occurring. The sensor is just giving an on–off signal. The kettle then automatically switches off. The heating element of the kettle is not continuously controlled but only given start and stop signals.

As another example, the filling of a container with water might have a sensor at the bottom which registers when the container is empty and so gives an input to the controller to switch the water flow on and a sensor at the top which registers when the container is full and so gives an input to the controller to switch off the flow of water (Figure 4.27). We have two sensors giving on–off signals in order to obtain the required sequence of events.

As an illustration of the type of control that might be used with a machine consider the system for a drill which is required to automatically drill a hole in a workpiece when it is placed on the work table (Figure 4.28). A switch sensor can be used to detect when the workpiece is on the work table, such a sensor being an on–off sensor. This then gives an on input signal to the controller and it then gives an output signal to actuate a motor to lower the drill head and start drilling. When the drill reaches the full extent of its movement in the workpiece, the drill head triggers another switch sensor. This provides an on input to the controller and it then reverse the direction of rotation of the drill head motor and the drill retracts.

See chapter 7 for a further discussion of sequential control systems and how they can be realised.

Problems

Questions 1 to 4 have four answer options: A, B, C and D. Choose the correct answer from the answer options.

1 Decide whether each of these statements is True (T) or False (F).

An open-loop control system:

(i) Has negative feedback.
(ii) Responds to changes in conditions.

A (i) T (ii) T
B (i) T (ii) F
C (i) F (ii) T
D (i) F (ii) F

2 Decide whether each of these statements is True (T) or False (F).

A closed-loop control system:
(i) Has negative feedback.
(ii) Responds to changes in conditions.

A (i) T (ii) T
B (i) T (ii) F
C (i) F (ii) T
D (i) F (ii) F

3 Decide whether each of these statements is True (T) or False (F).

A closed-loop control system:
(i) Has a measurement system which gives feedback of a signal which is a measure of the variable being controlled.
(ii) Has a controller which has an input based on the difference between the set value and the fed back value for the variable being controlled.

A (i) T (ii) T
B (i) T (ii) F
C (i) F (ii) T
D (i) F (ii) F

4 Decide whether each of these statements is True (T) or False (F).

Negative feedback with a control system is when:
(i) The fed back signal is added to the input signal.
(ii) The fed back signal is a measure of the output value of the control system.

A (i) T (ii) T
B (i) T (ii) F
C (i) F (ii) T
D (i) F (ii) F

5 Suggest the possible form control systems might take for the following situations:
(a) Controlling the thickness of sheet steel produced by a rolling mill.
(b) A conveyor belt is to be used to transport packages from a loading machine to a pick-up area. The control system must start the belt when a package is loaded onto the belt, run the belt until the

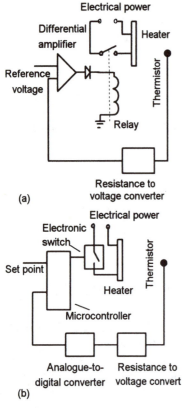

(a)

(b)

Figure 4.29 *Problem 6*

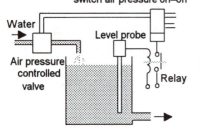

Figure 4.30 *Problem 8*

package arrives at the pick-up area, then stop the belt until the package is removed. Then the entire sequence can start again.

(c) Monitoring breathing in an intensive care unit, sounding an alarm if breathing stops.

(d) Controlling the amount of a chemical supplied by a hopper into sacks.

(e) Controlling the volume of water supplied to a tank in order to maintain a constant level.

(f) Controlling the illumination of the road in front of a car by switching on the lights.

(g) Controlling the temperature in a car by the driver manually selecting the heater controls, switching between them as necessary to obtain the required temperature.

6 Figure 4.29 shows two systems that might be used to control the temperature of a room. Explain how each operates.

7 Draw a block diagram of a domestic central heating system which has the following elements:

(a) A thermostat which has a dial which is set to the required temperature and has an input of the actual temperature in the house and which operates as a switch and gives an output electrical signal which is either on or off.

(b) A solenoid valve which has the input of the electrical signal from the thermostat and controls the flow of oil to the central heating furnace.

(c) A heating furnace where the input is the flow of oil and the output is heat to the rooms via water flowing through radiators in the house.

(d) The rooms in the house where the input is the heat from the radiators and the output is the temperature in the rooms.

8 Figure 4.30 shows a water level control system. Identify the basic functional elements of the system.

9 Draw a block diagram for a negative-feedback system that might be used to control the level of light in a room to a constant value.

10 Draw block diagrams which can be used to present the operation of a toaster when it is (a) an open-loop system, (b) a closed-loop system.

5 Process controllers

5.1 Introduction

Process controllers are control system components which basically have an input of the error signal, i.e. the difference between the required value signal and the feedback signal, and an output of a signal to modify the system output. The ways in which such controllers react to error changes are termed the *control laws*, or more often, the *control modes*. The simplest form of controller is an *on–off device* which switches on some correcting device when there is an error and switches it off when the error ceases. However, such a method of control has limitations and often more sophisticated controllers are used. While there are many ways a controller could be designed to react to an error signal, a form of controller which can give satisfactory control in a wide number of situations is the *three-term* or *PID controller*. The three basic control modes are *proportional* (P), *integral* (I) and *derivative* (D); the three-term controller is a combination of all three modes.

5.1.1 Direct and reverse action

In discussing the elements of a control system, the term *direct action* is used for an element that for an increase in its input gives an increase in its output, e.g. a domestic central heating furnace where an increase in the controlled input to the system results in an increase in temperature. The term *reverse action* is used when an increase in input gives a decrease in output, e.g. an air conditioner where an increase in the energy input to it results in a decrease in temperature.

5.1.2 Dead time

In any control system with feedback the system cannot respond instantly to any change and thus there are delays while the system takes time to accommodate the change. Such delays are referred to as *dead time* or *lags*. For example, in the control of the temperature in a room by means of a central heating system, if a window is suddenly opened and the temperature drops or the thermostat is suddenly set to a new value, a lag will occur before the control system responds, switches on the heater and gets the temperature back to its set value.

Transfer delays are a common event with control systems where flow is concerned, e.g. water flowing along a pipe from point A where the control valve is to point B where the rate of flow is required and monitored. Any change made at some point A will take some time before its affects are apparent at a point B, the *time delay* depending on the

Application

A source of dead time in a control system is the response time of the measurement sensor. Thus, for a system using a temperature sensor, a resistance temperature detector (RTD) has a slower response time than a thermocouple. A thermocouple has typically a response time of about 0.5 s while a RTD is a few seconds.

Application

An example of a transfer delay is where a hopper is loading material onto a conveyor belt moving with a velocity v. The rate at which the material leaves the hopper is controlled by a valve with feedback from a weight sensor. If the weight of deposited material per unit length of belt is monitored a distance L from the hopper discharge point, then there will be a time delay of L/v in the control system.

distance between A and B and the rate of flow between them. The term *distance–velocity lag* is sometimes used to describe such delays.

> Dead time effectively hides a disturbance from the control system until its well into the system and needs to be made as small as possible.

5.1.3 Capacitance

In the level control of the level of water in a tank, an important attribute of the system is its *capacitance*. If we have water leaving the tank and the control signal used to determine the rate of flow of water into the tank, then the greater the surface area of the water in the tank the longer it will take the controlled inflow of water to respond and restore a drop in level. We talk of the system having capacitance and the greater the capacitance the longer it takes to react to changes. If the capacitance were decreased then the system would react quicker to make the changes necessary to restore the required level.

Another example is a domestic heating system controlled by a thermostat. The larger the space being heated the longer it will take the controller to respond and restore a drop in temperature. Again we talk of the capacitance of the system.

> Capacitance has the tendency to dampen out disturbances.

Application

Consider an electrical circuit with a capacitor C and resistor R in series. When a voltage V is switched on, the voltage across the capacitor increases with time until it eventually reaches a steady-state value. The initial rate of change of voltage across the capacitor is V/RC. Thus the bigger the capacitance the smaller the rate of change and so the longer it takes for the capacitor to become fully charged.

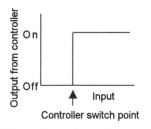

Figure 5.1   *On–off control*

5.2 On–off control

With on–off control, the controller is essentially a switch which is activated by the error signal and supplies just an on–off correcting signal (Figure 5.1). The controller output has just two possible values, equivalent to on and off. For this reason the controller is sometimes termed a *two-step controller*.

An example of such a controller is the bimetallic thermostat (Figure 5.2) used with a simple temperature control system. If the actual temperature is above the required temperature, the bimetallic strip is in an off position and the heater is switched off; if the actual temperature is below the required temperature, the bimetallic strip moves into the on position and the heater is switched on. The controller output is thus just on or off and so the correcting signal on or off.

Because the control action is discontinuous and there are time lags in the system, oscillations, i.e. cycling, of the controlled variable occur about the required condition. Thus, with temperature control using the bimetallic thermostat, when the room temperature drops below the required level there is a significant time before the heater begins to have an effect on the room temperature and, in the meantime, the temperature has fallen even more. When the temperature rises to the required temperature, since time elapses before the control system reacts and switches the heater off and it cools, the room temperature goes beyond

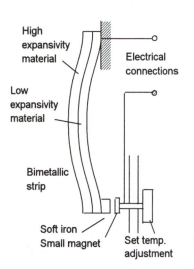

Figure 5.2   *Bimetallic thermostat*

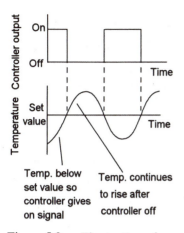

Figure 5.3 *Fluctuation of temperature about set value*

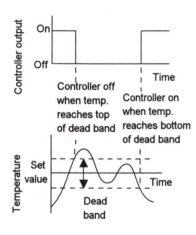

Figure 5.4 *On–off controller with a dead band*

Application
The normal domestic central heating system has an on–off controller, though a modern one is more likely to be an electronic on–off sensor rather than a bimetallic strip.

Another example of on–off control is the control of a car radiator cooling fan using a temperature-sensitive switch.

the required value. The result is that the room temperature oscillates above and below the required temperature (Figure 5.3).

There is also a problem with the simple on–off system in that when the room temperature is hovering about the set value the thermostat might be reacting to very slight changes in temperature and almost continually switching on or off. Thus, when it is at its set value a slight draught might cause it to operate. This problem can be reduced if the heater is switched on at a lower temperature than the one at which it is switched off (Figure 5.4). The term *dead band* or *neutral zone* is used for the values between the on and off values. For example, if the set value on a thermostat is 20°C, then a dead band might mean it switches on when the temperature falls to 19.5° and off when it is 20.5°. The temperature has thus to change by one degree for the controller to switch the heater on or off and thus smaller changes do not cause the thermostat to switch. A large dead band results in large fluctuations of the temperature about the set temperature; a small dead band will result in an increased frequency of switching. The bimetallic thermostat shown in Figure 5.2 has a permanent magnet on one switch contact and a small piece of soft iron on the other; this has the effect of producing a small dead band in that, when the switch is closed, a significant rise in temperature is needed for the bimetallic element to produce sufficient force to separate the contacts.

On–off control is not too bad at maintaining a constant value of the variable when the capacitance of the system is very large, e.g. a central heating system heating a large air volume, and so the effect of changes in, say, a heater output results in slow changes in the variable. It also involves simple devices and so is fairly cheap. On–off control can be implemented by mechanical switches such as bimetallic strips or relays, with more rapid switching being achieved with electronic circuits, e.g. thyristors or transistors used to control the speed of a motor.

> On–off control is simple and inexpensive and is often used where cycling can be reduced to an acceptable level.

5.2.1 Relays

A widely used form of on–off controller is a *relay*. Figure 5.5 shows the basic form of an electromagnetic relay. A small current at a low voltage applied to the solenoid produces a magnetic field and so an electromagnet. When the current is high enough, the electromagnet attracts the armature towards the pole piece and in doing so operates the relay contacts. A much larger current can then be switched on. When the current through the solenoid drops below the critical level, the springy nature of the strip on which the contacts are mounted pushes the armature back to the off position. Thus if the error signal is applied to the relay, it trips on when the error reaches a certain size and can then be used to switch on a much larger current in a correction element such as a heater or a motor.

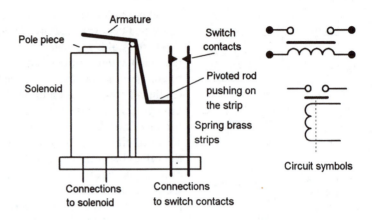

Figure 5.5 *Relay*

5.3 Proportional control

With the on–off method of control, the controller output is either an on or an off signal and so the output is not related to the size of the error. With *proportional control* the size of the controller output is proportional to the size of the error (Figure 5.6), i.e. the controller input. Thus we have: controller output ∝ controller input. We can write this as:

controller output = K_p × controller input

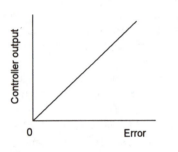

Figure 5.6 *Proportional control*

where K_p is a constant called the gain. This means the correction element of the control system will have an input of a signal which is proportional to the size of the correction required.

The float method of controlling the level of water in a cistern (Figure 5.7) is an example of the use of a proportional controller. The control mode is determined by the lever.

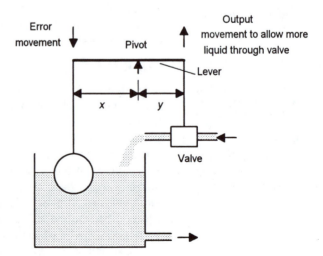

Figure 5.7 *The float-lever proportional controller*

The output is proportional to the error, the gain being x/y. The error signal is the input to the ball end of the lever, the output is the movement of the other end of the lever. Thus, we have output movement = $(x/y) \times$ the error.

Another example of a proportional mode controller is an amplifier which gives an output which is proportional to the size of the input. Figure 5.8 illustrates, for the control of temperature of the outflow of liquid from a tank, the use of a differential amplifier as a comparison element and another amplifier as supplying the proportional control mode.

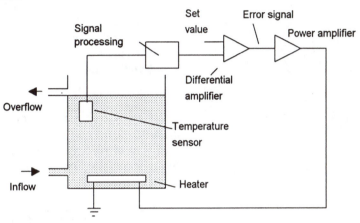

Figure 5.8 *Proportional controller for the control of temperature*

5.3.1 Proportional band

Note that it is customary to express the output of a controller as a percentage of the full range of output that it is capable of passing on to the correction element. Thus, with a valve as a correction element, as in the float operated control of level in Figure 5.9, we might require it to be completely closed when the output from the controller is 0% and fully open when it is 100% (Figure 5.10). Because the controller output is proportional to the error, these percentages correspond to a zero value for the error and the maximum possible error value. When the error is 50% of its maximum value then the controller output will be 50% of its full range.

Some terminology that is used in describing controllers:

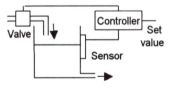

Figure 5.9 *Water level control system*

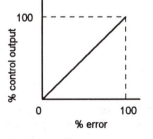

Figure 5.10 *Percentages*

1 *Range*
The range is the two extreme values between which the system operates. A common controller output range is 4 to 20 mA.

2 *Span*
The span is the difference between the two extreme values within which the system operates, e.g. a temperature control system might operate between 0°C and 30°C and so have a span of 30°C.

3   *Absolute deviation*

The set-point is compared to the measured value to give the error signal, this being generally termed the *deviation*. The term *absolute deviation* is used when the deviation is just quoted as the difference between the measured value and the set value, e.g. a temperature control system might operate between 0°C and 30°C and have an absolute deviation of 3°C.

4   *Fractional deviation*

The deviation is often quoted as a *fractional* or *percentage deviation*, this being the absolute deviation as a fraction or percentage of the span. Thus, a temperature control system operating between 0°C and 30°C with an error of 3°C has a percentage deviation of (3/30) × 100 = 10%. When there is no deviation then the percentage deviation is 0% and when the deviation is the maximum permitted by the span it is 100%.

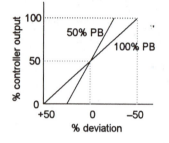

Figure 5.11 *Proportional band*

Generally with process controllers, the proportional gain is described in terms of its *proportional band* (PB). The proportional band is the fractional or percentage *deviation* that will produce a 100% change in controller output (Figure 5.11):

$$\%PB = \frac{\% \text{ deviation}}{\% \text{ change in controller output}} \times 100$$

The 100% controller output might be a signal that fully opens a valve, the 0% being when it fully closes it. A 50% proportional band means that a 50% error will produce a 100% change in controller output; 100% proportional band means that a 100% error will produce a 100% change in controller output.

Since the percentage deviation is the error e as a percentage of the span and the percentage change in the controller output is the controller output y_c as a percentage of the output span of the controller:

$$\%PB = \frac{e}{\text{measurement span}} \times \frac{\text{controller output span}}{y_c} \times 100$$

Since the controller gain K_p is y_c/e:

$$\%PB = \frac{1}{K_p} \frac{\text{controller output span}}{\text{measurement span}} \times 100$$

Example

What is the controller gain of a temperature controller with a 60% PB if its input range is 0°C to 50° and its output is 4 mA to 20 mA?

$$\%PB = \frac{1}{K_p} \frac{\text{controller output span}}{\text{measurement span}} \times 100$$

and so:

$$K_p = \frac{1}{60} \frac{20-4}{50-0} \times 100 = 0.53 \text{ mA/°C}$$

5.3.2 Limitations of proportional control

Proportional controllers have limitations. Consider the above example in Figure 5.8 of the amplifier as the proportional controller. Initially, take the temperature of the liquid in the bath to be at the set value. There is then no error signal and consequently no current to the heating element. Now suppose the temperature of the inflowing liquid changes to a constant lower value (Figure 5.12). The temperature sensor will, after a time lag, indicate a temperature value which differs from the set value. The greater the mass of the liquid in the tank, i.e. the capacitance, the longer will be the time taken for the sensor to react to the change. This is because it will take longer for the colder liquid to have mixed with the liquid in the tank and reached the sensor. The differential amplifier will then give an error signal and the power amplifier a signal to the heater which is proportional to the error. The current to the heater will be proportional to the error, the constant of proportionality being the gain of the amplifier. The higher the gain the larger will be the current to the heater for a particular error and thus the faster the system will respond to the temperature change. As indicated in Figure 5.12, the inflow is constantly at this lower temperature. Thus, when steady state conditions prevail, we always need current passing through the heater. Thus there must be a continuing error signal and so the temperature can never quite be the set value. This error signal which persists under steady state conditions is termed the *steady state error* or the *proportional offset*. The higher the gain of the amplifier the lower will be the steady state error because the system reacts more quickly.

In the above example, we could have obtained the same type of response if, instead of changing the temperature of the input liquid, we had made a sudden change of the set value to a new constant value. There would need to be a steady state error or proportional offset from the original value. We can also obtain steady state errors in the case of a control system which has to, say, give an output of an output shaft rotating at a constant rate, the error results in a velocity-lag.

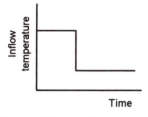

Figure 5.12 *Inflow change*

All proportional control systems have a steady state error. The proportional mode of control tends to be used in processes where the gain K_P can be made large enough to reduce the steady state error to an acceptable level. However, the larger the gain the greater the chance of the system oscillating. The oscillations occur because of time lags in the system, the higher the gain the bigger will be the controlling action for a particular error and so the greater the chance that the system will overshoot the set value and oscillations occur.

Example

A proportional controller has a gain of 4. What will be the percentage steady state error signal required to maintain an output from the controller of 20% when the normal set value is 0%?

With a proportional controller we have:

$$\% \text{ controller output} = \text{gain} \times \% \text{ error}$$

$$20 = 4 \times \% \text{ error}$$

Hence the percentage error is 5%.

Example

For the water level control system described in Figure 5.9, the water level is at the required height when the linear control valve has a flow rate of 5 m³/h and the outflow is 5 m³/h. The controller output is then 50% and operates as a proportional controller with a gain of 10. What will be the controller output and the offset when the outflow changes to 6 m³/h?

Since a controller output of 50% corresponds to 5 m³/h from the linear control valve, then 6 m³/h means that the controller output will need to be 60%. To give a change in output of 60 − 50 = 10% with a controller having a gain of 10 means that the error signal into the controller must be 1%. There is thus an offset of 1%.

5.4 Derivative control

With *derivative control* the change in controller output from the set point value is proportional to the rate of change with time of the error signal, i.e. controller output ∝ rate of change of error. Thus we can write:

$$\text{D controller output} = K_D \times \text{rate of change of error}$$

It is usual to express these controller outputs as a percentage of the full range of output and the error as a percentage of full range. K_D is the constant of proportionality and is commonly referred to as the *derivative time* since it has units of time.

Figure 5.13 illustrates the type of response that occurs when there is a steadily increasing error signal. Because the rate of change of the error with time is constant, the derivative controller gives a constant controller output signal to the correction element. With derivative control, as soon as the error signal begins to change there can be quite a large controller output since it is proportional to the rate of change of the error signal and not its value. Thus with this form of control there can be rapid corrective responses to error signals that occur.

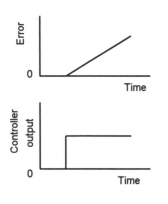

Figure 5.13 *Derivative control*

5.4.1 PD control

Derivative controllers give responses to changing error signals but do not, however, respond to constant error signals, since with a constant error the rate of change of error with time is zero. Because of this, derivative control D is combined with proportional control P. Then:

PD controller output = K_P × error + K_D × rate of change of error with time

Figure 5.14 shows how, with proportional plus derivative control, the controller output can vary when there is a constantly changing error. There is an initial quick change in controller output because of the derivative action followed by the gradual change due to proportional action. This form of control can thus deal with fast process changes better than just proportional control alone. It still, like proportional control alone, needs a steady state error in order to cope with a constant change in input conditions or a change in the set value.

The above equation for PD control is sometimes written as:

$$\text{PD controller output} = K_P\left(\text{error} + \frac{K_D}{K_P}\text{rate of change of error}\right)$$

K_D/K_P is called the *derivative action time* T_D and so:

PD controller output = K_P(error + T_D × rate of change of error)

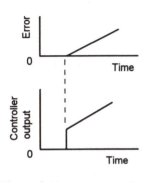

Figure 5.14 *PD control*

PD control can deal with fast process changes better than just proportional control alone. It still needs a steady state error in order to cope with a constant change in input conditions or a change in the set value.

Example

A derivative controller has a derivative constant K_D of 0.4 s. What will be the controller output when the error (a) changes at 2%/s, (b) is constant at 4%?

(a) Using the equation given above, i.e. controller output = K_D × rate of change of error, then we have:

controller output = 0.4 × 2 = 0.8%

This is a constant output.
(b) With a constant error there is no change of error with time and thus the controller output is zero.

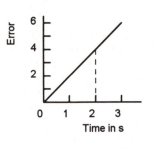

Figure 5.15 *Example*

Example

What will the controller output be for a proportional plus derivative controller (a) initially and (b) 2 s after the error begins to change from the zero error at the rate of 2%/s (Figure 5.15). The controller has $K_P = 4$ and $T_D = 0.4$ s.

(a) Initially the error is zero and so there is no controller output due to proportional action. There will, however, be an output due to derivative action since the error is changing at 2%/s. Since the output of the controller, even when giving a response due to derivative action alone, is multiplied by the proportional gain, we have:

controller output = T_D × rate of change of error

$$= 4 \times 0.4 \times 2\% = 3.2\%$$

(b) Because the rate of change is constant, after 2 s the error will have become 4%. Hence, then the controller output due to the proportional mode will be given by:

controller output = K_P × error

and so that part of the output is:

controller output = 4 × 4% = 16%

The error is still changing and so there will still be an output due to the derivative mode. This will be given by:

controller output = K_P T_D × rate of change of error

and so:

controller output = 4 × 0.4 × 2% = 3.2%

Hence the total controller output due to both modes is the sum of these two outputs and 16% + 3.2% = 19.2%.

5.5 Integral control

Integral control is the control mode where the controller output is proportional to the integral of the error with respect to time, i.e.:

controller output ∝ integral of error with time

and so we can write:

I controller output = K_I × integral of error with time

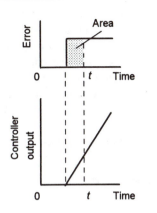

Figure 5.16 *Integral control*

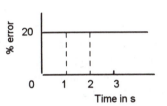

Figure 5.17 *Example*

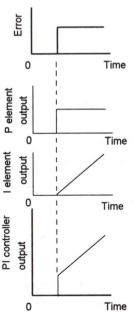

Figure 5.18 *PI control*

where K_I is the constant of proportionality and, when the controller output is expressed as a percentage and the error as a percentage, has units of s⁻¹.

To illustrate what is meant by the integral of the error with respect to time, consider a situation where the error varies with time in the way shown in Figure 5.16. The value of the integral at some time t is the area under the graph between $t = 0$ and t. Thus we have:

controller output ∝ area under the error graph between $t = 0$ and t

Thus as t increases, the area increases and so the controller output increases. Since, in this example, the area is proportional to t then the controller output is proportional to t and so increases at a constant rate. Note that this gives an alternative way of describing integral control as:

rate of change of controller output ∝ error

A constant error gives a constant rate of change of controller output.

Example

An integral controller has a value of K_I of 0.10 s⁻¹. What will be the output after times of (a) 1 s, (b) 2 s, if there is a sudden change to a constant error of 20%, as illustrated in Figure 5.17?

We can use the equation:

controller output = K_I × integral of error with time

(a) The area under the graph between a time of 0 and 1 s is 20%s. Thus the controller output is $0.10 \times 20 = 2\%$.
(b) The area under the graph between a time of 0 and 2 s is 40%s. Thus the controller output is $0.10 \times 40 = 4\%$.

5.5.1 PI control

The integral mode I of control is not usually used alone but generally in conjunction with the proportional mode P. When integral action is added to a proportional control system the controller output is given by:

PI controller output = $K_P$error + K_I integral of error with time

where K_P is the proportional control constant and K_I the integral control constant.

Figure 5.18 shows how a system with PI control reacts when there is an abrupt change to a constant error. The error gives rise to a proportional controller output which remains constant since the error does not change. There is then superimposed on this a steadily increasing controller output due to the integral action.

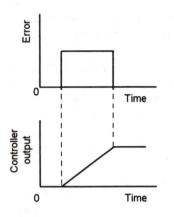

Figure 5.19 *Controller output when error becomes zero*

The combination of integral mode with proportional mode has one great advantage over the proportional mode alone: the steady state error can be eliminated. This is because the integral part of the control can provide a controller output even when the error is zero. The controller output is the sum of the area all the way back to time $t = 0$ and thus even when the error has become zero, the controller will give an output due to previous errors and can be used to maintain that condition. Figure 5.19 illustrates this.

The above equation for PI controller output is often written as:

$$\text{PI controller output} = K_P\left(\text{error} + \frac{K_I}{K_P}\text{integral of error}\right)$$

K_P/K_I is called the *integral action time* T_I and so:

$$\text{PI controller output} = K_P\left(\text{error} + \frac{1}{T_I}\text{integral of error}\right)$$

Because of the lack of a steady state error, a PI controller can be used where there are large changes in the process variable. However, because the integration part of the control takes time, the changes must be relatively slow to prevent oscillations.

5.6 PID control

Combining all three modes of control (proportional, integral and derivative) enables a controller to be produced which has no steady state error and reduces the tendency for oscillations. Such a controller is known as a *three-mode controller* or *PID controller*. The equation describing its action is:

$$\text{controller output} = K_P \times \text{error} + K_I \times \text{integral of error} + K_D \times \text{rate of change of error}$$

where K_P is the proportionality constant, K_I the integral constant and K_D the derivative constant. The above equation can be written as:

$$\text{PID controller output} = K_P\left(\text{error} + \frac{1}{T_I}\text{integral of error} + T_D\text{rate of change of error}\right)$$

A PID controller can be considered to be a proportional controller which has integral control to eliminate the offset error and derivative control to reduce time lags.

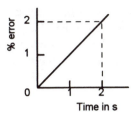

Figure 5.20 *Example*

Example

Determine the controller output of a three-mode controller having K_P as 4, T_I as 0.2 s, T_D as 0.5 s at time (a) $t = 0$ and (b) $t = 2$ s when there is an error input which starts at 0 at time $t = 0$ and increases at 1%/s (Figure 5.20).

(a) Using the equation:

controller output = K_P(error + (1/T_I) × integral of error + T_D × rate of change of error)

we have for time $t = 0$ an error of 0, a rate of change of error with time of 1 s^{-1}, and an area between this value of t and $t = 0$ of 0. Thus:

controller output = $4(0 + 0 + 0.5 \times 1) = 2.0\%$

(b) When $t = 2$ s, the error has become 1%, the rate of change of the error with time is 1%/s and the area under between $t = 2$ and $t = 0$ is 1%s. Thus:

controller output = $4(1 + (1/0.2) \times 1 + 0.5 \times 1) = 26\%$

5.6.1 PID process controller

Figure 5.21 shows the basic elements that tend to figure on the front face of a typical three-term process controller. The controller can be operated in three modes by pressing the relevant key:

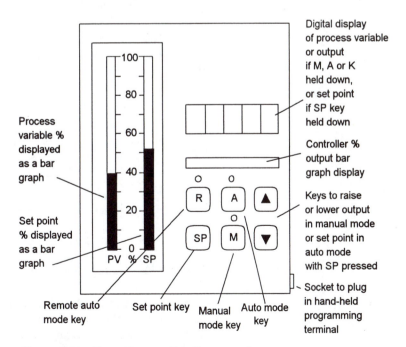

Figure 5.21 *Typical controller front panel*

1 *Manual mode*

The operator directly controls the operation and can increase or decrease the controller output signal by holding down the M key and pressing the up or down keys. A LED above the key shows when this mode has been selected. The output is shown on the digital display and on the bar graph display.

2 *Automatic mode*

The controller operates as a three-term controller with a set point specified by the operator. A LED above the key shows when this mode has been selected. The digital display shows the set point value when the SP key is depressed and the value changed by pressing the up or down keys. The digital display shows the set point value in units such as °C, the unit previously having been set up to give such values in the set up procedure. The set point is also displayed on the vertical bar graph as a percentage.

3 *Remote automatic mode*

The controller is operated in a similar manner to the automatic mode but with the set point established by an external signal. A LED above the key shows when this mode has been selected.

When no key is depressed, the process variable is shown on the digital display and on the vertical bar graph.

The procedure adopted when using the controller is to initially set the mode as manual. The set point is then set to the required value and the controller output manually adjusted until the deviation is zero and the plant thus operating at the required set point. Figure 5.22 shows the block diagram of the control system when it is being operated in manual mode and the operator adjusting the controller output by adding in a signal. The controller can then be switched to automatic control. When this happens, the manual input signal is held constant at the value that was set in manual mode.

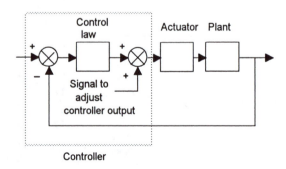

Figure 5.22 *Control system in manual mode*

Switching back to manual mode from automatic mode to make some adjustment and then back to automatic mode gain can present a problem. There can be a sudden change in controller output on the transition from manual to automatic modes, this being termed a 'bump' in the plant operation. This arises because of the integral element in the controller which bases its error on the duration of the error signal input to the controller and does not take account of any manually introduced signals. Thus, changing the manually introduced signal can lead to the output from the controller in the automatic mode not being the same as that in the manual mode. To avoid this 'bump' and give a *bumpless transfer*, modern controllers automatically adjust the contribution to the control law from the integral element.

Modern process controllers are likely to be microprocessor-based controllers, though operating as though they are conventional analogue controllers. They can be programmed by connecting a hand-held terminal to them so that the parameters of the PID controller can be set.

5.7 Tuning

The design of a controller for a particular situation involves selecting the control modes to be used and the control mode settings. This means determining whether proportional control, proportional plus derivative, proportional plus integral or proportional plus integral plus derivative is to be used and selecting the values of K_P, K_I and K_D. These determine how the system reacts to a disturbance or a change in the set value, how fast it will respond to changes, how long it will take to settle down after a disturbance or change to the set value, and whether there will be a steady state error.

Figure 5.23 illustrates the types of response that can occur with the different modes of control when subject to a step input, i.e. a sudden change to a different constant set value or perhaps a sudden constant disturbance. Proportional control gives a fast response with oscillations which die away to leave a steady state error. Proportional plus integral control has no steady state error but is likely to show more oscillations before settling down. Proportional but integral plus derivative control has also no steady state error, because of the integral element, and is likely to show less oscillations than the proportional plus integral control. The inclusion of derivative control reduces the oscillations.

The term *tuning* is used to describe methods used to select the best controller setting to obtain a particular form of performance, e.g. the component being where an error signal results in the controlled variable oscillating about the required value with an oscillation which decays quite rapidly so that each successive amplitude is a quarter of the preceding one. The following is a description of some of the methods used for tuning. Two methods that are widely used, are the process reaction method and the ultimate cycle method, both by *Ziegler and Nichols*.

5.7.1 Process reaction tuning method

This method uses certain measurements made from testing the system with the control loop open so that no control action occurs. Generally the break is made between the controller and the correction unit (Figure 5.24). A test input signal is then applied to the correction unit and the response of the controlled variable determined.

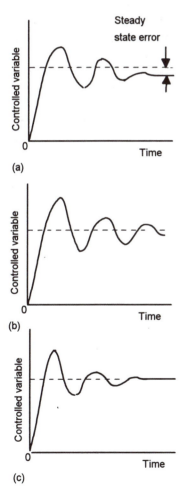

Figure 5.23 *Responses to (a) P, (b) PI, (c) PID control*

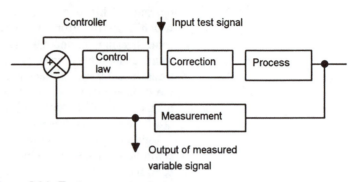

Figure 5.24 *Test arrangement*

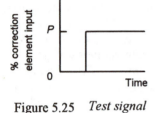

Figure 5.25   *Test signal*

The test signal is a step signal with a step size expressed as the percentage change P in the correction unit (Figure 5.25). The output response of the controlled variable, as a percentage of the full-scale range, to such an input is monitored and a graph (Figure 5.26) of the variable plotted against time. This graph is called the *process reaction curve*. A tangent is drawn to give the maximum gradient of the graph. The time between the start of the test signal and the point at which this tangent intersects the graph time axis is termed the lag L. If the value of the maximum gradient is M, expressed as the percentage change of the set value of the variable per minute, Table 5.1 shows the criteria given by Ziegler and Nichols to determine the controller settings. The basis behind these criteria is to give a closed-loop response for the system which exhibits a *quarter amplitude decay* (Figure 5.27), i.e. the amplitude of the response of the system shows oscillations which decay with time so that the amplitude decreases by a quarter on each oscillation.

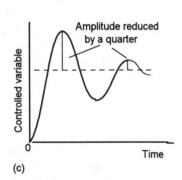

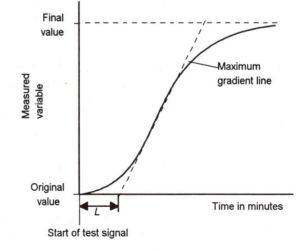

Figure 5.27   *Quarter amplitude decay*

Figure 5.26 *The process reaction graph*

Table 5.1 *Settings from the process reaction curve method*

Type of controller	K_p	T_i	T_d
P	*P/RL*		
PI	0.9*P/RL*	3.3*L*	
PID	1.2*P/RL*	2*L*	0.5*L*

Example

Determine the settings of K_P, K_I and K_D required for a three-mode controller which gave a process reaction curve shown in Figure 5.28 when the test signal was a 10% change in the control valve position.

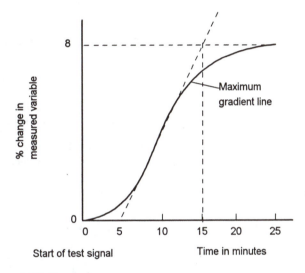

Figure 5.28 *Example*

Drawing a tangent to the maximum gradient part of the graph gives a lag L of 5 minutes and a gradient M of 8/10 = 0.8%/min. Hence

$$K_P = \frac{1.2P}{ML} = \frac{1.2 \times 10}{0.8 \times 5} = 3$$

$$T_1 = 2L = 2 \times 5 = 10 \text{ min}$$

$$T_D = 0.5L = 0.5 \times 5 = 2.5 \text{ min}$$

Since $K_I = K_P/T_I$, then $K_I = 0.3$ min^{-1}. Since $K_D = K_P T_D$, then $K_D = 7.5$ min.

5.7.2 Ultimate cycle tuning method

With this method, the integral and derivative actions are first reduced to their least effective values. The proportional constant K_P is then set low and gradually increased until oscillations in the controlled variable start to occur. The critical value of the proportional constant K_{Pc} at which this occurs is noted and the periodic time of the oscillations T_c measured. The procedure is thus:

1 Set the controller to manual operation and the plant near to its normal operating conditions.

2 Turn off all control modes but proportional.

3 Set K_p to a low value, i.e. the proportional band to a wide value.

4 Switch the controller to automatic mode and then introduce a small set-point change, e.g. 5 to 10%.

5 Observe the response.

6 Set K_p to a slightly higher value, i.e. make the proportional band narrower.

7 Introduce a small set-point change, e.g. 5 to 10%.

8 Observe the response.

9 Keep on repeating 6, 7 and 8, until the response shows sustained oscillations which neither grow nor decay. Note the value of K_p giving this condition (K_{pu}) and the period (T_u) of the oscillation.

10 The Ziegler and Nichols recommended criteria for controller settings for a system to have quarter amplitude decay are given by Table 5.2. For a PID system with some overshoot or with no overshoot, the criteria are given by Table 5.3.

Table 5.2 *Settings for the ultimate cycle method for quarter amplitude decay*

Type of controller	K_p	T_i	T_d
P	$0.5K_{pu}$		
PI	$0.45K_{pu}$	$T_u/1.2$	
PID	$0.6K_{pu}$	$T_u/2$	$T_u/8$

Table 5.3 *Settings for the ultimate cycle method for PI control*

Type of controller	K_p	T_i	T_d
PID, ¼ decay	$0.6K_{pu}$	$T_u/2$	$T_u/8$
PID, some overshoot	$0.33K_{pu}$	$T_u/2$	$T_u/3$
PID, no overshoot	$0.2K_{pu}$	$T_u/3$	$T_u/2$

Example

When tuning a three-mode control system by the ultimate cycle method it was found that, with derivative and integral control switched off, oscillations begin when the proportional gain is increased to 3.3. The oscillations have a periodic time of 500 s. What are the suitable values of K_P, K_I and K_D for quarter amplitude decay?

Using the equations given above:

$$K_P = 0.6K_{Pc} = 0.6 \times 3.3 = 1.98$$

$$T_I = 500/2 = 250 \text{ s and so } K_I = K_P/T_I = 1.98/200 = 0.0099 \text{ s}^{-1}$$

$$T_D = 500/8 = 62.5 \text{ s and so } K_D = K_PT_D = 1.98 \times 62.5 = 123.75 \text{ s}$$

5.7.3 Quarter amplitude decay

A variation on the ultimate cycle method involves no adjusting the control system for sustained oscillations but for oscillations which have a quarter amplitude decay. The controller is set to proportional only. Then, with a step input to the control system, the output is monitored and the amplitude decay determined. If the amplitude decay is less than a quarter the proportional gain is increased, if less than a quarter it is decreased. The step input is then repeated and the amplitude decay again determined. By a method of trial and error, the test input is repeated until a quarter wave amplitude decay is obtained. We then have the value for the proportional gain constant. The integral time constant is then set to be $T/1.5$ and the derivative time constant to $T/6$.

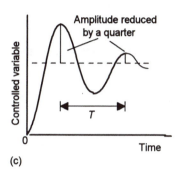

Figure 5.29 *Quarter amplitude decay*

5.8 Digital systems

The term *direct digital control* is used to describe the use of digital computers in the control system to calculate the control signal that is applied to the actuators to control the plant. Such a system is of the form shown in Figure 5.30. At each sample instant the computer samples, via the analogue-to-digital converter (ADC), the plant output to produce the sampled output value. This, together with the discrete input value is then processed by the computer according to the required control law to give the required correction signal which is then sent via the digital-to-analogue converter (DAC) to provide the correcting action to the plant to give the required control. Direct digital control laws are computer programs that take the set value and feedback signals and operate on them to give the output signal to the actuator. The program might thus be designed to implement PID control.

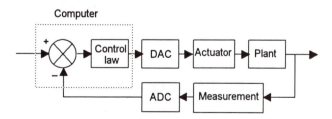

Figure 5.30 *Direct digital control*

The program involves the computer carrying out operations on the fed back measurement value occurring at the instant it is sampled and also using the values previously obtained. The program for proportional control thus takes the form of setting initial values to be used in the program and then a sequence of program instructions which are repeated every sampling period:

Initialise
Set the initial value of the error (this will be zero if the program is to
     start at the measurement value then occurring)
Set the value of the proportional gain
Loop
Input the error at the instant concerned
Calculate the output by multiplying the error by the set value of the
     proportional gain
Output the value of the calculated output
Wait for the end of the sampling period
Go back to Loop and repeat the program

For PD control the program is:

Initialise
Set the initial value of the error (this will be zero if the program is to
     start at the measurement value then occurring)
Set the initial value of the error that is assumed to have occurred in
     the previous sampling period
Set the value of the proportional gain
Set the value of the derivative gain
Loop
Input the error at the instant concerned
Calculate the proportional part of the output by multiplying the error
     by the set value of the proportional gain
Calculate the derivative part of the output by subtracting the value of
     the error at the previous sampling instant from the value at the
     current sampling instant (the difference is a measure of the rate
     of change of the error since the signals are sampled at regular
     intervals of time) and multiply it by the set value of the
     derivative gain.
Calculate the output by adding the proportional and derivative
     output elements
Output the value of the calculated output
Wait for the end of the sampling period
Go back to Loop and repeat the program

For PI control the program is:

Initialise
Set the initial value of the error (this will be zero if the program is to
     start at the measurement value then occurring)

Set the value of the output that is assumed to have occurred in the previous sampling period

Set the value of the proportional gain

Set the value of the integral gain

Loop

Input the error at the instant concerned

Calculate the proportional part of the output by multiplying the error by the set value of the proportional gain

Calculate the integral part of the output by multiplying the value of the error at the current sampling instant by the sampling period and the set value of the integral gain (this assumes that the output has remained constant over the previous sampling period and so multiplying its value by the sampling period gives the area under the output–time graph) and add to it the previous value of the output.

Calculate the output by adding the proportional and integral output elements

Output the value of the calculated output

Wait for the end of the sampling period

Go back to Loop and repeat the program

5.8.1 Embedded systems

The term *embedded system* is used for control systems involving a microprocessor being used as the controller and located as an integral element, i.e. embedded, in the system. Such a system is used with engine management control systems in modern cars, exposure and focus control in modern cameras, the controlling of the operation of modern washing machines and indeed is very widely used in modern consumer goods.

Problems   *Questions 1 to 20 have four answer options: A, B, C and D. Choose the correct answer from the answer options.*

1   Decide whether each of these statements is True (T) or False (F).

An on–off temperature controller must have:
(i)   An error signal input which switches the controller on or off.
(ii) An output signal to switch on or off the correction element.

A   (i) T   (ii) T
B   (i) T   (ii) F
C   (i) F   (ii) T
D   (i) F   (ii) F

2   Decide whether each of these statements is True (T) or False (F).

Oscillations of the variable being controlled occur with on–off temperature controller because:
(i)   There is a time delay in switching off the correction element when the variable reaches the set value.

(ii) There is a time delay in switching on the correction element when the variable falls below the set value.

A (i) T (ii) T
B (i) T (ii) F
C (i) F (ii) T
D (i) F (ii) F

3 Decide whether each of these statements is True (T) or False (F).

With a proportional controller:
(i)   The controller output is proportional to the error.
(ii) The controller gain is proportional to the error.

A (i) T (ii) T
B (i) T (ii) F
C (i) F (ii) T
D (i) F (ii) F

4 A steady state error will not occur when there is a change to the set value with a control system operating in the mode:

A Proportional
B Proportional plus derivative
C Derivative
D Proportional plus integral

Questions 5 to 8 concern the error input to a controller shown in Figure 5.31(a) and the possible controller outputs shown in Figure 5.31(b).

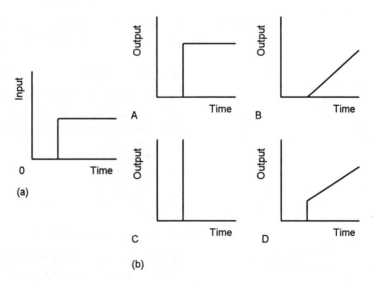

Figure 5.31 *Problems 5 to 8*

5 Which one of the outputs could be given by a proportional controller?
6 Which one of the outputs could be given by a derivative controller?
7 Which one of the outputs could be given by an integral controller?
8 Which one of the outputs could be given by a proportional plus integral controller?
9 Decide whether each of these statements is True (T) or False (F).

With a PID process control system tested at start-up using the ultimate cycle method with the derivative mode turned off and the integral mode set to its lowest setting, the period of oscillation was found to be 10 minutes with a proportional gain setting of 2. The optimum settings, using the criteria of Ziegler and Nichols, will be:
(i) An integral constant of 0.2 min^{-1}.
(ii) A proportional gain setting of 1.2.

A (i) T (ii) T
B (i) T (ii) F
C (i) F (ii) T
D (i) F (ii) F

10 Decide whether each of these statements is True (T) or False (F).

With a PI process control system tested by the process reaction method and the controller output changed by 10%, the response graph obtained was as shown in Figure 5.32. The optimum settings, using the criteria of Ziegler and Nichols, will be:
(i) Proportional gain constant 7.2.
(ii) Integral constant 0.15 min^{-1}.

A (i) T (ii) T
B (i) T (ii) F
C (i) F (ii) T
D (i) F (ii) F

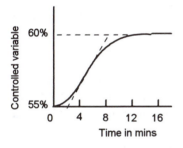

Figure 5.32 *Problem 10*

11 Decide whether each of these statements is True (T) or False (F).

With a PI process control system tested at start-up using the ultimate cycle method with the derivative mode turned off and the integral mode set to its lowest setting, the period of oscillation was found to be 20 minutes with a proportional gain setting of 1.2. The optimum settings, using the criteria of Ziegler and Nichols, will be:
(i) An integral constant of 0.06 min^{-1}.
(ii) A proportional gain setting of 1.2.

A (i) T (ii) T
B (i) T (ii) F
C (i) F (ii) T
D (i) F (ii) F

12 A control system is designed to control temperatures between $-10°$ and $+30°C$. What is (a) the range, (b) the span?

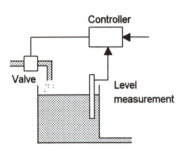

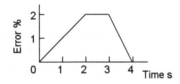

Figure 5.33   *Problem 16*

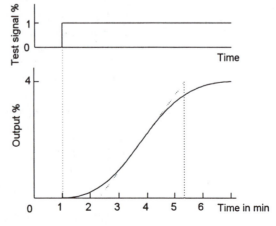

Figure 5.34   *Problem 17*

13 A temperature control system has a set point of 20°C and the measured value is 18°C. What is (a) the absolute deviation, (b) the percentage deviation?

14 What is the controller gain of a temperature controller with a 80% PB if its input range is 40°C to 90° and its output is 4 mA to 20 mA?

15 A controller gives an output in the range 4 to 20 mA to control the speed of a motor in the range 140 to 600 rev/min. If the motor speed is proportional to the controller output, what will be the motor speed when the controller output is (a) 8 mA, (b) 40%?

16 Figure 5.33 shows a control system designed to control the level of water in the container to a constant level. It uses a proportional controller with K_p equal to 10. The valve gives a flow rate of 10 m³/h per percent of controller output, its flow rate being proportional to the controller input. If the controller output is initially set to 50% what will be the outflow from the container? If the outflow increases to 600 m³/h, what will be the new controller output to maintain the water level constant?

17 Sketch graphs showing how the controller output will vary with time for the error signal shown in Figure 5.34 when the controller is set initially at 50% and operates as (a) just proportional with $K_p = 5$, (b) proportional plus derivative with $K_p = 5$ and $K_d = 1.0$ s, (c) proportional plus integral with $K_p = 5$ and $K_I = 0.5$ s⁻¹.

18 Using the Ziegler–Nichols ultimate cycle method for the determination of the optimum settings of a PID controller, oscillations began with a 30% proportional band and they had a period of 11 min. What would be the optimum settings for the PID controller?

19 Using the Ziegler–Nichols ultimate cycle method for the determination of the optimum settings of a PID controller, oscillations began with a gain of 2.2 with a period of 12 min. What would be the optimum settings for the PID controller?

20 Figure 5.35 shows the open-loop response of a system to a unit step in controller output. Using the Ziegler–Nichols data, determine the optimum settings of the PID controller.

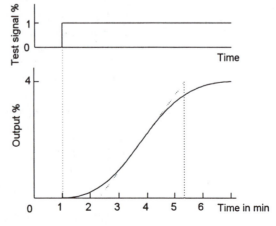

Figure 5.35   *Problem 20*

6 Correction elements

6.1 Introduction

The *correction element* or *final control element* is the element in a control system which is responsible for transforming the output of a controller into a change in the process which aims to correct the change in the controlled variable. Thus, for example, it might be a valve which is operated by the output from the controller and used to change the rate at which liquid passes along a pipe and so change the controlled level of the liquid in a cistern. It might be a motor which takes the electrical output from the controller and transforms it a rotary motion in order to move a load and so control its position. It might be a switch which is operated by the controller and so used to switch on a heater to control temperature.

The term *actuator* is used for the part of a correction/final control element that provides the power, i.e. the bit which moves, grips or applies forces to an object, to carry out the control action. Thus a valve might have an input from the controller and be used to vary the flow of a fluid along a pipe and so make a piston move in a cylinder and result in linear motion. The piston–cylinder system is termed an actuator.

In this chapter pneumatic/hydraulic and electric correction control elements, along with actuators, are discussed.

6.2 Pneumatic and hydraulic systems

Process control systems frequently require control of the flow of a fluid. The valves used as the correction elements in such situations are frequently pneumatically operated, even when the control system is otherwise electrical. This is because such pneumatic devices tend to be cheaper and more easily capable of controlling large rates of flow. The main drawback with pneumatic systems is, however, the compressibility of air. This makes it necessary to have a storage reservoir to avoid changes in pressure occurring as a result of loads being applied. Hydraulic signals do not have this problem and can be used for even higher power control devices. They are, however, expensive and there are hazards associated with oil leaks which do not occur with air leaks.

6.2.1 Current to pressure converter

Generally the signals required by a pneumatic correction element are in the region of 20 to 100 kPa gauge pressure, i.e. pressure above the atmospheric pressure. Figure 6.1 shows the principle of one form of a *current to pressure converter* that can be used to convert a current output from a controller, typically in the range 4 to 20 mA, to a pneumatic pressure signal of 20 to 100 kPa to operate a final control element. The current from the controller passes through coils mounted on a pivoted beam. As a consequence, the coils are then attracted towards a magnet, the extent of the attraction depending on the size of the current. The movement of the coils cause the lever to rotate about its pivot and so change the separation of a flapper from a nozzle. The position of the flapper in relation to the nozzle determines the size of the output pressure in the system.

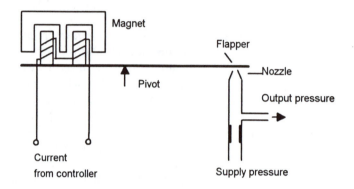

Figure 6.1 *Current to pressure converter*

6.2.2 Pressure sources

With a pneumatic system a source of pressurised air is required. This can be provided by an electric motor driving an air compressor (Figure 6.2). The air is drawn from the atmosphere via a filter. Since the air compressor increases the temperature of the air, a cooling system is likely to follow and, since air also contains a significant amount of moisture, a moisture separator to remove the moisture from the air. A storage reservoir is used to smooth out any pressure fluctuations due to the compressibility of air. A pressure relief valve provides protection against the pressure in the system rising above a safe level.

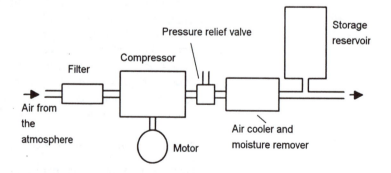

Figure 6.2 *A pressurised air source*

With a hydraulic system a source of pressurised oil is required. This can be provided by a pump driven by an electric motor. The pump pumps oil from a sump through a non-return valve and an accumulator and back to the sump (Figure 6.3). The non-return valve is to prevent the oil being back-driven to the pump. A pressure relief valve is included so that the pressure is released if it rises above a safe level. The accumulator is essentially just a container in which the oil is held under pressure against an external force and is there to smooth out any short-term fluctuations in the output oil pressure. If the oil pressure rises then the piston moves to increase the volume the oil can occupy and so reduces the pressure. If the oil pressure falls then the piston moves in to reduce the volume occupied by the oil and so increase its pressure.

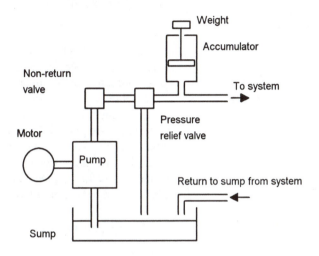

Figure 6.3 *A source of pressurised oil*

6.2.3 Control valves

Pneumatic and hydraulic systems use control valves to give direction to the flow of fluid through a system, control its pressure and control the rate of flow. These types of valve can be termed *directional control valves*, *pressure control valves* and *flow control valves*. Directional control valves, sometimes termed *finite position valves* because they are either completely open or completely closed, i.e. they are on/off devices, are used to direct fluid along one path or another. They are equivalent to electric switches which are either on or off. Pressure control valves, often termed pressure regulator valves, react to changes in pressure in switching a flow on or off, or varying it. Flow control valves, sometimes termed *infinite position valves*, vary the rate at which a fluid passes through a pipe and are used to regulate the flow of material in process control systems. Valves are discussed in more detail later in this chapter.

6.2.4 Actuators

Fluid power actuators can be classified in two groups: *linear actuators* which are used to move an object or apply a force in a straight line and *rotary actuators* which are used to move an object in a circular path.

The *hydraulic* or *pneumatic cylinder* is a linear actuator, the principles and form being the same for both versions with the differences being purely a matter of size as a consequence of the higher pressures

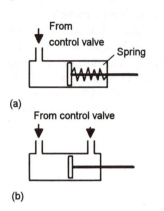

(a)

(b)

Figure 6.4   *(a) Single acting, (b) double acting cylinder*

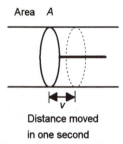

Figure 6.5   *Movement of a piston in a cylinder*

used with hydraulics. The hydraulic/pneumatic cylinder consists of a hollow cylindrical tube along which a piston can slide. Figure 6.4(a) shows the single acting form and Figure 6.4(b) the double acting form. The *single acting* form has the control pressure applied to just one side of the piston, a spring often being used to provide the opposition to the movement of the piston. The piston can only be moved in one direction along the cylinder by the signal from the controller. The *double acting* form has control pressures that can be applied to each side of the piston. When there is a difference in pressure between the two sides the piston moves, the piston being able to move in either direction along the cylinder.

The choice of cylinder is determined by the force required to move the load and the speed required. Hydraulic cylinders are capable of much larger forces than pneumatic cylinders. However, pneumatic cylinders are capable of greater speeds.

Since pressure is force per unit area, the force produced by a piston in a cylinder is equal to the cross-sectional area of the piston, this being effectively the same as the internal cross-sectional area of the cylinder, multiplied by the difference in pressure between the two sides of the piston. Thus for a pneumatic cylinder with a pressure difference of 500 kPa and having an internal diameter of 50 mm,

$$\text{force} = \text{pressure} \times \text{area} = 500 \times 10^3 \times \tfrac{1}{4}\pi \times 0.050^2 = 982 \text{ N}$$

A hydraulic cylinder with the same diameter and a pressure difference of 15 000 kPa, hydraulic cylinders being able to operate with higher pressures than pneumatic cylinders, will give a force of 29.5 kN. Note that the maximum force available is not related to the flow rate of hydraulic fluid or air into a cylinder but is determined solely by the pressure and piston area.

The speed with which the piston moves in a cylinder is determined by the rate at which fluid enters the cylinder. If the flow rate of hydraulic liquid into a cylinder is a volume of Q per second, then the piston must sweep out a volume of Q. If a piston moves with a velocity v then, in one second, it moves a distance of v (Figure 6.5). But for a piston of cross-sectional area A this must mean that the volume swept out by the piston in 1 s is Av. Thus we must have:

$$Q = Av$$

Thus the speed v of a hydraulic cylinder is equal to the flow rate of liquid Q through the cylinder divided by the cross-sectional area A of the cylinder. The speed is determined by just the piston area and the flow rate. For example, for a hydraulic cylinder of diameter 50 mm and a hydraulic fluid flow of 7.5×10^{-3} m³/s:

$$\text{speed } v = \frac{Q}{A} = \frac{7.5 \times 10^{-3}}{\tfrac{1}{4}\pi \times 0.050^2} = 3.8 \text{ m/s}$$

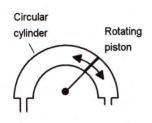

Figure 6.6 *Rotary actuator*

Rotary actuators give rotary motion as a result of the applied fluid pressure. Figure 6.6 shows a rotary actuator which gives partial rotary movement. Continuous rotation is possible with some forms and then they are the equivalent of electric motors. Figure 6.7 shows one form, known as a *vane motor*. The vanes are held out against the walls of the motor by springs or hydraulic pressure. Thus, when there is a pressure difference between the inlet and outlets of the motor, rotation occurs.

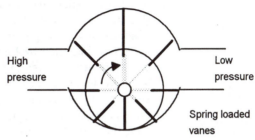

Figure 6.7 *Vane motor*

Example

A hydraulic cylinder is to be used in a manufacturing operation to move a workpiece through a distance of 250 mm in 20 s. If a force of 50 kN is required to move the workpiece, what is the required pressure difference and hydraulic liquid flow rate if a cylinder with a piston diameter of 150 mm is to be used?

As derived above, the force produced by the cylinder is equal to the product of the cross-sectional area of the cylinder and the working pressure. Thus the required pressure is:

$$\text{pressure} = \frac{F}{A} = \frac{50 \times 10^3}{\frac{1}{4}\pi \times 0.150^2} = 2.8 \times 10^6 \text{ Pa} = 2.8 \text{ MPa}$$

The average speed required is 250/20 = 12.5 mm/s. As derived above, the speed of a hydraulic cylinder is equal to the flow rate of liquid through the cylinder divided by the cross-sectional area of the cylinder. Thus the required flow rate is:

flow rate = speed × area

$$= 0.0125 \times \tfrac{1}{4}\pi \times 0.150^2 = 2.2 \times 10^{-4} \text{ m/s}$$

6.3 Directional control valves   Directional control valves are widely used in control systems as elements for switching on or off hydraulic or pneumatic pressures which can then, via some actuator, control the movement of some item. A *directional control valve* on the receipt of some external signal, which might be mechanical, electrical or a pressure signal, changes the direction of, or

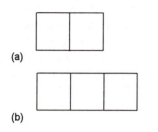

(a)

(b)

Figure 6.8 *(a) Two position, (b) three position valves*

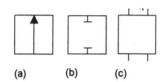

(a) **(b)** **(c)**

Figure 6.9 *(a) Flow path, (b) shut-off, (c) input connections*

stops, or starts the flow of fluid in some part of a pneumatic/hydraulic circuit.

The basic symbol for a control valve is a square. With a directional control valve two or more squares are used, with each square representing the positions to which the valve can be switched. Thus, Figure 6.8(a) represents a valve with two switching positions, Figure 6.8(b) a valve with three switching positions. Lines in the boxes are used to show the flow paths with arrows indicating the direction of flow (Figure 6.9(a)) and shut-off positions indicated by terminated lines (Figure 6.9(b)). The pipe connections, i.e. the inlet and outlet ports of the valve, are indicated by lines drawn on the outside of the box and are drawn for just the 'rest/initial/neutral position', i.e. when the valve is not actuated (Figure 6.9(c)). You can imagine each of the position boxes to be moved by the action of some actuator so that it connects up with the pipe positions to give the different connections between the ports. Directional control valves are described by the number of ports and the number of positions. Thus, a 2/2 valve has 2 ports and 2 positions, a 3/2 valve 3 ports and 2 positions, a 4/2 valve 4 ports and 2 positions, a 5/3 valve 5 ports and 3 positions. Figure 6.10 shows some commonly used examples and their switching options and Figure 6.11 the means by which valves can be switched between positions.

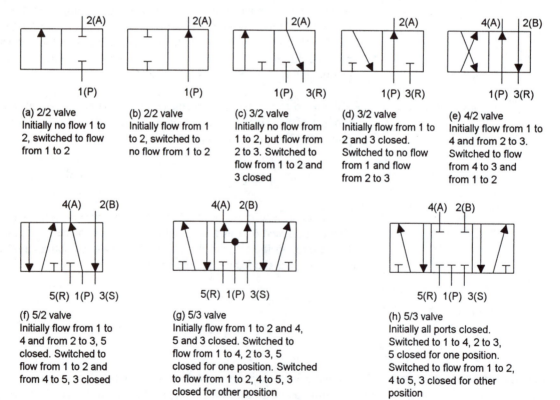

Figure 6.10 *Commonly used direction valves: P or 1 indicates the pressure supply ports, R and S or 3 and 5 the exhaust ports, A and B or 2 and 4 the signal output ports*

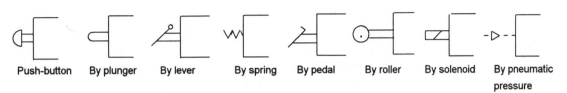

Figure 6.11 *Examples of valve actuation methods*

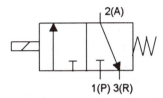

Figure 6.12 *Symbol for a solenoid-activated valve with return spring*

Application
In section 4.4.5 an antilock brake system (ABS) for a car is discussed and Figure 4.17 shows how valves are used.

As an illustration, Figure 6.12 shows the symbol for a 3/2 valve with solenoid activation and return by means of a spring. Thus, when the solenoid is not activated by a current through it, the signal port 2 is connected to the exhaust 3 and so is at atmospheric pressure. When the solenoid is activated, the pressure supply P is connected to the signal port 2 and thus the output is pressurised.

Figure 6.13 shows how such a valve might be used to cause the piston in a single-acting cylinder to move; the term single-acting is used when a pressure signal is applied to only one side of the piston. When the switch is closed and a current passes through the solenoid, the valve switches position and pressure is applied to extend the piston in the cylinder.

Figure 6.14 shows how a double-solenoid activated valve can be used to control a double-acting cylinder. Momentary closing switch S1 causes a current to flow through the solenoid at the left-hand end of the valve and so result in the piston extending. On opening S1 the valve remains in this extended position until a signal is received by the closure of switch S2 to activate the right-hand solenoid and return the piston.

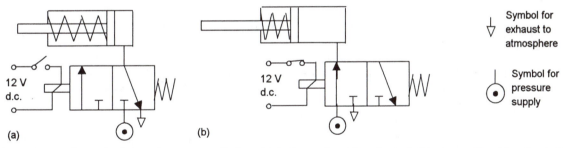

Figure 6.13 *Control of a single-acting cylinder: (a) before solenoid activated, (b) when solenoid activated*

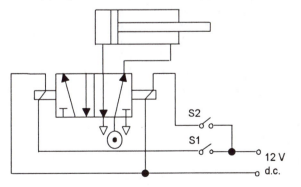

Figure 6.14 *Control of a double-acting cylinder*

6.3.1 Sequencing

Situations often occur where it is necessary to activate a number of cylinders in some sequence. Thus event 2 might have to start when event 1 is completed, event 3 when event 2 has been completed. For example, we might have: only when cylinder A is fully extended (event 1) can cylinder B start extending (event 2), and cylinder A can only start retracting (event 3) when cylinder B has fully extended (event 2). In discussions of sequential control it is common practice to give each cylinder a reference letter A, B, C, D, etc., and to indicate the state of each cylinder by using a + sign if it's extended or a − sign if retracted. Thus a sequence of operations might be shown as A+, B+, A−, B−. This indicates that the sequence of events is cylinder A extend, followed by cylinder B being extended, followed by cylinder A retracting, followed by cylinder B retracting. Figure 6.15 illustrates this with a displacement-step diagram. Figure 6.16 shows a circuit that could be used to generate this displacement-event diagram for two cylinders A and B.

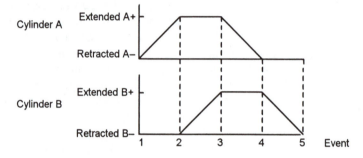

Figure 6.15 *Displacement-event diagram*

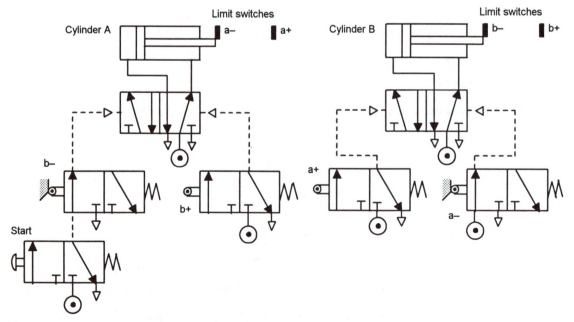

Figure 6.16 *Two-actuator sequential operation*

In order to generate the displacement-event diagram of Figure 6.15 the sequence of operations with Figure 6.16 is:

Event 1

1   Start push-button pressed.

2   Cylinder A extends, releasing limit switch a–.

Event 2

3   Cylinder A fully extended, limit switch a+ operated to start B extending.

4   Cylinder B extends, limit switch b– released.

Event 3

5   Cylinder B fully extended, limit switch b+ operated to start cylinder A retracting.

6   Cylinder A retracts, limit switch a+ released.

Event 4

7   Cylinder A fully retracted, limit switch a– operated to start cylinder B retracting.

8   Cylinder B retracts, limit switch b+ released.

Event 5

9   Cylinder B fully retracted, limit switch b– operated to complete the cycle.

The cycle can be started again by pushing the start button. If we wanted the system to run continuously then the last movement in the sequence would have to trigger the first movement.

As an illustration of the types of problems that sequential control can be used for, consider the control required with an automatic machine to perform a number of sequential actions such as positioning objects, operating clamps and then operating some machine tool (Figure 6.17). This requires the switching in sequence of a number of cylinders, the movements of the cylinder pistons being the mechanisms by which the actions are initiated.

6.3.2 Shuttle valve

The most common form of directional control valve is the *shuttle* or *spool valve*. Shuttle valves have a spool moving horizontally within the valve body. Raised areas, termed *lands*, block or open ports to give the required valve operation. Figure 6.18 illustrates these features with a 3/2 valve. In the first position, the shuttle is located so that its lands block off the 3 port and leave open, and connected, the 1 and 2 ports. In the second position, the shuttle is located so that it blocks of the 1 port and leaves open, and connected, the 2 and 3 ports. The shuttle can be made to move between these two positions by manual, mechanical, electrical or pressure signals applied to the two ends of the shuttle.

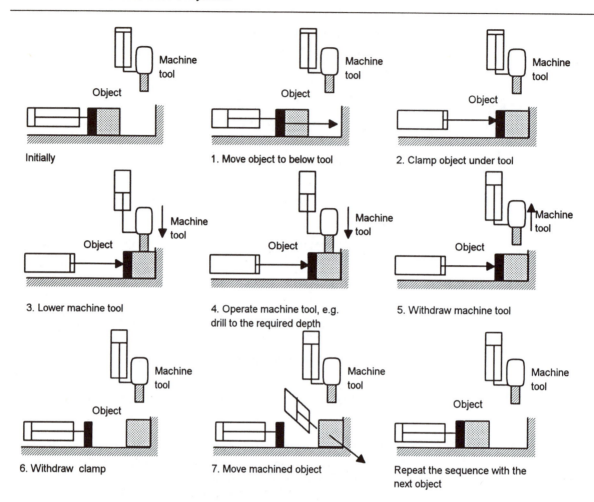

Figure 6.17 *Operations required for an automatic machine operation*

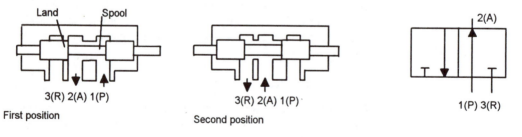

Figure 6.18 *3/2 shuttle valve*

Figure 6.19 shows an example of a 4/3 shuttle valve. It has the rest position with all ports closed. When the shuttle is moved from left to right, the pressure is applied to output port 2 and port 4 is connected to the exhaust port. When the shuttle moves from right to left, pressure is applied to output port 4 and port 2 is connected to the exhaust port.

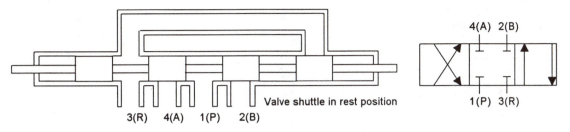

Valve shuttle in rest position

3(R)   4(A)   1(P)   2(B)

4(A) 2(B)

1(P) 3(R)

Figure 6.19 *4/3 shuttle valve*

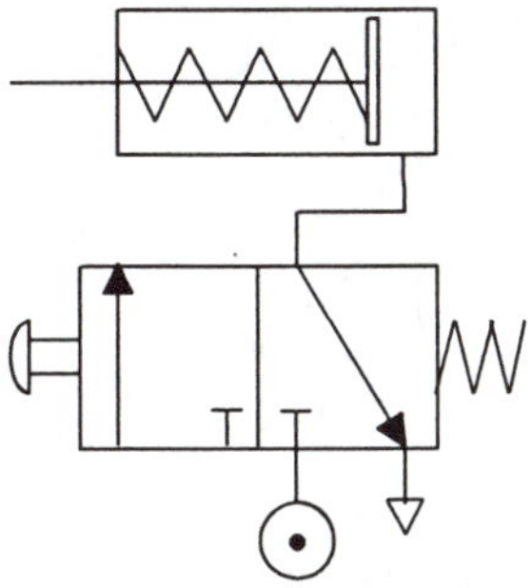

Example

State what happens for the pneumatic circuit shown in Figure 6.20 when the push-button is pressed and then released.

The right-hand box shows the initial position with the pressure source, i.e. the circle with the dot in the middle, connected to a closed port and the output from the right-hand end of the cylinder connected to the exhaust port, i.e. the open triangle. When the push-button is pressed the connections between the ports become those indicated in the left-hand box. The pressure source is then connected to the output port and hence to the right-hand end of the cylinder and forces the piston back against its spring and so from left to right. When the push-button is released, the connections between the ports become those in the right-hand box and the right-hand end of the cylinder is exhausted. The piston then moves back from left to right.

Figure 6.20 *Example*

6.4 Flow control valves

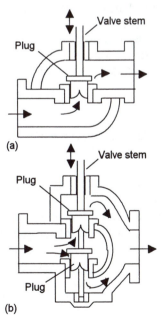

Valve stem

Plug

(a)

Valve stem

Plug

Plug

(b)

Figure 6.21 *Body globe valve:*
(a) single, (b) double seated

In many control systems the rate of flow of a fluid along a pipe is controlled by a valve which uses pneumatic action to move a valve stem and hence a plug or plugs into the flow path, so altering the size of the gap through which the fluid can flow (Figure 6.21). The term *single seated* is used where just one plus is involved and *double seated* where there are two. A single-seated valve has the advantage compared with the double-seated valve of being able to close more tightly but the disadvantages that the force on the plug is greater from the fluid and so a larger area diaphragm may be needed.

Figure 6.22 shows the basic elements of a common form of such a control valve. The movement of the stem, and hence the position of the plug or plugs in the fluid flow, results from the use of a diaphragm moving against a spring and controlled by air pressure (Figure 6.22). The air pressure from the controller exerts a force on one side of the diaphragm, the other side of the diaphragm being at atmospheric pressure, which is opposed by the force due to the spring on the other side. When the air pressure changes then the diaphragm moves until there is equilibrium between the forces resulting from the pressure and those from the spring. Thus the pressure signals from the controller result in the movement of the stem of the valve. There are two alternative forms, *direct* and *reverse action* forms (Figure 6.23) with the difference being the position of the spring. The valve body is joined to the diaphragm element by the *yoke*.

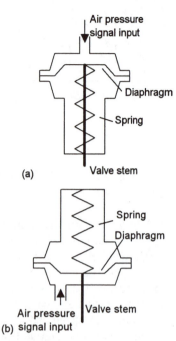

Figure 6.23 *(a) Direct action, (b) reverse action*

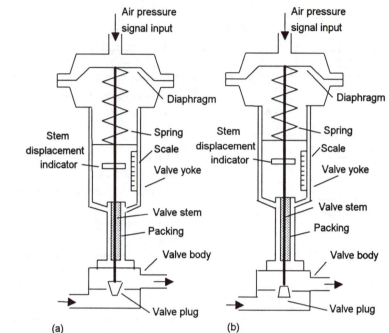

Figure 6.22 *Direct action: (a) air pressure increase to close, (b) air pressure increase to open*

6.4.1 Forms of plug

There are many forms of valve body and plug. The selection of the form of body and plug determine the characteristic of the control valve, i.e. the relationship between the valve stem position and the flow rate through it. For example, Figure 6.24 shows how the selection of plug can be used to determine whether the valve closes when the controller air pressure increases or opens when it increases and Figure 6.24 shows how the shape of the plug determines how the rate of flow is related to the displacement of the valve stem:

1 *Linear plug*

The change in flow rate is proportional to the change in valve stem displacement, i.e.:

change in flow rate = k (change in stem displacement)

where k is a constant. If Q is the flow rate at a valve stem displacement S and Q_{max} is the maximum flow rate at the maximum stem displacement S_{max}, then we have:

$$\frac{Q}{Q_{max}} = \frac{S}{S_{max}}$$

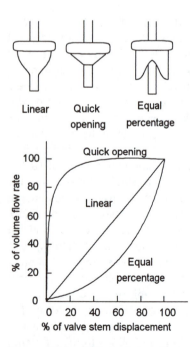

Figure 6.24 *Effect of plug shape on flow*

or the percentage change in the flow rate equals the percentage change in the stem displacement. Such valves are widely used for the control of liquids entering cisterns when the liquid level is being controlled.

2   *Quick-opening plug*
A large change in flow rate occurs for a small movement of the valve stem. This characteristic is used for on–off control systems where the valve has to move quickly from open to closed and vice versa.

3   *Equal percentage plug*
The amount by which the flow rate changes is proportional to the value of the flow rate when the change occurs. Thus, if the amount by which the flow rate changes is ΔQ for a change in valve stem position ΔS, then it is proportional to the value of the flow Q when the change occurs, i.e.:

$$\frac{\Delta Q}{\Delta S} \propto Q$$

Hence we can write

$$\frac{\Delta Q}{\Delta S} = kQ$$

where k is a constant. Generally this type of valve does not cut off completely when at the limit of its stem travel, thus when $S = 0$ we have $Q = Q_{min}$. If we write this expression for small changes and then integrate it we obtain:

$$\int_{Q_{min}}^{Q} \frac{1}{Q}\,dQ = k \int_{0}^{S} dS$$

Hence:

$$\ln Q - \ln Q_{min} = kS$$

If we consider the flow rate Q_{max} is given by S_{max} then:

$$\ln Q_{max} - \ln Q_{min} = kS_{max}$$

Eliminating k from these two equations gives:

$$\frac{\ln Q - \ln Q_{min}}{\ln Q_{max} - \ln Q_{min}} = \frac{S}{S_{max}}$$

$$\ln \frac{Q}{Q_{min}} = \frac{S}{S_{max}} \ln \frac{Q_{max}}{Q_{min}}$$

and so:

$$\frac{Q}{Q_{min}} = \left(\frac{Q_{max}}{Q_{min}}\right)^{S/S_{max}}$$

Example

A valve has a stem movement at full travel of 30 mm and has a linear plug which has a minimum flow rate of 0 and a maximum flow rate of 20 m³/s. What will be the flow rate when the stem movement is 15 mm?

The percentage change in the stem position from the zero setting is $(15/30) \times 100 = 50\%$. Since the percentage flow rate is the same as the percentage stem displacement, then a percentage stem displacement of 50% gives a percentage flow rate of 50%, i.e. 10 m³/s.

Example

A valve has a stem movement at full travel of 30 mm and an equal percentage plug. This gives a flow rate of 2 m³/s when the stem position is 0. When the stem is at full travel there is a maximum flow rate of 20 m³/s. What will be the flow rate when the stem movement is 15 mm?

Using the equation:

$$\frac{Q}{Q_{min}} = \left(\frac{Q_{max}}{Q_{min}}\right)^{S/S_{max}}$$

$$\frac{Q}{2} = \left(\frac{20}{2}\right)^{15/30}$$

gives $Q = 6.3$ m³/s.

6.4.2 Rangeability and turndown

The term *rangeability* R is used for the ratio Q_{max}/Q_{min}, i.e. the ratio of the maximum to minimum rates of controlled flow. Thus, if the minimum controllable flow is 2.0% of the maximum controllable flow, then the rangeability is $100/2.0 = 50$. Valves are often not required to handle the maximum possible flow and the term *turndown* is used for the ratio:

$$turndown = \frac{normal\ maximum\ flow}{minimum\ controllable\ flow}$$

For example, a valve might be required to handle a maximum flow which is 70% of that possible. With a minimum flow rate of 2.0% of the maximum flow possible, then the turndown is $70/2.0 = 35$.

6.4.3 Control valve sizing

The term *control valve sizing* is used for the procedure of determining the correct size, i.e. diameter, of the valve body. A control valve changes the flow rate by introducing a constriction in the flow path. But introducing such a constriction introduces a pressure difference between the two sides of the constriction. The basic equation (from an application of Bernoulli's equation) relating the rate of flow and pressure drop is:

$$\text{rate of flow } Q = K \sqrt{\text{pressure drop}}$$

where K is a constant which depends on the size of the constriction produced by the presence of the valve. The equations used for determining valve sizes are based on this equation. For a liquid, this equation is written as:

$$Q = A_V \sqrt{\frac{\Delta p}{\rho}} \text{ m}^3/\text{s}$$

where A_V is the *valve flow coefficient*, Δp the pressure drop in Pa across the valve and ρ the density in kg/m^3 of the fluid. Because the equation was originally specified with pressure in pounds per square inch and flow rate in American gallons per minute, another coefficient C_V based on these units is widely quoted. With such a coefficient and the quantities in SI units, we have:

$$Q = 2.37 \times 10^{-5} C_V \sqrt{\frac{\Delta p}{\rho}} \text{ m}^3/\text{s}$$

or

$$Q = 0.75 \times 10^{-6} C_V \sqrt{\frac{\Delta p}{G}} \text{ m}^3/\text{s}$$

G is the specific gravity (relative density) and Δp is the pressure difference. Other equations are available for gases and steam. For gases:

$$Q = 6.15 \times 10^{-4} C_V \sqrt{\frac{\Delta p \times p}{TG}} \text{ mm}^3/\text{s}$$

where T is the temperature on the Kelvin scale and p the inlet pressure. For steam:

$$Q = 27.5 \times 10^{-6} C_V \sqrt{\frac{\Delta p}{V}} \text{ kg/s}$$

where V is the specific volume of the steam in m³/kg, the specific volume being the volume occupied by 1 kg. Table 6.1 shows some typical values of A_V, C_V and the related valve sizes.

Table 6.1 *Flow coefficients and valve size*

Flow coefficients	Valve size in mm							
	480	640	800	960	1260	1600	1920	2560
C_V	8	14	22	30	50	75	110	200
$A_V \times 10^{-5}$	19	33	52	71	119	178	261	474

Example

Determine the valve size for a valve that is required to control the flow of water when the maximum flow rate required is 0.012 m³/s and the permissible pressure drop across the valve at this flow rate is 300 kPa.

Taking the density of water as 1000 kg/m³, we have:

$$A_V = Q\sqrt{\frac{\rho}{\Delta p}} = 0.012\sqrt{\frac{1000}{300 \times 10^3}} = 69.3 \times 10^{-5}$$

Thus, using Table 6.1, this value of coefficient indicates that the required valve size is 960 mm.

6.4.4 Valve positioners

Frictional forces and unbalanced forces on the plug may prevent the diaphragm from positioning the plug accurately. In order to overcome this, *valve positioners* may be fitted to the control valve stem. They position the valve stem more accurately and also provide extra power to operate the valve and so increase the speed of valve movement. Figure 6.25 shows the basic elements of a positioner.

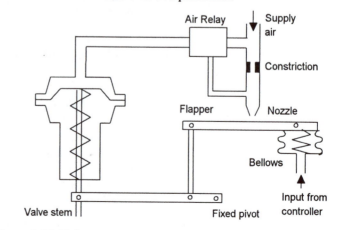

Figure 6.25 *Valve positioner*

The output from the controller is applied to a spring-loaded bellows. A flapper is attached to the bellows and is moved by pressure applied to the bellows. An increase in this pressure brings the flapper closer to the nozzle and so cuts down the air escaping from it. As a consequence, the pressure applied to the diaphragm is increased. The resulting valve stem displacement takes the flapper away from the nozzle until the air leakage from the nozzle is just sufficient to maintain the correct pressure on the diaphragm.

6.4.5 Other forms of flow control valves

The type of control valve described in the earlier parts of this section is basically the *split-body globe* valve body with a plug or plugs. This is the most commonly used form. There are, however, other forms. Figure 6.26(a) shows a 3-way globe. Other valve types are the *gate* (Figure 6.26(b)), the *ball* (Figure 6.26(c)), the *butterfly* (Figure 6.26(d)) and the *louvre* (Figure 6.26(e)). All excise control by restricting the fluid flow. Ball valves use a ball with a through-hole which is rotated; they have excellent shut-off capability. Butterfly valves rotate a vane to restrict the air flow and, as a consequence, suffer from the problem of requiring significant force to move from the full-open position and so can 'stick' in that position.

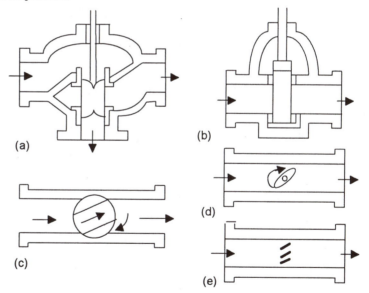

Figure 6.26 *(a) 3-way globe, (b) gate, (c) ball, (d) butterfly, (e) louvre*

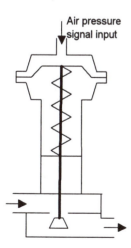

Figure 6.27 *Shut down if air pressure fails*

6.4.6 Fail-safe design

Fail-safe design means that the design of a plant has to take account of what will happen if the power or air supply fails so that a safe shut-down occurs. Thus, in the case of a fuel valve, the valve should close if failure occurs, while for a cooling water valve the failure should leave the valve open. Figure 6.27 shows a direct acting valve which shuts down the fluid flow if the air supply to the diaphragm fails.

6.5 Motors

Electric motors are frequently used as the final control element in position or speed-control systems. The basic principle on which motors are based is that a force is exerted on a conductor in a magnetic field when a current passes through it. For a conductor of length L carrying a current I in a magnetic field of flux density B at right angles to the conductor, the force F equals BIL.

There are many different types of motor. In the following, discussion is restricted to those types of motor that are commonly used in control systems, this including d.c. motors and the stepper motor. A *stepper motor* is a form of motor that is used to give a fixed and consistent angular movement by rotating an object through a specified number of revolutions or fraction of a revolution.

6.5.1 D.c. motors

In the d.c. motor, coils of wire are mounted in slots on a cylinder of magnetic material called the *armature*. The armature is mounted on bearings and is free to rotate. It is mounted in the magnetic field produced by *field poles*. This magnetic field might be produced by permanent magnets or an electromagnet with its magnetism produced by a current passing through the, so-termed, *field coils*. Whether permanent magnet or electromagnet, these generally form the outer casing of the motor and are termed the *stator*. Figure 6.28 shows the basic elements of d.c. motor with the magnetic field of the stator being produced by a current through coils of wire. In practice there will be more than one armature coil and more than one set of stator poles. The ends of the armature coil are connected to adjacent segments of a segmented ring called the *commutator* which rotates with the armature. Brushes in fixed positions make contact with the rotating commutator contacts. They carry direct current to the armature coil. As the armature rotates, the commutator reverses the current in each coil as it moves between the field poles. This is necessary if the forces acting on the coil are to remain acting in the same direction and so continue the rotation.

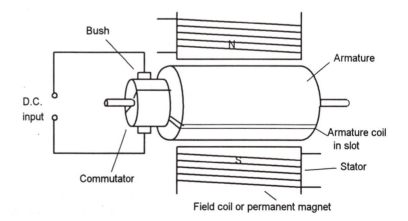

Figure 6.28 *Basic elements of a d.c. motor*

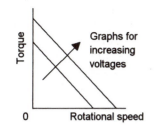

Figure 6.29 *Permanent magnet motor characteristic*

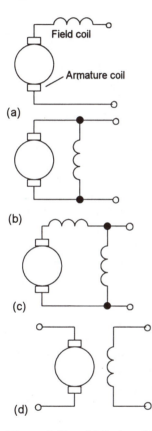

Figure 6.30 *(a) Series, (b) shunt, (c) compound, (d) separately wound*

Application
Section 4.4.4 shows a closed-loop arrangement that could be used to control the speed of a motor shaft.

For a d.c. motor with the field provided by a permanent magnet, the speed of rotation can be changed by changing the size of the current to the armature coil, the direction of rotation of the motor being changed by reversing the current in the armature coil. Figure 6.29 shows how, for a permanent magnet motor, the torque developed varies with the rotational speed for different applied voltages. The starting torque is proportional to the applied voltage and the developed torque decreases with increasing speed.

D.c. motors with field coils are classified as series, shunt, compound and separately excited according to how the field windings and armature windings are connected.

1 *Series-wound motor*
With the *series-wound motor* the armature and field coils are in series (Figure 6.30(a)). Such a motor exerts the highest starting torque and has the greatest no-load speed. However, with light loads there is a danger that a series-wound motor might run at too high a speed. Reversing the polarity of the supply to the coils has no effect on the direction of rotation of the motor, since both the current in the armature and the field coils are reversed.

2 *Shunt-wound motor*
With the *shunt-wound motor* (Figure 6.30(b)) the armature and field coils are in parallel. It provides the lowest starting torque, a much lower no-load speed and has good speed regulation. It gives almost constant speed regardless of load and thus shunt wound motors are very widely used. To reverse the direction of rotation, either the armature or field current can be reversed.

3 *Compound motor*
The *compound motor* (Figure 6.30(c)) has two field windings, one in series with the armature and one in parallel. Compound-wound motors aim to get the best features of the series and shunt-wound motors, namely a high starting torque and good speed regulation.

4 *Separately excited motor*
The *separately excited motor* (Figure 6.30(d)) has separate control of the armature and field currents. The direction of rotation of the motor can be obtained by reversing either the armature or the field current.

Figure 6.31 indicates the general form of the torque–speed characteristics of the above motors. The separately excited motor has a torque–speed characteristic similar to the shunt wound motor. The speed of such d.c. motors can be changed by either changing the armature current or the field current. Generally it is the armature current that is varied. The choice of d.c. motor will depend on what it is to be used for. Thus, for example, with a robot manipulator the robot wrist might use a series-wound motor because the speed decreases as the load increases. A shunt-wound motor might be used if a constant speed was required, regardless of the load.

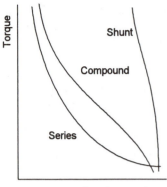

Figure 6.31 *Torque–speed characteristics of d.c. motors*

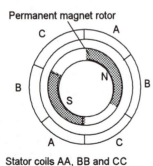

Figure 6.32 *PWM: (a) principle of PWM circuit, (b) varying the average armature voltage by chopping the constant d.c. voltage*

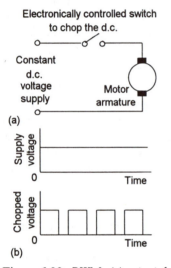

Figure 6.33 *Brushless permanent magnet d.c. motor*

The speed of a permanent magnet motor can be controlled by varying the current through the armature coil, with a field coil motor by either varying the armature current or the field current though generally it is the armature current that is varied. Thus speed control can be obtained by controlling the voltage applied to the armature. Rather than just try to directly vary the input voltage, a more convenient method is to use *pulse width modulation* (PWM). This basically involves taking a constant d.c. supply voltage and using an electronic circuit to chop it so that the average value is varied (Figure 6.32).

6.5.2 Brushless permanent magnet d.c. motor

A problem with the d.c. motors described in the previous section, is that they require a commutator and brushes in order to periodically reverse the current through each armature coil. Brushes have to be periodically changed and the commutator resurfaced because the brushes make sliding contacts with the commutator and suffer wear. Brushless d.c. motors do not have this problem.

A current-carrying conductor in a magnetic field experiences a force and with the conventional d.c. motor the magnet is fixed and the current-carrying conductors consequently made to move. However, as a consequence of Newton's third law of motion, the magnet will experience an opposite and equal force to that acting on the current-carrying conductors and so, with the brushless permanent magnet d.c. motor, the current carrying conductors are fixed and the magnet moves. With just one current carrying coil, the resulting force on the magnet would just cause it to deflect. In order to keep the magnet moving, a sequence of current carrying coils have to be used and each in turn switched on.

Figure 6.33 shows the basic form of such a motor. The rotor is a ferrite or ceramic permanent magnet. The current to the stator coils AA, BB and CC is electronically switched by transistors in sequence round them, the switching being controlled by the position of the rotor so that there are always forces acting on the magnet causing it to rotate in the same direction. Hall sensors (a magnetic field input to the sensor gives a voltage output) are generally used to sense the position of the rotor and initiate the switching by the transistors, the sensors being positioned around the stator. Figure 6.34 shows the transistor switching circuits that might be used with the motor shown in Figure 6.33.

To switch the coils in sequence we need to supply signals to switch the transistors on in the right sequence. This is provided by the outputs from the three sensors operating through a decoder circuit to give the appropriate base currents. Thus when the rotor is in the vertical position, i.e. 0°, there is an output from sensor c but none from a and b and this is used to switch on transistors A+ and B−. When the rotor is in the 60° position there are signals from the sensors b and c and transistors A+ and C− are switched on. Table 6.2 shows the entire switching sequence. The entire circuit for controlling such a motor is available as a single integrated circuit.

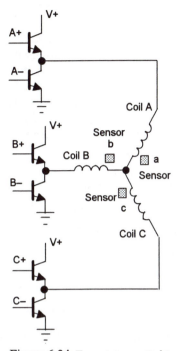

Figure 6.34 *Transistor switching*

Table 6.2 *Switching sequence*

Rotor position	Sensor signals			Transistors on	
	a	b	c		
0°	0	0	1	A+	B–
60°	0	1	1	A+	C–
120°	0	1	0	B+	C–
180°	1	1	0	B+	A–
240°	1	0	0	C+	A–
360°	1	0	1	C+	B–

6.5.3 Stepper motor

The *stepper or stepping motor* produces rotation through equal angles, the so-called *steps*, for each digital pulse supplied to its input. For example, if with such a motor 1 input pulse produces a rotation of 1.8° then 20 input pulses will produce a rotation through 36.0° , 200 input pulses a rotation through one complete revolution of 360°. It can thus be used for accurate angular positioning. By using the motor to drive a continuous belt, the angular rotation of the motor is transformed into linear motion of the belt and so accurate linear positioning can be achieved. Such a motor is used with computer printers, *x-y* plotters, robots, machine tools and a wide variety of instruments for accurate positioning.

There are two basic forms of stepper motor, the *permanent magnet* type with a permanent magnet rotor and the *variable reluctance* type with a soft steel rotor. Figure 6.35 shows the basic elements of the permanent magnet type with two pairs of stator poles.

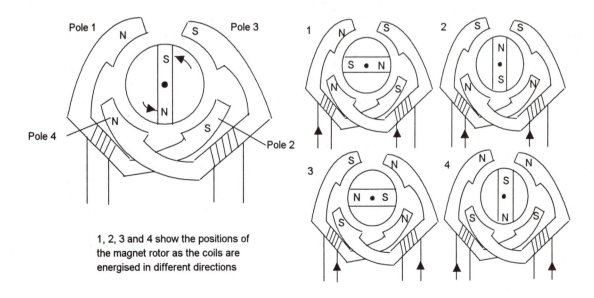

1, 2, 3 and 4 show the positions of the magnet rotor as the coils are energised in different directions

Figure 6.35 *The basic principles of the permanent magnet stepper motor (2-phase) with 90° steps*

Each pole is activated by a current being passed through the appropriate field winding, the coils being such that opposite poles are produced on opposite coils. The current is supplied from a d.c. source to the windings through switches. With the currents switched through the coils such that the poles are as shown in Figure 6.35, the rotor will move to line up with the next pair of poles and stop there. This would be, for Figure 6.35, an angle of 45°. If the current is then switched so that the polarities are reversed, the rotor will move a step to line up with the next pair of poles, at angle 135° and stop there. The polarities associated with each step are:

Step	Pole 1	Pole 2	Pole 3	Pole 4
1	North	South	South	North
2	South	North	South	North
3	South	North	North	South
4	North	South	North	South
5	Repeat of steps 1 to 4			

There are thus, in this case, four possible rotor positions: 45°, 135°, 225° and 315°.

Figure 6.36 shows the basic form of the *variable reluctance* type of stepper motor. With this form the rotor is made of soft steel and is not a permanent magnet. The rotor has a number of teeth, the number being less than the number of poles on the stator. When an opposite pair of windings on stator poles has current switched to them, a magnetic field is produced with lines of force which pass from the stator poles through the nearest set of teeth on the rotor. Since lines of force can be considered to be rather like elastic thread and always trying to shorten themselves, the rotor will move until the rotor teeth and stator poles line up. This is termed the position of minimum reluctance. Thus by switching the current to successive pairs of stator poles, the rotor can be made to rotate in steps. With the number of poles and rotor teeth shown in Figure 6.36, the angle between each successive step will be 30°. The angle can be made smaller by increasing the number of teeth on the rotor.

There is another version of the stepper motor and that is a *hybrid stepper*. This combines features of both the permanent magnet and variable reluctance motors. They have a permanent magnet rotor encased in iron caps which are cut to have teeth. The rotor sets itself in the minimum reluctance position in response to a pair of stator coils being energised.

The following are some of the terms commonly used in specifying stepper motors:

1 *Phase*

This is the number of independent windings on the stator, e.g. a four-phase motor. The current required per phase and its resistance

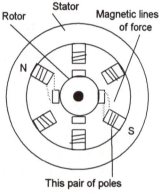

Figure 6.36 *Basic principles of a 3-phase variable reluctance stepper motor*

Application
Section 4.4.2 shows the application of a stepper motor to the control of the position of a tool.

and inductance will be specified so that the controller switching output is specified. Figure 6.35 is an example of a two-phase motor, such motors tending to be used in light-duty applications. Figure 6.36 is an example of a three-phase motor. Four-phase motors tend to be used for higher power applications.

2 *Step angle*
This is the angle through which the rotor rotates for one switching change for the stator coils.

3 *Holding torque*
This is the maximum torque that can be applied to a powered motor without moving it from its rest position and causing spindle rotation.

4 *Pull-in torque*
This is the maximum torque against which a motor will start, for a given pulse rate, and reach synchronism without losing a step.

5 *Pull-out torque*
This is the maximum torque that can be applied to a motor, running at a given stepping rate, without losing synchronism.

6 *Pull-in rate*
This is the maximum switching rate or speed at which a loaded motor can start without losing a step.

7 *Pull-out rate*
This is the switching rate or speed at which a loaded motor will remain in synchronism as the switching rate is reduced.

8 *Slew range*
This is the range of switching rates between pull-in and pull-out within which the motor runs in synchronism but cannot start up or reverse.

Figure 6.37 shows the general characteristics of a stepper motor.

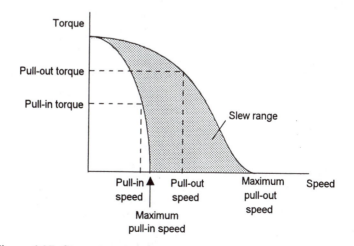

Figure 6.37 *Stepper motor characteristics*

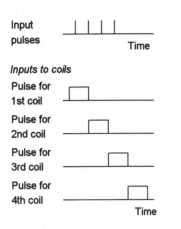

Input pulses — Time

Inputs to coils

Pulse for 1st coil

Pulse for 2nd coil

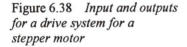

Pulse for 3rd coil

Pulse for 4th coil — Time

Figure 6.38 Input and outputs for a drive system for a stepper motor

To drive a stepper motor, so that it proceeds step-by-step to provide rotation, requires each pair of stator coils to be switched on and off in the required sequence when the input is a sequence of pulses (Figure 6.38). Driver circuits are available to give the correct sequencing and Figure 6.39 shows an example, the SAA 1027 for a four-phase unipolar stepper. Motors are termed *unipolar* if they are wired so that the current can only flow in one direction through any particular motor terminal, *bipolar* if the current can flow in either direction through any particular motor terminal. The stepper motor will rotate through one step each time the trigger input goes from low to high. The motor runs clockwise when the rotation input is low and anticlockwise when high. When the set pin is made low the output resets. In a control system, these input pulses might be supplied by a microprocessor.

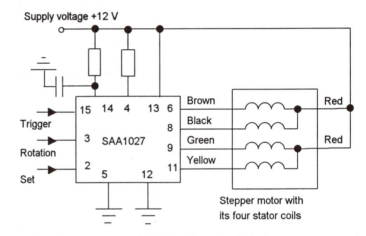

Figure 6.39 Driver circuit SAA1027 for a 12 V 4-phase stepper motor

Application

A manufacturer's data for a stepper motor includes:

12 V 4-phase, unipolar
Step angle 7.5°
Suitable driver SAA1027

Some applications require very small step angles. Though the step angle can be made small by increasing the number of rotor teeth and/or the number of phases, generally more than four phases and 50 to 100 teeth are not used. Instead a technique known as *mini-stepping* is used with each step being divided into a number of equal size sub-steps by using different currents to the coils so that the rotor moves to intermediate positions between normal step positions. For example, this method might be used so that a step of 1.8° is subdivided into 10 equal steps.

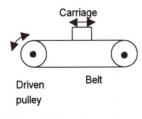

Figure 6.40 Example

Example

A stepper motor is to be used to drive, through a belt and pulley system (Figure 6.40), the carriage of a printer. The belt has to move a mass of 500 g which has to be brought up to a velocity of 0.2 m/s in a time of 0.1 s. Friction in the system means that movement of the carriage requires a constant force of 2 N. The pulleys have an effective diameter of 40 mm. Determine the required pull-in torque.

The force F required to accelerate the mass is:

$$F = ma = 0.500 \times (0.2/0.1) = 1.0 \text{ N.}$$

The total force that has to be overcome is the sum of the above force and that due to friction. Thus the total force that has to be overcome is $1.0 + 2 = 3$ N.

This force acts at a radius of 0.020 m and so the torque that has to be overcome to start, i.e. the pull-in torque, is

$$\text{torque} = \text{force} \times \text{radius} = 3 \times 0.020 = 0.06 \text{ N m}$$

6.6 Case studies   The following are case studies designed to illustrate the use of correction elements discussed in this chapter.

6.6.1 A liquid level process control system

Figure 6.41 one method of how a flow control valve can be used to control the level of a liquid in a container. Because there may be surface turbulence as a result of liquid entering the container or stirring of the liquid or perhaps boiling, such high frequency 'noise' in the system is often filtered out by the use of a *stilling well*, as shown in Figure 6.41. However, it must be recognised that the stilling well constitutes a U-tube in which low frequency oscillations of the liquid level can occur.

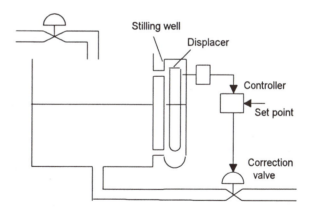

Figure 6.41 *Liquid level control*

6.6.2 A robot control system

Figure 6.42 shows how directional control valves can be used for a control system of a robot. When there is an input to solenoid A of valve 1, the piston moves to the right and causes the gripper to close. If solenoid B is energised, with A de-energised, the piston moves to the left and the gripper opens. When both solenoids are de-energised, no air passes to either side of the piston in the cylinder and the piston keeps its position without change. Likewise, inputs to the solenoids of valve 2 are used to extend or retract the arm. Inputs to the solenoids of valve 3 are used to move the arm up or down. Inputs to the solenoids of valve 4 are used to rotate the base in either a clockwise or anticlockwise direction.

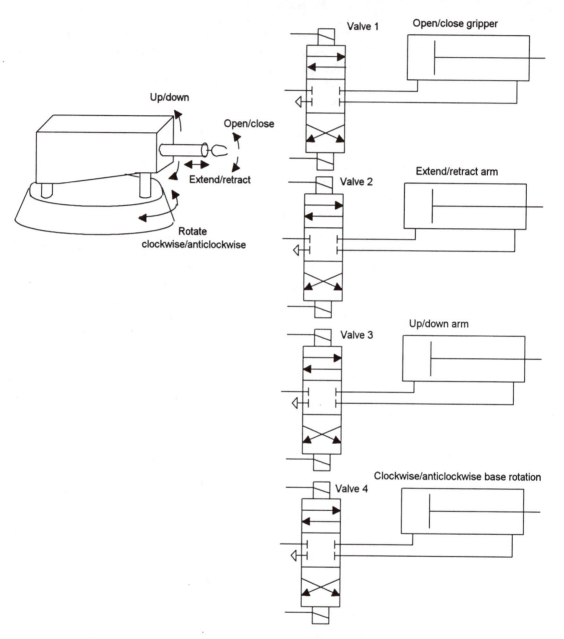

Figure 6.42 *Robot controls*

6.6.3 Milling machine control system

Figure 6.43 shows how a stepper motor can be used to control the movement of the workpiece in an automatic milling machine. The stepping motor rotates by controlled steps and gives, via a lead screw and gears, controlled displacements of the worktable in the xx direction. A similar arrangement is used for displacement in the yy direction. The system is open-loop control with no feedback of the work table position. The system relies on the accuracy with which the stepper motor can set the position of the work table.

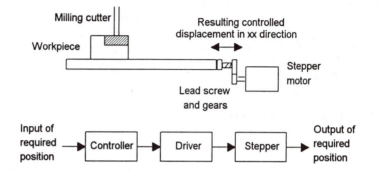

Figure 6.43 *Automatic milling machine*

Problems   *Questions 1 to 19 have four answer options: A, B, C and D. Choose the correct answer from the answer options.*

1 Decide whether each of these statements is True (T) or False (F).

For a hydraulic cylinder:
(i) The force that can be exerted by the piston is determined solely by the product of the pressure exerted on it and its cross-sectional area.
(ii) The speed with which the piston moves is determined solely by the product of the rate at which fluid enters the cylinder and the cross-sectional area of the piston.

A (i) T (ii) T
B (i) T (ii) F
C (i) F (ii) T
D (i) F (ii) F

2 A pneumatic cylinder has a piston of cross-sectional area 0.02 m². The force exerted by the piston when the working pressure applied to the cylinder is 2 MPa will be:

A 100 MN
B   40 MN
C 40 kN
D 20 kN

3 A hydraulic cylinder with a piston having a cross-sectional area of 0.01 m² is required to give a workpiece an average velocity of 20 mm/s. The rate at which hydraulic fluid should enter the cylinder is:

A 4×10^{-6} m³/s
B 2×10^{-4} m³/s
C 0.2 m³/s
D 2 m³/s

Questions 4 to 6 refer to Figure 6.44 which shows a valve symbol.

Figure 6.44   *Problems 4 to 6*

4 Decide whether each of these statements is True (T) or False (F).
The valve has:
(i)   2 ports
(ii) 4 positions

A (i) T   (ii) T
B (i) T   (ii) F
C (i) F   (ii) T
D (i) F   (ii) F

5 Decide whether each of these statements is True (T) or False (F).

When the push button is pressed:
(i) Hydraulic fluid from the supply is transmitted through port B.
(ii) The hydraulic fluid in the line to port A is returned to the sump.

A (i) T   (ii) T
B (i) T   (ii) F
C (i) F   (ii) T
D (i) F   (ii) F

6 Decide whether each of these statements is True (T) or False (F).

When the press button is released:
(i) Hydraulic fluid from the supply is transmitted through port A.
(ii) The hydraulic fluid in the line to port B is returned to the sump.

A (i) T   (ii) T
B (i) T   (ii) F
C (i) F   (ii) T
D (i) F   (ii) F

Questions 7 to 10 refer to Figure 6.45 which shows a pneumatic circuit involving two valves and a single acting cylinder.

7 Decide whether each of these statements is True (T) or False (F).

When push button 1 is pressed:
(i) The load is lifted.
(ii) Port A is closed.

A (i) T   (ii) T
B (i) T   (ii) F
C (i) F   (ii) T
D (i) F   (ii) F

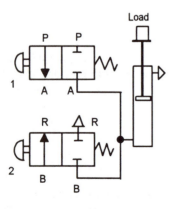

Figure 6.45   *Problems 7 to 10*

8 Decide whether each of these statements is True (T) or False (F).

When push button 1, after being pressed, is released:
(i) The load descends.
(ii) Port A is closed.

A (i) T (ii) T
B (i) T (ii) F
C (i) F (ii) T
D (i) F (ii) F

9 Decide whether each of these statements is True (T) or False (F).

When push button 2 is pressed:
(i) The load is lifted.
(ii) Port B is vented to the atmosphere.

A (i) T (ii) T
B (i) T (ii) F
C (i) F (ii) T
D (i) F (ii) F

10 Decide whether each of these statements is True (T) or False (F).

When push button:
(i) 1 is pressed the load is lifted.
(ii) 2 is pressed the load descends.

A (i) T (ii) T
B (i) T (ii) F
C (i) F (ii) T
D (i) F (ii) F

11 Decide whether each of these statements is True (T) or False (F).

Figure 6.46 shows a two-way spool valve. For this valve, movement of the shuttle from left to right:
(i) Closes port A.
(ii) Connects port P to port B.

A (i) T (ii) T
B (i) T (ii) F
C (i) F (ii) T
D (i) F (ii) F

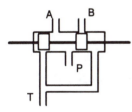

Figure 6.46   *Problem 11*

12 A flow control valve has a diaphragm actuator. The air pressure signals from the controller to give 0 to 100% correction vary from 0.02 MPa to 0.1 MPa above the atmospheric pressure. The diaphragm area needed to 100% open the control valve if a force of 400 N has to be applied to the stem to fully open the valve is:

A 0.02 m^3
B 0.016 m^3
C 0.004 m^3
D 0.005 m^3

13 Decide whether each of these statements is True (T) or False (F).

A quick-opening flow control valve has a plug shaped so that:
(i) A small change in the flow rate occurs for a large movement of the valve stem.
(ii) The change in the flow rate is proportional to the change in the displacement of the valve stem.

A (i) T (ii) T
B (i) T (ii) F
C (i) F (ii) T
D (i) F (ii) F

14 A flow control valve with a linear plug gives a minimum flow rate of 0 and a maximum flow rate of 10 m³/s. It has a stem displacement at full travel of 20 mm and so the flow rate when the stem displacement is 5 mm is:

A 0 m³/s
B 2.5 m³/s
C 5.0 m³/s
D 7.5 m³/s

15 A flow control valve with an equal percentage plug gives a flow rate of 0.1 m³/s when the stem displacement is 0 and 1.0 m³/s when it is at full travel. The stem displacement at full travel is 30 mm. The flow rate with a stem displacement of 15 mm is:

A 0.32 m³/s
B 0.45 m³/s
C 1.41 m³/s
D 3.16 m³/s

16 Decide whether each of these statements is True (T) or False (F).

A flow control valve has a minimum flow rate which is 1.0% of the maximum controllable flow. Such a valve is said to have a:
(i) Rangeability of 100.
(ii) Turndown of 100.

A (i) T (ii) T
B (i) T (ii) F
C (i) F (ii) T
D (i) F (ii) F

17 Decide whether each of these statements is True (T) or False (F).

A stepper motor is specified as having a step angle of 7.5°. This means that:
(i) The shaft takes 1 s to rotate through 7.5°.
(ii) Each pulse input to the motor rotates the motor shaft by 7.5°.

A (i) T (ii) T
B (i) T (ii) F
C (i) F (ii) T
D (i) F (ii) F

18 Decide whether each of these statements is True (T) or False (F).

For a series wound d.c. motor:
(i) The direction of rotation can be reversed by reversing the direction of the supplied current.
(ii) The speed of rotation of the motor can be controlled by changing the supplied current.

A (i) T (ii) T
B (i) T (ii) F
C (i) F (ii) T
D (i) F (ii) F

19 Decide whether each of these statements is True (T) or False (F).

With a shunt wound d.c. motor:
(i) The direction of rotation can be changed by reversing the direction of the armature current.
(ii) The direction of rotation can be changed by reversing the direction of the current to the field coils.

A (i) T (ii) T
B (i) T (ii) F
C (i) F (ii) T
D (i) F (ii) F

20 A force of 400 N is required to fully open a pneumatic flow control valve having a diaphragm actuator. What diaphragm area is required if the gauge pressure from the controller is 100 kPa?

21 An equal percentage flow control valve has a rangeability of 25. If the maximum flow rate is 50 m³/s, what will be the flow rate when the valve is one-third open?

22 A stepper motor has a step angle of 7.5°. What digital input rate is required to produce a rotation of 10.5 rev/s?

23 A control valve is to be selected to control the rate of flow of water into a tank requiring a maximum flow of 0.012 m³/s. The permissible pressure drop across the valve at maximum flow is 200 kPa. What valve size is required ? Use Table 6.1. The density of water is 1000 kg/m³.

24 A control valve is to be selected to control the flow of steam to a process, the maximum flow rate required being 0.125 kg/s. The permissible pressure drop across the valve at maximum flow is 40 kPa. What valve size is required? Use Table 6.1. The specific volume of the steam is 0.6 m³/s.

7 PLC systems

7.1 Introduction

In Section 4.7 the principle of *discrete-time control systems* was introduced, these being control systems in which one or more inputs can change only at discrete instants of time and involve logic control functions. For example, the control system for an automatic drilling machine (Figure 7.1) might be required to start lowering the drill when the workpiece is in position, start drilling when the drill reaches the workpiece and the workpiece is in position, stop drilling when the drill has produced the required depth of hole, retract the drill and then switch off and wait for the next workpiece to be put in position before repeating the operation.

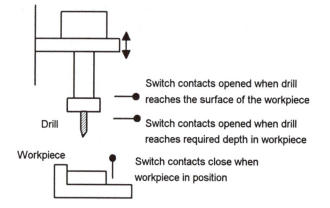

Figure 7.1 *An automatic drilling machine*

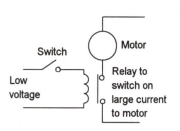

Figure 7.2 *Control circuit using a relay*

For such control operations, what form might a controller have? We could wire up electrical circuits in which the closing or opening of switches would result in motors being switched on or valves being actuated. Thus we might have the closing of a switch activating a relay which, in turn, switches on the current to a motor and causes the drill to rotate (Figure 7.2). Another switch might be used to activate a relay and switch on the current to a pneumatic or hydraulic valve which results in pressure being switched to drive a piston in a cylinder and so results in the workpiece being pushed into the required position. Such electrical circuits would have to be specific to the automatic drilling machine. However, instead of hardwiring each control circuit for each control situation we can use the same basic system for all situations if we use a microprocessor-based system and write a program to instruct the microprocessor how to react to each input signal from, say, switches and give the required outputs to, say, motors and valves. Thus we might have a program of the form:

> If switch A closes
>> Output to motor circuit
> If switch B and C closed
>> Output to valve circuit

By changing the instructions in the program we can use the same microprocessor system to control a wide variety of situations.

This chapter, after considering the functions of logic gates, takes a look at the *programmable logic controller* (PLC).

A PLC is a microprocessor-based system that uses a programmable memory to store instructions and implement functions such as logic, sequencing, timing, counting and arithmetic in order to control machines and processes and is designed to be operated by engineers with perhaps a limited knowledge of computers and computing languages.

Thus, the designers of the PLC have pre-programmed it so that the control program can be entered using a simple, rather intuitive, form of language. Input devices, e.g. sensors such as switches, and output devices in the system being controlled, e.g. motors, valves, etc., are connected to the PLC. The operator then enters a sequence of instructions, i.e. a program, into the memory of the PLC. The controller then monitors the inputs and outputs according to this program and carries out the control rules for which it has been programmed.

PLCs have the great advantage that the same basic controller can be used with a wide range of control systems. To modify a control system and the rules that are to be used, all that is necessary is for an operator to key in a different set of instructions. There is no need to rewire. The result is a flexible, cost effective, system which can be used with control systems which vary quite widely in their nature and complexity. The first PLC was developed in 1969. They are now widely used and extend from small self-contained units for use with perhaps 20 digital inputs/outputs to modular systems which can be used for large numbers of inputs/outputs, handle digital or analogue inputs/outputs, and also carry out proportional-integral-derivative control modes.

7.2 Logic gates

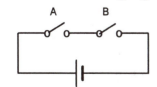

Figure 7.3   *An AND gate*

Consider the control system (Figure 7.3) where the water input valve to the domestic washing machine switched on if we have both the door to the machine closed, input signal A, and a particular time in the operating cycle has been reached, input signal B. There are two input signals which can be either yes or no signals and an output signal which can be a yes or no signal. The controller is programmed to give a yes output if both the input signals are yes and a no output when one or both of them are no.

Table 7.1 *AND gate*

Inputs		Output
A	B	Q
0	0	0
0	1	0
1	0	0
1	1	1

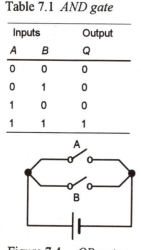

Figure 7.4   OR gate

Table 7.2 *OR gate*

Inputs		Output
A	B	Q
0	0	0
0	1	1
1	0	1
1	1	1

Figure 7.5   NOT gate

Table 7.3 *NOT gate*

Input	Output
A	Q
0	1
1	0

These two levels may be represented by the *binary* number system with the on level being represented by 1 and the off level by 0. Thus, for Figure 7.3, if input *A* and input *B* are both 1 then there is an output of 1. If either input A or input B or both are 0 then the output is 0. Such an operation is said to be controlled by a *logic gate*, in this case an AND gate.

The relationships between inputs to a logic gate and the outputs can be tabulated in a form known as a *truth table*. Thus for an AND gate with inputs *A* and *B* and a single output *Q*, we will have a 1 output when, and only when, *A* = 1 *and B* = 1. All other combinations of *A* and *B* will generate a 0 output. The truth table is given in Table 7.1.

An example of an AND gate is an interlock control system for a machine tool such that if the safety guard is in place, giving a 1 signal, and the power is on, giving a 1 signal, then there can be a 1 output and the machine will operate. If either of the inputs is 0 then the machine will not operate.

An OR gate is a system which with inputs *A* and *B* gives an output of a 1 when *A or B* is 1. We can visualise the OR gate as an electrical circuit which has two switches in parallel (Figure 7.4). When switch *A or B* is closed then there is a current. Table 7.2 is the truth table. An example of an OR gate is a conveyor belt system transporting finished bottled products to packaging where an arm is required to deflect bottles off the belt if either the weight is not within certain tolerances or there is no cap on a bottle.

A NOT gate has just one input and one output, giving a 1 output when the input is 0 and a 0 output when the input is 1. The NOT gate gives an output which is the inversion of the input and is thus often called an *inverter*. We can visualise such a gate as being an electrical circuit (Figure 7.5) with a switch which is normally allowing current to pass but when pressed switches the current off. Table 7.3 is the truth table. An example of a situation where a NOT gate might be used is where a light has to come on when the light level falls below a set value. This might be a light which comes on at night. When there is an input there is not an output.

Two forms of *standard circuit symbols* are in use for logic gates, one having originated in the United States and the other being an international form (IEEE/ANSI) which uses a rectangle with a symbol for the logic function inside it. Figure 7.6 shows the symbols for the AND, OR and NOT gates.

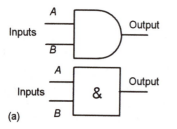

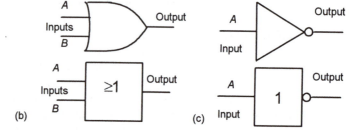

Figure 7.6 *Logic gate symbols: (a) AND, (b) OR, (c) NOT*

Gates can be combined to produce other relationships between inputs and outputs. Combining an AND gate with a NOT gate (Figure 7.7) gives what is termed a NAND gate. Table 7.4 is the truth table.

Table 7.4 *NAND gate*

Inputs		Output from AND gate	Output from NOT gate
A	B		Q
0	0	0	1
0	1	0	1
1	0	0	1
1	1	1	0

Table 7.5 *NOR gate*

Inputs		
A	B	Output
0	0	1
0	1	0
1	0	0
1	1	0

Table 7.6 *XOR gate*

Inputs		
A	B	Output
0	0	0
0	1	1
1	0	1
1	1	0

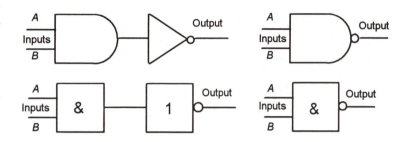

Figure 7.7 *NAND gate*

Likewise, if an OR gate is combined with a NOT gate we obtain a NOR gate (Figure 7.8) (Table 7.5), an exclusive OR (XOR) gate by a combination of OR and AND gates (Figure 7.9) (Table 7.6).

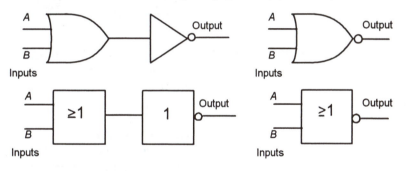

Figure 7.8 *NOR gate*

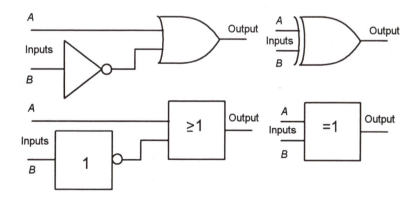

Figure 7.9 *XOR gate*

Application
Logic gates are available as integrated circuit chips. For example, the integrated circuit 74LS21 has two 4-input AND gates in one 14 pin package (Figure 7.10). The integrated circuit 74LS08 has four 2-input AND gates in one 14 pin package. The integrated circuit 74LS32 has four 2-input OR gates in one 14 pin package.

Example

What types of logic gates might be needed in the following control situations: (a) part of a chemical plant where an alarm is to be

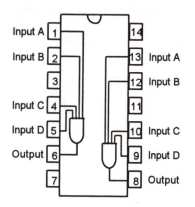

Figure 7.10 *74LS21 integrated circuit*

activated if the temperature falls below a certain level, (b) an automatic door is to open if a person approaches from either side?

(a) This requires a NOT gate in that the conditions required are:

Temperature	Alarm
Low (0)	On (1)
High (1)	Off (0)

(b) This requires an OR gate in that the conditions required are:

Person on side A	Person on side B	Door
Yes (1)	No (0)	Open (1)
No (0)	Yes (1)	Open (1)
No (0)	No (0)	Closed (0)
Yes (1)	Yes (1)	Open (1)

7.3 PLC system

Typically a PLC system has five basic components. These are the processor unit, memory, the power supply unit, input/output interface section and the programming device. Figure 7.11 shows the basic arrangement.

1. The *processor unit* or *central processing unit (CPU)* is the unit containing the microprocessor and this interprets the input signals and carries out the control actions, according to the program stored in its memory, communicating the decisions as action signals to the outputs.

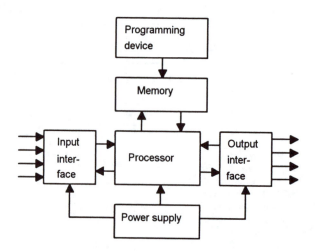

Figure 7.11 *The PLC system*

2 The *power supply unit* is needed to convert the mains a.c. voltage to the low d.c. voltage (5 V) necessary for the processor and the circuits in the input and output interface modules.

3 The *programming device* is used to enter the required program into the memory of the processor. The program is developed in the device and then transferred to the memory unit of the PLC.

4 The *memory unit* is where the program is stored that is to be used for the control actions to be exercised by the microprocessor.

5 The *input and output sections* are where the processor receives information from external devices and communicates information to external devices. Every input/output point has a unique address in the system. The inputs might thus be from switches, as illustrated in Figure 7.1 with the automatic drill, or other sensors such as photo-electric cells, temperature sensors, flow sensors, etc. The outputs might be to motor starter coils, solenoid valves, etc. The input/output channels provide isolation and signal conditioning functions so that sensors and actuators can often be directly connected to them without the need for other circuitry. Electrical isolation from the external world is usually by means of *optoisolators* (the term *optocoupler* is also often used) (Figure 7.12). When a digital pulse passes through the light-emitting diode, a pulse of infrared radiation is produced which is detected by the phototransistor and gives rise to a voltage in that circuit. The gap between the light-emitting diode and the phototransistor gives electrical isolation but the arrangement still allows for a digital pulse in one circuit to give rise to a digital pulse in another circuit. Outputs are often specified as being of relay type, transistor type or triac type. With the *relay type*, the signal from the PLC output is used to operate a relay and so is able to switch currents of the order of a few amperes in an external circuit. The relay not only allows small currents to switch much larger currents but also isolates the PLC from the external circuit. Relays are, however, relatively slow to operate. Relay outputs are suitable for a.c. and d.c. switching. They can withstand high surge currents and voltage transients. The *transistor type* of output uses a transistor to switch current through the external circuit. This gives a considerably faster switching action. It is, however, strictly for d.c. switching and is destroyed by overcurrent and high reverse voltage. As a protection, either a fuse or built-in electronic protection are used. Optoisolators are used to provide isolation. *Triac* outputs, with optoisolators for isolation, can be used to control external loads which are connected to the a.c. power supply. It is strictly for a.c. operation and is very easily destroyed by overcurrent. Fuses are virtually always included to protect such outputs.

There are a wide range of PLCs commercially available. Small PLCs are generally in a single box (or, as sometimes termed, brick) and typical commercial forms might have 6, 8, 12 or 24 inputs and 4, 8 or 16

Figure 7.12 *Optocoupler*

Application
The following are some examples of small PLCs.

With the Mitsubishi FX family of PLCs, the model FX1S is available with twelve 24 V d.c. outputs and relay type outputs, transistor type outputs, or triac type outputs. Thus, FX1S-10 MT-ESS/UL has six 24 V d.c. inputs and four transistor outputs, FX1S-10 MR-ES/UL has six 24 V d.c. inputs and four relay outputs.

The Toshiba T1 family of PLCs is likewise available in a wide variety of forms, e.g. with eight 24 D d.c. inputs, six relay outputs and two transistor outputs.

The Siemens Sismatic S5 range of PLCs includes a ten input, six relay output model.

outputs. They are designed to be used close to the equipment being controlled. Systems with larger numbers of inputs and outputs are likely to be modular and designed to fit in racks. The number of inputs and outputs of a system can then be readily increased by adding more modules.

7.4 PLC programming

The basic form of programming used with PLCs is *ladder programming*. As an introduction to ladder diagrams, consider the simple wiring diagram for an electrical circuit in Figure 7.13(a). The diagram shows the circuit for switching on or off an electric motor. We can redraw this diagram in a different way, using two vertical lines to represent the input power rails and stringing the rest of the circuit between them. Figure 7.13(b) shows the result. Both circuits have the switch in series with the motor and supplied with electrical power when the switch is closed. The circuit shown in Figure 7.13(b) is termed a *ladder diagram*.

With the ladder diagram the power supply for the circuits is always shown as two vertical lines with the rest of the circuit as horizontal lines. The power lines, or rails as they are often termed, are like the vertical sides of a ladder with the horizontal circuit lines like the rungs of the ladder. The horizontal rungs show only the control portion of the circuit, in the case of Figure 7.13 it is just the switch in series with the motor. Circuit diagrams often show the relative physical location of the circuit components and how they are actually wired. With ladder diagrams no attempt is made to show the actual physical locations and the emphasis is on clearly showing how the control is exercised.

Writing a program for a PLC using ladder programming is equivalent to drawing a switching circuit of the form shown in Figure 7.13(b). Figure 7.14 shows a simple ladder program. In drawing a ladder diagram, certain conventions are adopted:

1   The vertical lines of the diagram represent the power rails between which circuits are connected.

2   Each rung on the ladder defines one operation in the control process.

3   A ladder diagram is read from left to right and from top to bottom. Thus, the top rung is read from left to right. Then the second rung down is read from left to right and so on. When the PLC is in its run mode, it goes through the entire ladder program to the end, the end rung of the program being clearly denoted, and then promptly resumes at the start. This procedure of going through all the rungs of the program is termed a *cycle*.

4   Each rung must start with an input or inputs and must end with at least one output, input devices being represented by two short parallel lines to represent switching contacts and output devices being represented by circles. The term input is used for a control action, such as closing the contacts of a switch, used as an input to the PLC. The term output is used for a device connected to the output of a PLC, e.g. a motor.

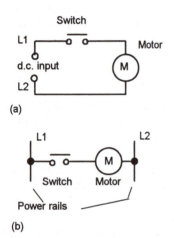

(a)

(b)

Figure 7.13 *Ways of drawing the same electrical circuit*

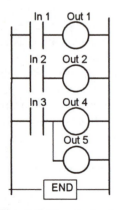

Figure 7.14 *A simple ladder program*

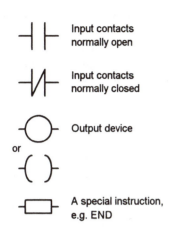

Input contacts
normally open

Input contacts
normally closed

Output device

or

A special instruction,
e.g. END

Figure 7.15 *Standard symbols*

X400   Y430

Figure 7.16 *An example of a ladder rung*

X400   X401   Y430

Figure 7.17 *An AND gate*

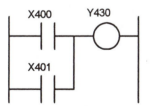

X400   Y430

X401

Figure 7.18 *An OR gate*

X400   Y430

Figure 7.19 *NOT gate*

5 Electrical devices are shown in their normal condition. Thus a switch which is normally open until some object closes it, is shown as open on the ladder diagram. A switch that is normally closed is shown closed.

6 A particular device can appear in more than one rung of a ladder. For example, we might have a relay which switches on one or more devices. The same letters and/or numbers are used to label the device in each situation.

7 The inputs and outputs are all identified by their addresses, the notation used depending on the PLC manufacturer. This is the address of the input or output in the memory of the PLC. For example, *Mitsubishi* PLCs precede input elements by an X and output elements by a Y and thus we have numbers such as X400 and X401 for inputs and Y430 and Y431 for outputs.

Figure 7.15 shows some of the standard symbols used in ladder diagrams.

To illustrate the drawing of a ladder diagram rung, consider a situation where the starting of a motor output device depends on a normally open start switch being closed. Starting with the input, we represent the normally open switch by the symbol | |. This might be labelled with the address X400. The ladder rung terminates with the output, the motor, which is designated by the symbol O. This might be labelled with the address Y430. We thus have the ladder diagram shown in Figure 7.16. When the switch is closed the motor is activated.

7.4.1 Logic gates

Ladder rungs are frequently written to carry out logic functions. For example, consider a situation where a motor is started when two, normally open, switches have both to be closed. This might represent a machine tool motor which will not start until the power switch is on and the switch indicating the closure of the safety guard is on. This describes an AND logic gate situation. The ladder diagram starts with | |, labelled X400, to represent switch A and in series with it | |, labelled X401, to represent switch B. The line then terminates with O, labelled Y430, to represent the output. Figure 7.17 shows the ladder rung.

Figure 7.18 shows a situation where a motor is not switched on until either, normally open, switch A or switch B is closed. The situation is an OR logic gate. The ladder diagram starts with | |, labelled X400, to represent switch A and in parallel with it | |, labelled X401, to represent switch B. The line then terminates with O, labelled Y430, to represent the output.

Figure 7.19 shows a NOT gate system on a ladder diagram. The input X400 contacts are shown as being normally closed. This is in series with the output Y430. With no input to X400, the contacts are closed and so there is an output. When there is an input, the contacts open and there is then no output. An example of a NOT gate control system is a light that

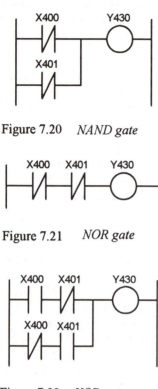

Figure 7.20   *NAND gate*

Figure 7.21   *NOR gate*

Figure 7.22   *XOR gate*

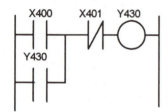

Figure 7.23   *Latch circuit*

comes on when it becomes dark, i.e. when there is no light input to the light sensor there is an output.

Figure 7.20 shows a ladder diagram which gives a NAND gate. When the inputs to input X400 and input X401 are both 0 then the output is 1. When the inputs to input X400 and input X401 are both 1, or one is 0 and the other 1, then the output is 0. An example of a NAND gate control system is a warning light that comes on if, with a machine tool, the safety guard switch has not been activated and the limit switch signalling the presence of the workpiece has not been activated.

Figure 7.21 shows a ladder diagram of a NOR system. When input X400 and input X401 are both not activated, there is a 1 output. When either X400 or X401 are 1 there is a 0 output.

Figure 7.22 shows a ladder diagram for an XOR gate system. When input X400 and input X401 are not activated then there is 0 output. When just input X400 is activated, then the upper branch results in the output being 1. When just input X401 is activated, then the lower branch results in the output being 1. When both input X400 and input X401 are activated, there is no output. In this example of a logic gate, input X400 and input X401 have two sets of contacts in the circuits, one set being normally open and the other normally closed. With PLC programming, each input may have as many sets of contacts as necessary.

7.4.2 Latching

There are often situations where it is necessary to hold an output energised, even when the input ceases. A simple example of such a situation is a motor which is started by pressing a push button switch. Though the switch contacts do not remain closed, the motor is required to continue running until a stop push button switch is pressed. The term *latch circuit* is used for the circuit used to carry out such an operation. It is a self-maintaining circuit in that, after being energised, it maintains that state until another input is received.

An example of a latch circuit is shown in Figure 7.23. When the input X400 contacts close, there is an output at Y430. However, when there is an output, another set of contacts | | Y430 associated with the output closes. These contacts form an OR logic gate system with the input contacts. Thus, even if the input X400 opens, the circuit will still maintain the output energised. The only way to release the output is by operating the normally closed contact X401.

As an illustration of the application of a latching circuit, consider a motor controlled by stop and start push button switches and for which one signal light must be illuminated when the power is applied to the motor and another when it is not applied. Figure 7.24 shows the ladder diagram. When X400 is momentarily closed, Y430 is energised and its contacts close. This results in latching and also the switching off of Y431 and the switching on of Y432. To switch the motor off, X401 is pressed and opens. Y430 contacts open in the top rung and third rung, but close in the second rung. Thus Y431 comes on and Y432 off.

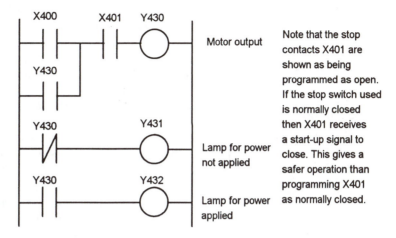

Motor output

Note that the stop contacts X401 are shown as being programmed as open. If the stop switch used is normally closed then X401 receives a start-up signal to close. This gives a safer operation than programming X401 as normally closed.

Lamp for power not applied

Lamp for power applied

Figure 7.24 *Motor on–off, with signal lamps, ladder diagram*

7.4.3 Internal relays

In PLCs there are elements that behave like relays, being able to be switched on or off and, in turn, switch other devices on or off. Hence the term *internal relay*. Such internal relays do not exist as real-world switching devices but are merely bits in the storage memory that behave in the same way as relays. For programming, they can be treated in the same way as an external relay output and input. Thus we might have (Figure 7.25):

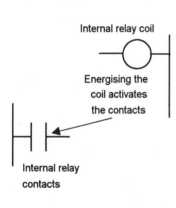

Internal relay coil

Energising the coil activates the contacts

Internal relay contacts

Figure 7.25 *Internal relay*

> *On one rung of the program*:
>   Inputs to external inputs activate the internal relay output.

> *On a later rung of the program*:
>   As a consequence of the internal relay output:
>   internal relay contacts are activated and so control some output.

In using an internal relay, it has to be activated on one rung of a program and then its output used to operate switching contacts on another rung, or rungs, of the program. Internal relays can be programmed with as many sets of associated contacts as desired.

To distinguish internal relay outputs from external relay outputs, they are given different types of addresses. Different manufacturers tend to use different terms for internal relays and different ways of expressing their addresses. For example, Mitsubishi uses the term *auxiliary relay* or *marker* and the notation M100, M101, etc. , the M indicating that it is an internal relay or marker rather than an external device. Siemens use the term *flag* and notation F0.0, F0.1, etc. The internal relay switching contacts are designated with the symbol for an input device, namely ||, and given the same address as the internal relay output, e.g. M100.

As an illustration of the use that can be made of internal relays, consider the following situation. A system is to be activated when two different sets of input conditions are realised. We might just program this as an AND logic gate system; however, if a number of inputs have to be checked in order that each of the input conditions can be realised, it may be simpler to use an internal relay. The first input conditions then are used to give an output to an internal relay with the associated contacts becoming part of the input conditions with the second input.

Figure 7.26 shows a ladder program for such a task. For the first rung: when input X400 or input X402 is closed and input X401 closed, then internal relay M100 is activated. This results in the contacts M100 closing. If input X403 is then activated, there is an output from output Y430. Such a task might be involved in the automatic lifting of a barrier when someone approaches from either side. Input X400 and input X402 are inputs from photoelectric sensors that detect the presence of a person, input X400 being activated from one side of it and input X402 from the other. Input X401 is an enabling switch to enable the system as a whole to be switched on. Thus when input X400 or input X402, and input X401, are activated, there is an output from the internal relay M100. This will close the internal relay contacts M100. If input X403, perhaps a limit switch, detects that the barrier is closed then it is activated and closes. The result is then an output from Y430, a motor which lifts the barrier. If the limit switch detects that the barrier is already open, the person having passed through it, then it remains open and so output Y430 is no longer energised and a counterweight might then close the barrier. The internal relay has enabled two parts of the program to be linked, one part being the detection of the presence of a person and the second part the detection of whether the barrier is already up or down.

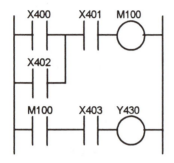

Figure 7.26 *Program*

7.4.4 Timers

In many control tasks there is a need to control time. For example, a motor or a pump might need to be controlled to operate for a particular interval of time, or perhaps be switched on after some time interval. PLCs thus have *timers* as built-in devices. Timers count fractions of seconds or seconds using the internal CPU clock.

PLC manufacturers differ on how timers should be programmed and hence how they can be considered. A common approach is to consider timers to behave like relays with coils which when energised result in the closure or opening of contacts after some pre-set time. The timer is thus treated as an output for a rung with control being exercised over pairs of contacts elsewhere (Figure 7.27). There are a number of different forms of timers that can be found with PLCs. With small PLCs there is likely to be just one form, the *on-delay timers*. These are timers which come on after a particular time delay (Figure 7.28). The time duration for which a timer has been set is termed the pre-set and is set in multiples of the time base used. Typical time bases are 10 ms, 100 ms, 1 s, 10 s and 100 s. Thus a pre-set value of 5 with a time base of 100 ms is a time of 500 ms.

Figure 7.29 shows a ladder rung diagram involving a delay-on timer. The timer is like a relay with a coil which is energised when the input X400 occurs (rung 1). It then closes, after some pre-set time delay, its contacts on rung 2. Thus the output from Y430 occurs some pre-set time after the input X400 occurs.

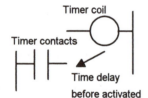

Figure 7.27 *Timer*

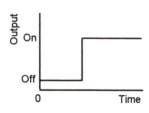

Figure 7.28 *On-delay timer*

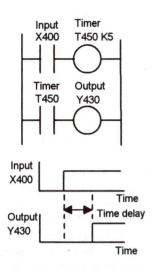

Figure 7.29 *Ladder program with a delay-on timer*

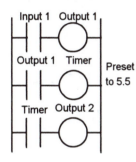

Figure 7.30 *Sequenced outputs*

As an illustration of the use of a timer, consider the ladder diagram shown in Figure 7.30. When the input 1 is on, the output 1 is switched on. The contacts associated with this output then start the timer. The contacts of the timer will close after the pre-set time delay. In this case with $K = 5.5$ and using a time base of 1 s the time delay is 5.5 s. When this happens, output 2 is switched on. Thus, following the input 1, output 1 is switched on and followed 5.5 s later by output 2. This illustrates how timed sequence of outputs can be achieved.

Figure 7.31 shows how timers can be used to start three outputs, e.g. three motors, in sequence following a single start button being pressed. When the start push button is pressed there is an output from internal relay IR1. This latches the start input. It also starts both the timers, T1 and T2, and motor 1. When the pre-set time for timer 1 has elapsed then its contacts close and motor 2 starts. When the pre-set time for timer 2 has elapsed then its contacts close and motor 3 starts. The three motors are all stopped by pressing the stop push button. Since this is seen as a complete program, the end instruction has been used.

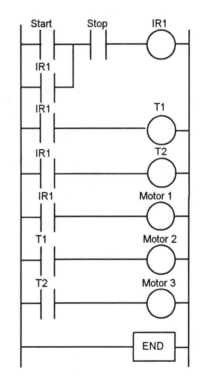

Figure 7.31 *Motor sequence*

7.4.5 Counters

Counters are provided as built-in elements in PLCs. A *counter* allows a number of occurrences of input signals to be counted. This might be counting the number of revolutions of a shaft, or perhaps the number of people passing through a door.

A counter is set to some pre-set number value and, when this value of input pulses has been received, it will operate its contacts. Thus normally open contacts would be closed, normally closed contacts opened. There are two types of counter, though PLCs may not include both types. These are down-counters and up-counters. *Down-counters* count down from the pre-set value to zero, i.e. events are subtracted from the set value. When the counter reaches the zero value, its contacts change state. Most PLCs offer down counting. *Up-counters* count from zero up to the pre-set value, i.e. events are added until the number reaches the pre-set value. When the counter reaches the set value, its contacts change state.

Different PLC manufacturers deal with counters in slightly different ways. Some treat the counter as two basic elements: one relay coil to count input pulses and one (RST) to reset the counter, the associated contacts of the counter being used in other rungs. Figure 7.32 illustrates this with a basic counting circuit. When there is a pulse input to input 1, the counter is reset. When there is an input to input 2, the counter starts counting. If the counter is set for, say, 10 pulses, then when 10 pulse inputs have been received at input 2, the counter's contacts will close and there will be an output from output 1. If at any time during the counting there is an input to input 1, the counter will be reset and start all over again and count for 10 pulses.

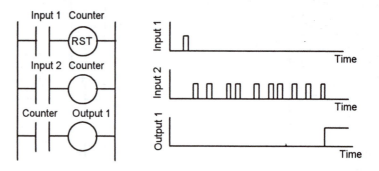

Figure 7.32 *Basic counter program*

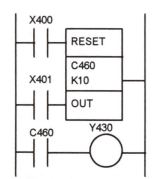

Figure 7.33 *Mitsubishi program*

Figure 7.33 shows how the above program would appear with a Mitsubishi PLC. The reset and counting elements are combined in a single box spanning the two rungs. You can consider the rectangle to be enclosing the two counter outputs in Figure 7.32.

Counters can be used to ensure that a particular part of a sequence is repeated a known number of times. This is illustrated by the following program which is designed to enable a three-cylinder, double solenoid-controlled arrangement to give the sequence A+, A−, A+, A−, A+, A−, B+, C+, B−, C−. The A+, A− sequence is repeated three times before B+, C+, B−, C− occur. We can use a counter to enable this repetition. Figure 7.34 shows a possible program. The counter only allows B+ to occur after it has received three inputs corresponding to three a− signals.

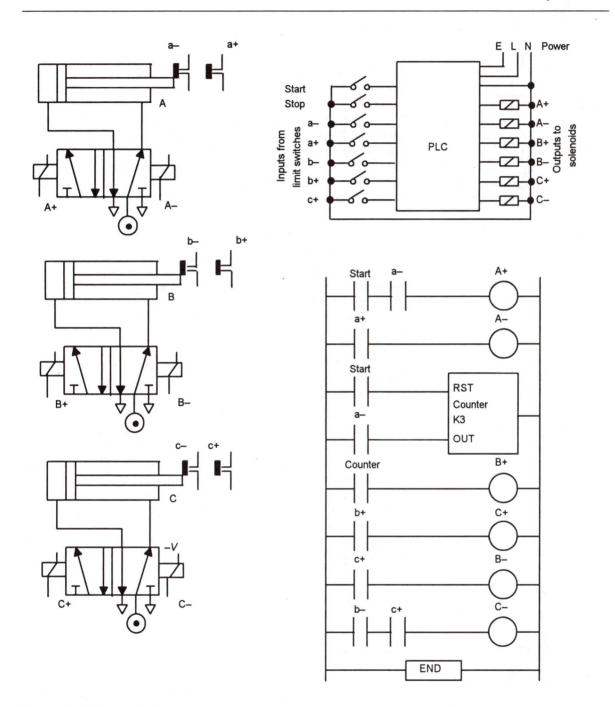

Figure 7.34 *Three-cylinder system*

7.5 Case studies The following case studies are intended to illustrate the application of the PLC programming techniques given in this chapter.

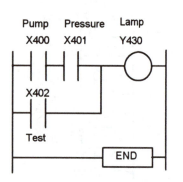

Figure 7.35 *Signal lamp*

7.5.1 Signal lamp to monitor operations

Consider a basic situation where a signal lamp is required to be switched on if a pump is running and the pressure is satisfactory, or if the lamp test switch is closed. For the inputs from the pump and the pressure sensors we have an AND logic situation since both are required if there is to be an output from the lamp. We, however, have an OR logic situation with the test switch in that it is required to give an output of lamp on regardless of whether there is a signal from the AND system. The ladder program is thus of the form shown in Figure 7.35. Note that we tell the PLC when it has reached the end of the program by the use of the END instruction.

7.5.2 Cyclic movement of a piston

Consider the task of obtaining cyclic movement of a piston in a cylinder. This might be to periodically push workpieces into position in a machine tool with another similar, but out of phase, arrangement being used to remove completed workpieces. Figure 7.36 shows the valve and piston arrangement that might be used, a possible ladder program and chart indicating the timing of each output.

Consider both timers set for 10 s. When the start contacts X400 are closed, timer T450 starts. Also there is an output from Y431. The output Y431 is one of the solenoids used to actuate the valve. When it is energised it causes the pressure supply P to be applied to the right-hand end of the cylinder and the left-hand side to be connected to the vent to the atmosphere. The piston thus moves to the left. After 10 s, the normally open T450 contacts close and the normally closed T450 contacts open. This stops the output Y431, starts the timer T451 and energises the output Y430. As a result, the pressure supply P is applied to the left-hand side of the piston and the right-hand side connected to the vent to the atmosphere. The piston now moves to the right. After 10 s, the T451 normally closed contacts are opened. This causes the normally closed contacts of T450 to close and so Y431 is energised. Thus the sequence repeats itself.

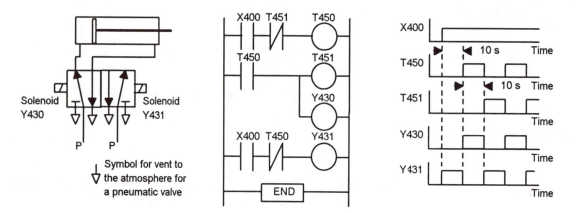

Figure 7.36 *Cyclic movement of a piston*

7.5.3 Sequential movement of pistons

Consider another task involving three pistons A, B and C that have to be actuated in the sequence: A to the right, A to the left, B to the right, B to the left, C to the right, C to the left (such a sequence is often written A+, A–, B+, B–, C+, C–). Figure 7.37 shows the valves and a ladder program using timers that might be used.

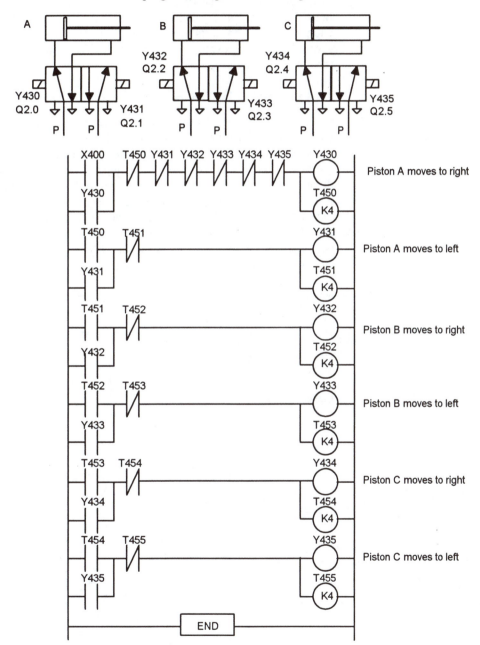

Figure 7.37 *Sequential movement of pistons*

X400 is the start switch. When it is closed there is an output from Y430 and the timer T450 starts. The start switch is latched by the output. Piston A moves to the right. After the set time, K = 4, the normally closed timer T450 contacts open and the normally open timer T450 contacts close. This switches off Y430 and energises Y431 and starts timer T451. Piston A moves to the left. In rung 2, the T450 contacts are latched and so the output remains on until the set time has been reached. When this occurs the normally closed timer T451 contacts open and the normally open T451 contacts close. This switches off Y431 and energises Y432 and starts timer T452. Piston B moves to the right. Each succeeding rung activates the next solenoid. Thus in sequence, each of the outputs is energised.

7.5.4 Central heating system

Consider a domestic central heating system where the central heating boiler is to be thermostatically controlled and supply hot water to the radiator system in the house and also to the hot water tank to provide hot water from the taps in the house. Pump motors have to be switched on to direct the hot water from the boiler to either, or both the radiator and hot water systems according to whether the temperature sensors for the room temperature and the hot water tank indicate that the radiators or tank need heating. The entire system is to be controlled by a clock so that it only operates for certain hours of the day. Figure 7.38 shows the system and how a Mitsubishi PLC might be used.

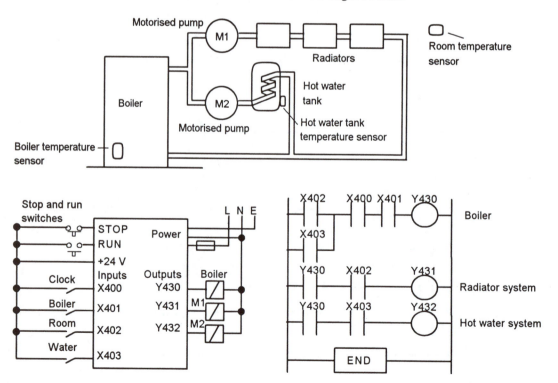

Figure 7.38 *Central heating system*

The boiler, output Y430, is switched on if X400 and X401 and either X402 or X403 are switched on. This means if the clock is switched on, the boiler temperature sensor gives an on input, and either the room temperature sensor or the water temperature sensor give on inputs. The motorised valve M1, output Y431, is switched on if the boiler, Y430, is on and if the room temperature sensor X402 gives an on input. The motorised valve M2, output Y432, is switched on if the boiler, Y430, is on and if the water temperature sensor gives an on input.

Problems *Questions 1 to 6 have four answer options: A, B, C or D. Choose the correct answer from the answer options.*

1 Decide whether each of these statements is True (T) or False (F).

Figure 7.39 shows a ladder diagram rung for which:
(i) The input contacts are normally open.
(ii) There is an output when there is an input to the contacts.

A (i) T (ii) T
B (i) T (ii) F
C (i) F (ii) T
D (i) F (ii) F

Figure 7.39   *Problem 1*

2 Decide whether each of these statements is True (T) or False (F).

Figure 7.40 shows a ladder diagram rung for which:
(i) The input contacts are normally open.
(ii) There is an output when there is an input to the contacts.

A (i) T (ii) T
B (i) T (ii) F
C (i) F (ii) T
D (i) F (ii) F

Figure 7.40   *Problem 2*

3 Decide whether each of these statements is True (T) or False (F).

Figure 7.41 shows a ladder diagram rung for which:
(i) When only input 1 contacts are activated, there is an output.
(ii) When only input 2 contacts are activated, there is an output.

A (i) T (ii) T
B (i) T (ii) F
C (i) F (ii) T
D (i) F (ii) F

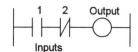

Figure 7.41   *Problem 3*

4 Decide whether each of these statements is True (T) or False (F).

Figure 7.42 shows a ladder diagram rung for which there is an output when:
(i) Inputs 1 and 2 are both activated.
(ii) Either one of inputs 1 and 2 is not activated.

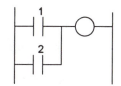

Figure 7.42   *Problem 4*

A (i) T (ii) T
B (i) T (ii) F
C (i) F (ii) T
D (i) F (ii) F

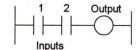

Inputs

Figure 7.43   *Problem 5*

5 Decide whether each of these statements is True (T) or False (F).

Figure 7.43 shows a ladder diagram rung with an output when:
(i) Inputs 1 and 2 are both activated.
(ii) Input 1 or 2 is activated.

A (i) T (ii) T
B (i) T (ii) F
C (i) F (ii) T
D (i) F (ii) F

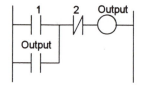

Figure 7.44   *Problem 6*

6 Decide whether each of these statements is True (T) or False (F).

Figure 7.44 shows a ladder diagram rung for which there is an output when:
(i) Input 1 is momentarily activated before reverting to its normally open state.
(ii) Input 2 is activated.

A (i) T (ii) T
B (i) T (ii) F
C (i) F (ii) T
D (i) F (ii) F

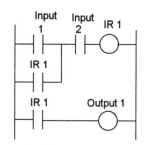

Figure 7.45   *Problems 7, 8 and 9*

Problems 7 to 9 refer to Figure 7.45 which shows a ladder diagram with an internal relay, designated IR 1, two inputs Input 1 and Input 2, and an output Output 1.

7 Decide whether each of these statements is True (T) or False (F).

For the ladder diagram shown in Figure 7.45, there is an output from Output 1 when:
(i) There is just an input to Input 1.
(ii) There is just an input to Input 2.

A (i) T (ii) T
B (i) T (ii) F
C (i) F (ii) T
D (i) F (ii) F

8 Decide whether each of these statements is True (T) or False (F).

For the ladder diagram shown in Figure 7.45, there is an output from Output 1 when:
(i) There is an input to Input 2 and a momentary input to Input 1.
(ii) There is an input to Input 1 or an input to Input 2.

A (i) T (ii) T
B (i) T (ii) F
C (i) F (ii) T
D (i) F (ii) F

9 Decide whether each of these statements is True (T) or False (F).

For the ladder diagram shown in Figure 7.45, the internal relay:
(i)   Switches on when there is just an input to Input 1.
(ii) Switches on when there is an input to Input 1 and to Input 2.

A (i) T (ii) T
B (i) T (ii) F
C (i) F (ii) T
D (i) F (ii) F

Problems 10 to 12 refer to Figure 7.46 which shows a ladder diagram involving internal relays IR 1 and IR 2, inputs Input 1, Input 2 and Input 3, and output Output 1.

10 Decide whether each of these statements is True (T) or False (F).

For the ladder diagram shown in Figure 7.46, the internal relay IR 1 is energised when:
(i)   There is an input to Input 1.
(ii) There is an input to Input 3.

A (i) T (ii) T
B (i) T (ii) F
C (i) F (ii) T
D (i) F (ii) F

11 Decide whether each of these statements is True (T) or False (F).

For the ladder diagram shown in Figure 7.46, the internal relay IR 2 is energised when:
(i)   Internal relay IR 1 is energised.
(ii) Input 4 is energised.

A (i) T (ii) T
B (i) T (ii) F
C (i) F (ii) T
D (i) F (ii) F

12 Decide whether each of these statements is True (T) or False (F).

For the ladder diagram shown in Figure 7.46, there is an output from Output 1 when:
(i)   There are inputs to just Input 1, Input 2 and Input 4.
(ii) There are inputs to just Input 3 and Input 4.

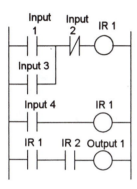

Figure 7.46 *Problems 10, 11 and 12*

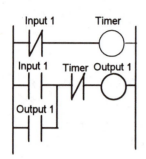

Figure 7.47   *Problems 13 to 16*

A (i) T (ii) T
B (i) T (ii) F
C (i) F (ii) T
D (i) F (ii) F

Problems 13 to 16 refer to Figure 7.47 which shows a ladder diagram with an on-delay timer, an input Input 1 and an output Output 1.

13 Decide whether each of these statements is True (T) or False (F).

When there is an input to Input 1 in Figure 7.47:
(i) The timer starts.
(ii) There is an output from Output 1.

A (i) T (ii) T
B (i) T (ii) F
C (i) F (ii) T
D (i) F (ii) F

14 Decide whether each of these statements is True (T) or False (F).

The timer in Figure 7.47 starts when:
(i) There is an output.
(ii) The input ceases.

A (i) T (ii) T
B (i) T (ii) F
C (i) F (ii) T
D (i) F (ii) F

15 Decide whether each of these statements is True (T) or False (F).

When there is an input to Input 1 in Figure 7.47, the output is switched:
(i) On for the time for which the timer was pre-set.
(ii) Off for the time for which the timer was pre-set.

A (i) T (ii) T
B (i) T (ii) F
C (i) F (ii) T
D (i) F (ii) F

Problems 16 to 18 refer to Figure 7.48 which shows a ladder diagram with a counter, two inputs Input 1 and Input 2 and an output Output 1.

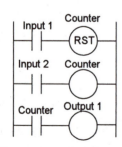

Figure 7.48   *Problems 16 to 18*

16 Decide whether each of these statements is True (T) or False (F).

For the ladder diagram shown in Figure 7.48, when the counter is set to 5, there is an output from Output 1 every time:
(i) Input 1 has closed 5 times.

(ii) Input 2 has closed 5 times.

A (i) T (ii) T
B (i) T (ii) F
C (i) F (ii) T
D (i) F (ii) F

17 Decide whether each of these statements is True (T) or False (F).

For the ladder diagram shown in Figure 7.48:
(i) The first rung gives the condition required to reset the counter.
(ii) The second rung gives the condition required to generate pulses to be counted.

A (i) T (ii) T
B (i) T (ii) F
C (i) F (ii) T
D (i) F (ii) F

18 Decide whether each of these statements is True (T) or False (F).

When there is an input to Input 1 in Figure 7.48:
(i) The counter contacts in the third rung close.
(ii) The counter is ready to start counting the pulses from Input 2.

A (i) T (ii) T
B (i) T (ii) F
C (i) F (ii) T
D (i) F (ii) F

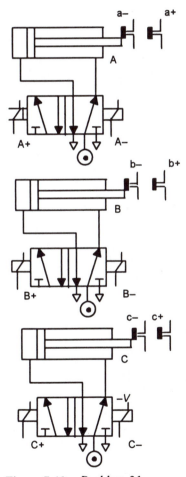

Figure 7.49   *Problem 21*

19 This problem is essentially concerned with part of the domestic washing machine program. Devise a ladder program to switch on a pump for 100 s. It is then to be switched off and a heater switched on for 50 s. Then the heater is switched off and another pump is used to empty the water.

20 Devise a ladder program that can be used with a solenoid valve controlled double-acting cylinder, i.e. a cylinder with a piston which can be moved either way by means of solenoids for each of its two positions, and which moves the piston to the right, holds it there for 2 s and then returns it to the left.

21 The inputs from the limit switches and the start switch and the outputs to the solenoids of the valves shown in Figure 7.49 are connected to a PLC which has the ladder program shown in Figure 7.50. What is the sequence of the cylinders?

22 The inputs from the limit switches and the start switch and the outputs to the solenoids of the valves shown in Figure 7.51(a) are connected to a PLC which has the ladder program shown in Figure 7.51(b). What is the sequence of the cylinders?

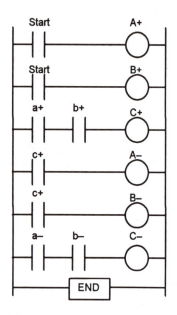

Figure 7.50   *Problem 21*

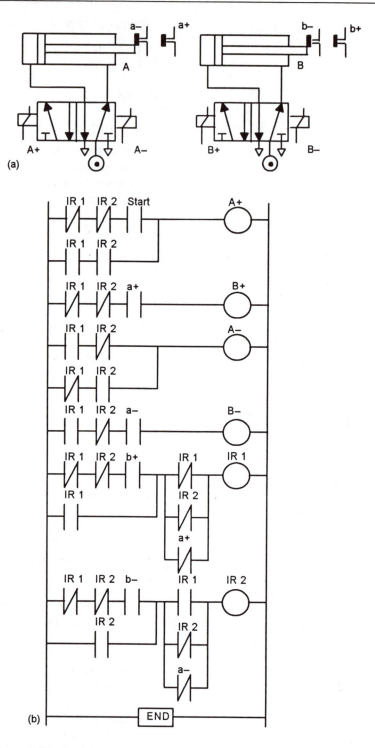

Figure 7.51   *Problem 22*

8 System models

Suppose we have a control system for the temperature in a room. What will happen to the temperature when the thermostat has its set value increased from, say, 20°C to 22°C? In order to determine how the output of a control system will react to different inputs, we need a mathematical model of the system so that we have an equation describing how the output of the system is related to its input. Then we can use the equation to make a prediction of what will happen. We are concerned with not only what temperature the room ends up at but how it varies with time as it changes to the required value.

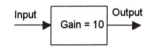

Figure 8.1 *Amplifier system with the output ten times the input*

> We use the term *static response* to describe the response of a system to an input without any reference to the time taken to reach that response and the term *dynamic response* when we also consider how it varies with time.

8.1.1 Static response

In the case of an amplifier system (Figure 8.1) we might be able to use the simple relationship that the output is always 10 times the input. If we have an input of a 1 V signal we can calculate that the output will be 10 V. This is a simple model of a system where the input is just multiplied by a gain of 10 in order to give the output and we are looking at just the static response. This chapter starts off with a discussion of this simple model of a system.

8.1.2 Dynamic response

If we consider a system representing a spring balance with an input of a load signal and an output of a deflection (Figure 8.2) then, when we have an input to the system and put a fixed load on the balance (this type of input is known as a *step input* because the input variation with time looks like a step), it is likely that it will not instantaneously give the weight but the pointer on the spring balance will oscillate for a little time before settling down to the weight value. The static response is just the weight value the spring balance eventually settles down to. But, for a complete picture of the behaviour of the spring balance with time, we cannot just state, for an input of some constant load, that the output is just the input multiplied by some constant number but need some way of describing an output which varies with time.

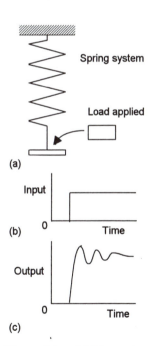

Figure 8.2 *(a) The spring system with a constant load applied at some instant of time, (b) the step showing how the input varies with time, (c) the output showing how it varies with time for the step input*

With an electrical system of a circuit with capacitance and resistance, when the voltage to such a circuit is switched on, i.e. there is a step voltage input to the system, then the current changes with time before eventually settling down to a steady value. With a temperature control system, such as that used for the central heating system for a house, when the thermostat is changed from 20°C to 22°C, the output does not immediately become 22°C but there is a change with time and eventually it may become 22°C. In general, the mathematical model describing the relationship between input and output for a system is likely to involve terms which give values which change with time and are described by a differential equation (see Appendix B). A *differential equation* is an equation involving derivatives of a function, i.e. terms such as dy/dt and d^2y/dt^2. In this chapter we look at how such differential equation relationships arise. In chapter 9 we look at a way of simplifying the maths and making life much easier.

8.2 Gain

In the case of an amplifier system we might have the output directly proportional to the input and, with a gain of 10, if we have an input of a 1 V signal we can calculate that the output will be ten times greater and so 10 V. In general, for such a system where the output is directly proportional to the input, we can write:

$$\text{output} = G \times \text{input}$$

with G being the gain.

Example

A motor has an output speed which is directly proportional to the voltage applied to its armature. If the output is 5 rev/s when the input voltage is 2 V, what is the system gain?

With output = $G \times$ input, then $G = 5/2 = 2.5$ (rev/s)/V.

8.2.1 Gain of systems in series

Consider two systems, e.g. amplifiers, in series with the first having a gain G_1 and the second a gain G_2 (Figure 8.3(a)). The first system has an input of x_1 and an output of y_2 and thus:

$$y_1 = G_1x_1$$

The second system has an input of y_1 and an output of y_2 and thus:

$$y_2 = G_2y_1 = G_2 \times G_1x_1$$

The overall system has an input of x_1 and an output of y_2 and thus, if we represent the overall system as having a gain of G:

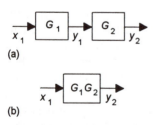

(a)

(b)

Figure 8.3   *(a) Two systems in series, (b) the equivalent system with a gain equal to the product of the gains of the two constituent systems*

$$y_2 = Gx_1$$

and so:

$$G = G_1 \times G_2$$

> For series-connected systems, the overall gain is the product of the gains of the constituent systems.

Example

A system consists of an amplifier with a gain of 10 providing the armature voltage for a motor which gives an output speed which is proportional to the armature voltage, the constant of proportionality being 5 (rev/s)/V. What is the relationship between the input voltage to the system and the output motor speed?

The overall gain $G = G_1 \times G_2 = 10 \times 5 = 50$ (rev/s)/V.

8.2.2 Feedback loops

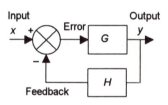

Figure 8.4 *System with negative feedback: the fed back signal subtracts from the input*

Consider a system with *negative feedback* (Figure 8.4). The output of the system is fed back via a measurement system with a gain H to subtract from the input to a system with gain G.

The input to the feedback system is y and thus its output, i.e. the feedback signal, is Hy. The error is $x - Hy$. Hence, the input to the G system is $x - Hy$ and its output y. Thus:

$$y = G(x - Hy)$$

and so:

$$(1 + GH)y = Gx$$

The overall input of the system is y for an input x and so the overall gain G of the system is y/x. Hence:

> $$\text{system gain} = \frac{y}{x} = \frac{G}{1 + GH}$$

The term *forward path* is used for the path from the error signal to the output and so, in this case, the forward path gain is G. The *feedback path* is the path from the output back to the comparison element and so, in this case, the feedback path gain is H.

Application

Figure 8.6 shows the circuit of a basic feedback amplifier. It consists of an operational amplifier with a potential divider of two resistors R_1 and R_2 connected across its output. The output from this potential divider is fed back to the inverting input of the amplifier. The input to the amplifier is via its non-inverting input. Thus the sum of the inverted feedback input and the non-inverted input is the error signal. The op amp has a very high voltage gain G. Thus GH, H being the gain of the feedback loop, is very large compared with 1 and so the overall system gain is:

$$\text{system gain} = \frac{G}{1+GH} \approx \frac{G}{GH} \approx \frac{1}{H}$$

Since the gain G of the op amp can be affected by changes in temperature, ageing, etc. and thus can vary, the use of the op amp with a feedback loop means that, since H is just made up of resistances which are likely to be more stable, a more stable amplifier system is produced. The feedback loop gain H is the fraction of the output signal fed back and so is $R_1/(R_1 + R_2)$. Hence, the overall gain of the system is:

$$\text{system gain} = \frac{R_2+R_1}{R_1}$$

Thus, if the op amp has a voltage gain of 200 000, $R_1 = 1$ kΩ and $R_2 = 49$ kΩ, the overall system gain is independent of the voltage gain of the op amp and is given by $(R_1 + R_2)/R_1 = 50/1 = 50$.

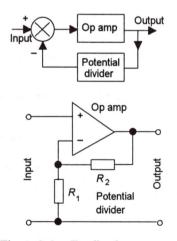

Figure 8.6 *Feedback amplifier*

For a system with a negative feedback, the overall gain is the forward path gain divided by one plus the product of the forward path and feedback path gains.

For a system with *positive feedback* (Figure 8.5), i.e. the fed back signal adds to the input signal, the feedback signal is Hy and thus the input to the G system is $x + Hy$. Hence:

$$y = G(x + Hy)$$

and so:

$$(1 - GH)y = Gx$$

$$\text{system gain} = \frac{y}{x} = \frac{G}{1-GH}$$

For a system with a positive feedback, the overall gain is the forward path gain divided by one minus the product of the forward path and feedback path gains.

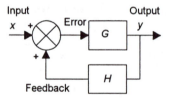

Figure 8.5 *System with positive feedback: the fed back signal adds to the input and so the error signal is the input signal plus the fed back signal*

Example

A negative feedback system has a forward path gain of 12 and a feedback path gain of 0.1. What is the overall gain of the system?

Using the equation derived above for the system gain:

$$\text{system gain} = \frac{G}{1+GH} = \frac{12}{1+0.1\times 12} = 5.45$$

The overall gain is thus 5.45.

8.3 Dynamic systems

To be able to describe how the output of a system depends on its input and how the output changes with time when the input changes, we need a mathematical equation relating the input and output. The following describes how we can arrive at the input–output relationships for systems by considering them to be composed of just a few simple basic elements. Thus, if we want to develop a model for a car suspension we need to consider how easy it is to extend or compress it, i.e. its stiffness, the forces damping out any motion of the suspension and the mass of the system and so its resistance of the system to acceleration, i.e. its inertia. So we think of the model as having the separate elements of stiffness, damping and inertia which we can represent by a spring, a dashpot and a mass (Figure 8.7) and then write down equations for the behaviour of each element using the fundamental physical laws governing the behaviour of each element. This way of modelling a system is known as *lumped-parameter modelling*.

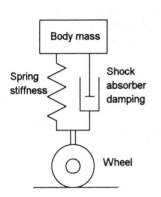

Figure 8.7   *Model for a car suspension system*

8.3.1 Mechanical systems

Mechanical systems, however complex, have stiffness (or springiness), damping and inertia and can be considered to be composed of basic elements which can be represented by springs, dashpots and masses.

1   *Spring*
   The 'springiness' or 'stiffness' of a system can be represented by an ideal spring (ideal because it has only springiness and no other properties). For a linear spring (Figure 8.8(a)), the extension y is proportional to the applied extending force F and we have:

$$F = ky$$

where k is a constant termed the *stiffness*.

2   *Dashpot*
   The 'damping' of a mechanical system can be represented by a dashpot. This is a piston moving in a viscous medium, e.g. oil, in a cylinder (Figure 8.8(b)). Movement of the piston inwards requires the trapped fluid to flow out past edges of the piston; movement outwards requires fluid to flow past the piston and into the enclosed space. For such a system, the resistive force F which has to be overcome is proportional to the velocity of the piston and hence the rate of change of displacement y with time, i.e. dy/dt. Thus we can write:

$$F = c\frac{dy}{dt}$$

where c is a constant.

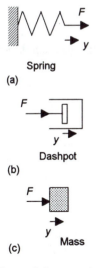

Figure 8.8   *Mechanical system building blocks*

3 *Mass*

The 'inertia' of a system, i.e. how much it resists being accelerated can be represented by mass. For a mass m (Figure 8.8(c)), the relationship between the applied force F and its acceleration a is given by Newton's second law as $F = ma$. But acceleration is the rate of change of velocity v with time t, i.e. $a = dv/dt$, and velocity is the rate of change of displacement y with time, i.e. $v = dy/dt$. Thus $a = d(dy/dt)/dt$ and so we can write:

$$F = m\frac{d^2y}{dt^2}$$

The following example illustrates how we can arrive at a model for a mechanical system.

Example

Derive a model for the mechanical system represented by the system of mass, spring and dashpot given in Figure 8.9(a). The input to the system is the force F and the output is the displacement y.

To obtain the system model we draw *free-body diagrams*, these being diagrams of masses showing just the external forces acting on each mass. For the system in Figure 8.9(a) we have just one mass and so just one free-body diagram and that is shown in Figure 8.9(b). As the free-body diagram indicates, the net force acting on the mass is the applied force minus the forces exerted by the spring and by the dashpot:

$$\text{net force} = F - ky - c\frac{dy}{dt}$$

Then applying Newton's second law, this force must be equal to ma, where a is the acceleration, and so:

$$m\frac{d^2y}{dt^2} = F - ky - c\frac{dy}{dt}$$

The relationship between the input F to the system and the output y is thus described by the second-order differential equation:

$$m\frac{d^2y}{dt^2} + c\frac{dy}{dt} + ky = F$$

The term *second-order* is used because the equation includes as its highest derivative d^2y/dt^2.

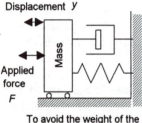

Displacement y

Applied force

F

To avoid the weight of the mass becoming involved, the forces and displacement are horizontal. The rollers enable us to ignore friction.

(a)

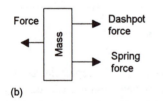

Force

Dashpot force

Spring force

(b)

Figure 8.9 *(a) Mechanical system with mass, damping and stiffness, (b) the free-body diagram for the forces acting on the mass*

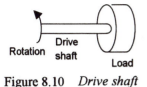

Figure 8.10   *Drive shaft*

(a) Torsional spring or
elastic twisting of a shaft

(b) Rotational dashpot

(c) Moment of inertia

Figure 8.11 *Rotational system
elements: (a) torsional spring,
(b) rotational dashpot, (c)
moment of inertia*

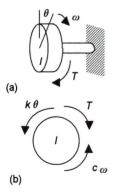

(a)

(b)

Figure 8.12   *Example*

8.3.2 Rotational systems

In control systems we are often concerned with rotational systems, e.g. we might want a model for the behaviour of a motor drive shaft (Figure 8.10) and how the driven load rotation will be related to the rotational twisting input to the drive shaft. For rotational systems the basic building blocks are a torsion spring, a rotary damper and the moment of inertia (Figure 8.11).

1   *Torsional spring*
The 'springiness' or 'stiffness' of a rotational spring is represented by a torsional spring. For a torsional spring, the angle θ rotated is proportional to the torque T:

$$T = k\theta$$

where k is a measure of the stiffness of the spring.

2   *Rotational dashpot*
The damping inherent in rotational motion is represented by a rotational dashpot. For a rotational dashpot, i.e. effectively a disk rotating in a fluid, the resistive torque T is proportional to the angular velocity ω and thus:

$$T = c\omega = c\frac{\mathrm{d}\theta}{\mathrm{d}t}$$

where c is the damping constant.

3   *Inertia*
The inertia of a rotational system is represented by the moment of inertia of a mass. A torque T applied to a mass with a moment of inertia I results in an angular acceleration a and thus, since angular acceleration is the rate of change of angular velocity ω with time, i.e. $\mathrm{d}\omega/\mathrm{d}t$, and angular velocity ω is the rate of change of angle with time, i.e. $\mathrm{d}\theta/\mathrm{d}t$, then the angular acceleration is $\mathrm{d}(\mathrm{d}\theta/\mathrm{d}t)/\mathrm{d}t$ and so:

$$T = Ia = I\frac{\mathrm{d}^2\theta}{\mathrm{d}t^2}$$

The following example illustrates how we can arrive at a model for a rotational system.

Example

Develop a model for the system shown in Figure 8.12(a) of the rotation of a disk as a result of twisting a shaft.

Figure 8.12(b) shows the free-body diagram for the system. The torques acting on the disk are the applied torque T, the spring torque $k\theta$ and the damping torque $c\omega$. Hence:

$$T - k\theta - c\frac{d\theta}{dt} = I\frac{d^2\theta}{dt^2}$$

We thus have the second-order differential equation relating the input of the torque to the output of the angle of twist:

$$I\frac{d^2\theta}{dt^2} + c\frac{d\theta}{dt} + k\theta = T$$

8.3.3 Electrical systems

The basic elements of electrical systems are the pure components of resistor, inductor and capacitor (Figure 8.13), the term pure is used to indicate that the resistor only possesses the property of resistance, the inductor only inductance and the capacitor only capacitance.

Resistor

Inductor

Capacitor

Figure 8.13 *Electrical system building blocks*

1 *Resistor*
   For a *resistor*, resistance R, the potential difference v across it when there is a current i through it is given by:

$$v = Ri$$

2 *Inductor*
   For an *inductor*, inductance L, the potential difference v across it at any instant depends on the rate of change of current i and is:

$$v = L\frac{di}{dt}$$

3 *Capacitor*
   For a *capacitor*, the potential difference v across it depends on the charge q on the capacitor plates with $v = q/C$, where C is the capacitance. Thus:

$$v = \frac{1}{C}q$$

$$\frac{dv}{dt} = \frac{1}{C}\frac{dq}{dt}$$

Since current i is the rate of movement of charge:

$$\frac{dv}{dt} = \frac{1}{C}\frac{dq}{dt} = \frac{1}{C}i$$

and so we can write:

$$i = C\frac{\mathrm{d}v}{\mathrm{d}t}$$

To develop the models for electrical circuits we use Kirchhoff's laws. These can be stated as:

1   *Kirchhoff's current law*
    The total current flowing into any circuit junction is equal to the total current leaving that junction, i.e. the algebraic sum of the currents at a junction is zero.

2   *Kirchhoff's voltage law*
    In a closed circuit path, termed a loop, the algebraic sum of the voltages across the elements that make up the loop is zero. This is the same as saying that for a loop containing a source of e.m.f., the sum of the potential drops across each circuit element is equal to the sum of the applied e.m.f.s. provided we take account of their directions.

The following examples illustrate the development of models for electrical systems.

Example

Develop a model for the electrical system described by the circuit shown in Figure 8.14. The input is the voltage v when the switch is closed and the output is the voltage v_C across the capacitor.

Using Kirchhoff's voltage law gives:

$$v = v_R + v_C$$

and, since $v_R = Ri$ and $i = C(\mathrm{d}v_C/\mathrm{d}t)$ we obtain the equation:

$$v = RC\frac{\mathrm{d}v_C}{\mathrm{d}t} + v_C$$

The relationship between an input v and the output v_C is a first-order differential equation. The term *first-order* is used because it includes as its highest derivative $\mathrm{d}v_C/\mathrm{d}t$.

Example

Develop a model for the circuit shown in Figure 8.15 when we have an input voltage v when the switch is closed and take an output as the voltage v_C across the capacitor.

Applying Kirchhoff's voltage law gives:

$$v = v_R + v_L + v_C$$

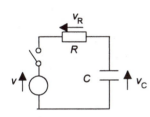

Figure 8.14   *Electrical system with resistance and capacitance*

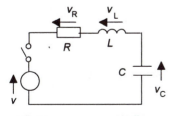

Figure 8.15   *Electrical system with resistance, inductance and capacitance*

and so:

$$v = Ri + L\frac{di}{dt} + v_C$$

Since $i = C(dv_C/dt)$, then $di/dt = C(d^2v_C/dt^2)$ and thus we can write:

$$v = RC\frac{dv_C}{dt} + LC\frac{d^2v_C}{dt^2} + v_C$$

The relationship between an input v and output v_C is described by a second-order differential equation.

Example

Develop a model for a d.c. permanent magnet motor relating the current through the armature to the applied voltage.

When a current flows through the armature coil, forces act on it as a result of its current carrying conductors being in the magnetic field provided by the permanent magnet and so cause the armature coil to rotate. For a coil rotating in a magnetic field, a voltage is induced in it in such a direction as to oppose the change producing it and so there is a back e.m.f. Thus, the electrical circuit we can use to describe the motor has two sources of e.m.f., that applied to produce the armature current and the back e.m.f. The electrical circuit model is thus as shown in Figure 8.16. Just two elements, an inductor and a resistor, are used to represent the armature coil. The equation is thus:

$$v_a - v_b = Ri_a + L\frac{di_a}{dt}$$

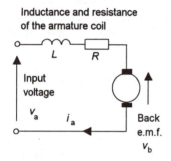

Inductance and resistance of the armature coil

Input voltage v_a　i_a　Back e.m.f. v_b

Figure 8.16 *Model for a d.c. motor*

8.3.4 Thermal systems

In considering a control system for the domestic central heating system we need a model for how the output of the system, i.e. the temperature of a room, depends on a change to the heat input to room. Likewise, for a process control system where we have a heater used to change the temperature of a liquid, we need a model relating the temperature of the liquid to the heat input. Thermal systems have two basic building blocks, resistance and capacitance (Figure 8.17).

1　*Thermal resistance*
  The thermal resistance R is the resistance offered to the rate of flow of heat q (Figure 8.17(a)) and is defined by:

$$q = \frac{T_1 - T_2}{R}$$

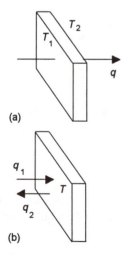

(a)

(b)

Figure 8.17　*(a) Thermal resistance, (b) thermal capacitance* where $T_1 - T_2$ is the temperature difference through which the heat flows.

For heat conduction through a solid we have the rate of flow of heat proportional to the cross-sectional area and the temperature gradient. Thus for two points at temperatures T_1 and T_2 and a distance L apart:

$$q = Ak\frac{T_1 - T_2}{L}$$

with k being the thermal conductivity. Thus with this mode of heat transfer, the thermal resistance R is L/Ak. For heat transfer by convection between two points, Newton's law of cooling gives:

$$q = Ah(T_2 - T_1)$$

where $(T_2 - T_1)$ is the temperature difference, h the coefficient of heat transfer and A the surface area across which the temperature difference is. The thermal resistance with this mode of heat transfer is thus $1/Ah$.

2   *Thermal capacitance*
The thermal capacitance (Figure 8.17(b)) is a measure of the store of internal energy in a system. If the rate of flow of heat into a system is q_1 and the rate of flow out q_2 then the rate of change of internal energy of the system is $q_1 - q_2$. An increase in internal energy can result in a change in temperature:

$$\text{change in internal energy} = mc \times \text{change in temperature}$$

where m is the mass and c the specific heat capacity. Thus the rate of change of internal energy is equal to mc times the rate of change of temperature. Hence:

$$q_1 - q_2 = mc\frac{dT}{dt}$$

This equation can be written as:

$$q_1 - q_2 = C\frac{dT}{dt}$$

where the capacitance $C = mc$.

The following examples illustrate the development of models for thermal systems.

Example

Develop a model for the simple thermal system of a thermometer at temperature T being used to measure the temperature of a liquid

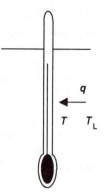

Figure 8.18 *Example*

when it suddenly changes to the higher temperature of T_L (Figure 8.18).

When the temperature changes there is heat flow q from the liquid to the thermometer. The thermal resistance to heat flow from the liquid to the thermometer is:

$$q = \frac{T_L - T}{R}$$

Since there is only a net flow of heat from the liquid to the thermometer the thermal capacitance of the thermometer is:

$$q = C\frac{dT}{dt}$$

Substituting for q gives:

$$C\frac{dT}{dt} = \frac{T_L - T}{R}$$

which, when rearranged gives:

$$RC\frac{dT}{dt} + T = T_L$$

This is a first-order differential equation.

Example

Determine a model for the temperature of a room (Figure 8.19) containing a heater which supplies heat at the rate q_1 and the room loses heat at the rate q_2.

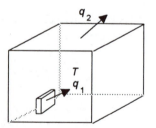

Figure 8.19 *Example*

We will assume that the air in the room is at a uniform temperature T. If the air and furniture in the room have a combined thermal capacity C, since the energy rate to heat the room is $q_1 - q_2$, we have:

$$q_1 - q_2 = C\frac{dT}{dt}$$

If the temperature inside the room is T and that outside the room T_0 then

$$q_2 = \frac{T - T_0}{R}$$

where R is the thermal resistance of the walls. Substituting for q_2 gives:

$$q_1 - \frac{T - T_0}{R} = C\frac{dT}{dt}$$

Hence:

$$RC\frac{dT}{dt} + T = Rq_1 + T_0$$

This is a first-order differential equation.

8.3.5 Hydraulic systems

A common fluid control system involves liquid flowing into a container and out of it through a valve, the requirement being to control the level of the liquid in the container. For such a system we need a model which indicates how the height of liquid in the container is related to the rates of inflow and outflow.

For a fluid system the three building blocks are resistance, capacitance and inertance; these are the equivalents of electrical resistance, capacitance and inductance. The equivalent of electrical current is the volumetric rate of flow and of potential difference is pressure difference. Hydraulic fluid systems are assumed to involve an incompressible liquid; pneumatic systems, however, involve compressible gases and consequently there will be density changes when the pressure changes. Here we will just consider the simpler case of hydraulic systems. Figure 8.20 shows the basic form of building blocks for hydraulic systems.

1 *Hydraulic resistance*

   Hydraulic resistance R is the resistance to flow which occurs when a liquid flows from one diameter pipe to another (Figure 8.20(a)) and is defined as being given by the hydraulic equivalent of Ohm's law:

$$p_1 - p_2 = Rq$$

2 *Hydraulic capacitance*

   Hydraulic capacitance C is the term used to describe energy storage where the hydraulic liquid is stored in the form of potential energy (Figure 8.20(b)). The rate of change of volume V of liquid stored is equal to the difference between the volumetric rate at which liquid enters the container q_1 and the rate at which it leaves q_2, i.e.

$$q_1 - q_2 = \frac{dV}{dt}$$

But $V = Ah$ and so:

$$q_1 - q_2 = A\frac{dh}{dt}$$

The pressure difference between the input and output is:

$$p_1 - p_2 = p = h\rho g$$

Hence, substituting for h gives:

Rate of flow q

Pressure p_1 Pressure p_2

(a) Resistance

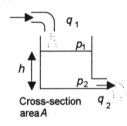

q_1

p_1

h

p_2

Cross-section area A q_2

(b) Capacitance

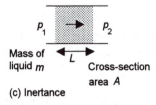

p_1 p_2

Mass of liquid m L Cross-section area A

(c) Inertance

Figure 8.20 *Hydraulic building blocks*

$$q_1 - q_2 = \frac{A}{\rho g}\frac{dp}{dt}$$

The hydraulic capacitance C is defined as:

$$C = \frac{A}{\rho g}$$

and thus we can write:

$$q_1 - q_2 = C\frac{dp}{dt}$$

Application

A control system used to control the height of a liquid in a container will have a response which depends on the hydraulic capacitance of the system. Thus a container which has a large surface area for the liquid will have a larger capacitance than one with a smaller surface area. As a consequence, this larger capacitance means for the same rates of inflow and outflow, the rate of change of the pressure with time will be smaller. Since the pressure is proportional to the height of the liquid, then the larger capacitance means a slower rate of response to an input or output change.

3   *Hydraulic inertance*

Hydraulic inertance is the equivalent of inductance in electrical systems. To accelerate a fluid a net force is required and this is provided by the pressure difference (Figure 8.20(c)). Thus:

$$(p_1 - p_2)A = ma = m\frac{dv}{dt}$$

where a is the acceleration and so the rate of change of velocity v. The mass of fluid being accelerated is $m = AL\rho$ and the rate of flow $q = Av$ and so:

$$(p_1 - p_2)A = L\rho\frac{dq}{dt}$$

$$p_1 - p_2 = I\frac{dq}{dt}$$

where the inertance I is given by $I = L\rho/A$.

The following example illustrates the development of a model for a hydraulic system.

Example

Develop a model for the hydraulic system shown in Figure 8.21 where there is a liquid entering a container at one rate q_1 and leaving through a valve at another rate q_2.

We can neglect the inertance since flow rates can be assumed to change only very slowly. For the capacitance term we have:

$$q_1 - q_2 = C\frac{dp}{dt} = \frac{A}{\rho g}\frac{dp}{dt}$$

For the resistance of the valve we have:

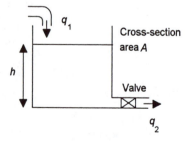

Figure 8.21   *Example*

$$p_1 - p_2 = Rq_1$$

Thus, substituting for q_2, and recognising that the pressure difference is $h\rho g$, gives:

$$q_1 = A\frac{dh}{dt} + \frac{h\rho g}{R}$$

$$A\frac{dh}{dt} + \frac{\rho g}{R}h = q_1$$

This is a first-order differential equation.

8.4 Differential equations

As the previous section indicates, the relationship between the input and output for systems is often in the form of a differential equation which shows how, when there is some input, the output varies with time. A differential equation is an equation involving derivatives, i.e. rates of change, of a function. For example, we might have the rate of change of displacement y with time t, written as dy/dt, or the rate of change with time of dy/dt, written as d^2y/dt^2. See Appendix A for a discussion of differential equations and how the outputs can be derived, here we consider only the results of such derivations.

8.4.1 First-order differential equations

Many systems have input–output relationships which can be described by a first-order differential equation and have an output y related to an input x by an equation of the form:

$$\tau\frac{dy}{dt} + y = kx$$

where τ and k are constants, τ being known as the *time constant*. The term first-order is used to describe a differential equation for which the highest derivative is of the form dy/dt.

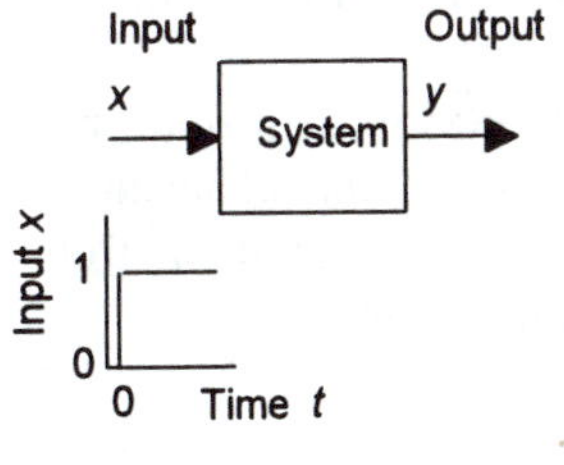

Consider the response of such a system when subject to a unit step input, i.e. an input which suddenly changes from 0 to a constant value of 1 (Figure 8.22). When we reach the time at which the output x is no longer changing with time, i.e. we have steady-state conditions, then $dy/dt = 0$ and so we have, for the above equation, output $y = kx$ and k is the *steady-state gain*. Thus, with a unit step input the steady-state output is $1k$.

Figure 8.22 *Step input to a system*

Steady-state output for unit step input = $1k$

If we consider how the output will vary with time in attaining the steady-state value, then the output is related to the input by an equation of the form:

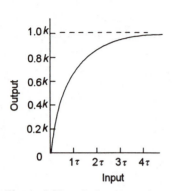

Figure 8.23 *Behaviour of a first-order system when subject to a unit step input*

Table 8.1 *Growth of output y with time*

Time	y/(steady-state value)
0	0
0.7T	0.5
1.4T	0.75
2.1T	0.875
2.8T	0.938
3.5T	0.969

$$y = \text{steady-state value} \times (1 - e^{-t/\tau})$$

τ is called the *time constant*. Figure 8.23 shows how such first-order systems behave when subject to a unit step input.

Using the above equation, the time taken for y to reach half the steady-state value is given by:

$$0.5 = (1 - e^{-t/\tau})$$

$$e^{-t/\tau} = 0.5$$

$$-\frac{t}{\tau} = \ln 0.5 = -0.693$$

In a time of 0.693τ y reaches half its steady-state value. The time taken to reach three-quarters of the steady-state value is given by:

$$0.75V = V(1 - e^{-t/\tau})$$

$$e^{-t/\tau} = 0.25$$

$$-\frac{t}{\tau} = \ln 0.25 = -1.386$$

Thus in a time of 1.386τ y will reach three-quarters of its steady-state value. This is twice the time taken to reach half the steady-state value. This is a characteristic of exponential graphs: if t is the time taken to reach half the steady-state value, then in $2t$ it will reach three-quarters, in $3t$ it will reach seven-eighths, etc. (Table 8.1).

When $t = 1\tau$ then y/(steady-state value) = $(1 - e^{-1})$ = 0.632. Thus in a time equal to the time constant the output rises to 63.2% of the steady-state value. When $t = 2\tau$ then y/(steady-state value) = $(1 - e^{-2})$ = 0.865. Thus the output rises to 86.5% of the steady-state value. When $t = 3\tau$ then y/(steady-state value) = $(1 - e^{-3})$ = 0.950. Thus the output rises to 95.0% of the steady-state value.

Examples of first-order systems are an electrical system having capacitance and resistance, an electrical system having inductance and resistance and a thermal system of a room with a heat input from an electrical heater and an output of the room temperature.

Example

Determine the time constant for a circuit having a capacitance of 8 μF in series with a resistance of 1 MΩ if the differential equation relating the input V and the output voltage across the capacitor v_C is:

$$v = RC\frac{dv_C}{dt} + v_C$$

Note that this differential equation was derived in an earlier example in this chapter.

In general with such problems, we put the equation in the standard form, i.e. $\tau(dy/dt) + y = kx$ with just a 1 in front of the y term, and then by comparison we can obtain the time constant and the steady-state gain. Thus, the time constant is RC and so $1 \times 10^6 \times 8 \times 10^{-6} = 8$ s.

Example

A bead thermistor temperature sensor is specified as having a thermal time constant of 11 s (see Section 2.7.4). When there is a step input temperature change, how long will it take for the thermistor to reach 95% of its steady-state value.

It will take about three times the time constant and so 33 s.

8.4.2 Second-order differential equations

The term second-order differential equation is used when the highest derivative is of the form d^2y/dt^2, i.e. they are concerned with the rate of change of a rate of change. Many systems have input–output relationships which can be described by second-order differential equations with output y related to input x by an equation of the form:

$$\frac{d^2y}{dt^2} + 2\zeta\omega_n\frac{dy}{dt} + \omega_n^2 y = k\omega_n^2 x$$

where k, ζ and ω_n are constants for the systems. The constant ζ is known as the *damping ratio* or *factor* and ω_n as the *undamped natural angular frequency*.

If we have steady-state conditions and the output y is no longer changing with time, then $d^2y/dt^2 = 0$ and $dy/dt = 0$ and so we have output $y = kx$ and k is the *steady-state gain*. Thus, with a unit step input:

steady-state output for unit step input = $1k$

Figure 8.24 shows how a second order system behaves when subject to a unit step input. The general form of the response varies with the damping factor.

Systems with damping factors less than 1 are said to be *underdamped*, with damping factors greater than 1 as *over-damped* and for a damping factor of 1 as *critically damped*.

1   With *no damping*, i.e. $\zeta = 0$, the system output oscillates with a constant amplitude and a frequency of ω_n (since $\omega_n = 2\pi f_n$, where f_n is the undamped natural frequency, and $f_n = 1/T$, where T_n is the time for one undamped oscillation, then $T_n = 2\pi/\omega_n = 6.3/\omega_n$).

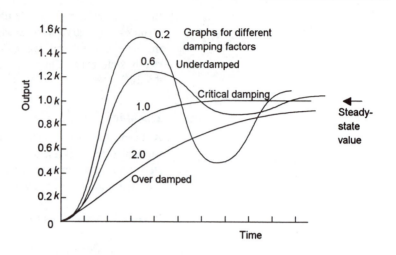

Figure 8.24 *Behaviour of a second-order system with a unit step input. With no damping the output is just a continuous oscillation following a step input. As the damping increases, so the oscillations become damped out and with a damping factor of 1.0 there are no oscillations and the output just rises over time to the steady-state output value. Further increases in damping mean that the output takes longer to reach the steady-state value.*

2   With *underdamping* i.e. $\zeta < 0$, the output oscillates but the closer the damping factor is to 1 the faster the amplitude of the oscillations diminishes.

3   With *critical damping*, i.e. $\zeta = 1$, there are no oscillations and the output just gradually approaches the steady-state value.

4   With *overdamping*, i.e. $\zeta > 1$, the output takes longer than critical damping to reach the steady-state value.

A mechanical system which can be modelled by a spring, dashpot and mass is an example of a second-order system. When we apply a load to the system then oscillations occur which have amplitudes which die away with time. This was illustrated in the opening section and Figure 8.2. Likewise with the second-order system of an electrical circuit having resistance, inductance and capacitance; when there is a step voltage input, i.e. a switch is closed and applies a constant voltage to the circuit, then the voltage across the capacitor will be described by a second-order differential equation and so can oscillate with amplitudes which die away with time.

Example

A mechanical system has an output of a rotation through an angle θ related to an input torque T by the differential equation:

$$I\frac{d^2\theta}{dt^2} + c\frac{d\theta}{dt} + k\theta = T$$

What are the natural angular frequency and the damping constant of the system?

The equation can be put into the form:

$$\frac{d^2y}{dt^2} + 2\zeta\omega_n\frac{dy}{dt} + \omega_n^2 y = k\omega_n^2 x$$

as:

$$\frac{d^2\theta}{dt^2} + \frac{c}{I}\frac{d\theta}{dt} + \frac{k}{I}\theta = \frac{1}{I}T$$

Hence, comparing the terms in front of y and θ gives $\omega_n = \sqrt{(k/I)}$. Comparing the terms in front of dy/dt and $d\theta/dt$ gives $2\zeta\omega_n = c/I$ and hence, since $\omega_n = \sqrt{(k/I)}$, gives $\zeta = c/[2\sqrt{(kI)}]$.

Example

A system has an input and output related by a second-order differential equation. How would you expect the system to behave if it has a damping factor of 0.6 and is subject to a unit step input?

The damping factor indicates that the system is under damped and so the output will oscillate about the steady-state value of $1k$ with decreasing amplitude oscillations before eventually settling down to the steady-state value.

8.4.3 System identification

The differential equations describe the input/output relationship when we consider the input and output to be functions of time. We can use the model building techniques described in the previous section to arrive at differential equations, alternatively we can find the response of a system to, say, a step input and by examining the response determine the form of the differential equation which described its behaviour. In Chapter 12 we consider the response of systems to sinusoidal inputs and use the response to determine the form of differential equation which describes its behaviour.

Problems   *Questions 1 to 7 have four answer options: A, B, C or D. Choose the correct answer from the answer options.*

1 Decide whether each of these statements is True (T) or False (F).

   For a system with negative feedback, the overall static gain is:
   (i) The steady-state output divided by the input.
   (ii) The forward path gain divided by one minus the product of forward path and feedback path gains.

A   (i) T (ii) T
B   (i) T (ii) F
C   (i) F (ii) T
D   (i) F (ii) F

2 Decide whether each of these statements is True (T) or False (F).

For two systems in series, the overall static gain is:
(i) The overall steady-state output divided by the input.
(ii) The sum of the gains of the two series elements.

A   (i) T (ii) T
B   (i) T (ii) F
C   (i) F (ii) T
D   (i) F (ii) F

Problems 3 and 4 refer to the mechanical system model shown in Figure 8.25.

3 Decide whether each of these statements is True (T) or False (F).

For the model in Figure 8.25:
(i) The resistive force which has to be overcome for the dashpot is proportional to the acceleration experienced by it.
(ii) The resistive force which has to be overcome for the spring is proportional to its extension y.

A   (i) T (ii) T
B   (i) T (ii) F
C   (i) F (ii) T
D   (i) F (ii) F

4 Decide whether each of these statements is True (T) or False (F).

For the model in Figure 8.25:
(i) The equation for dynamic conditions relating the output y and the input F is a second-order differential equation.
(ii) The steady-state gain of the system depends only on the constant of proportionality relating the force acting on the spring and its extension, i.e. the spring stiffness.

A   (i) T (ii) T
B   (i) T (ii) F
C   (i) F (ii) T
D   (i) F (ii) F

5 Decide whether each of these statements is True (T) or False (F).

For a hydraulic system model, the input rate of flow q is related to the output of the liquid level h in a container by the differential equation:

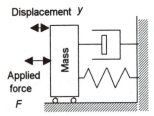

Displacement y

Mass

Applied force

F

Figure 8.25 *Problems 3 to 5*

$$A\frac{dh}{dt} + \frac{\rho g}{R}h = q$$

When there is a step input to the system:
(i) The steady-state gain will be 1.
(ii) The time constant for the system will be A.

A   (i) T (ii) T
B   (i) T (ii) F
C   (i) F (ii) T
D   (i) F (ii) F

6 For a rotational system, the output θ is related to the input T by the differential equation:

$$I\frac{d^2\theta}{dt^2} + c\frac{d\theta}{dt} + k\theta = T$$

For the system to be critically damped, we must have:

A $c = I$
B $c = k$
C $c = kI$
D $c = 2\sqrt{(kI)}$

7 For a system which can be represented by a second-order differential equation relating its input and output, for a step input to give an output which rises to the steady-state value with no oscillations about the steady-state value and take the minimum amount of time, the damping constant has to be:

A   Zero
B   Less than 1
C   1
D   More than 1

8 If a system has a gain of 5, what will be the output for an input voltage of 2 V?

9 An open-loop system consists of three elements in series, the elements having gains of 2, 5 and 10. What is the overall gain of the system?

10 A closed-loop control system has a forward loop with a gain of 6 and a feedback loop with a gain of 2. What will be the overall steady-state gain of the system if the feedback is (a) positive, (b) negative?

11 Derive a differential equation relating the input and output for each of the systems shown in Figure 8.26.

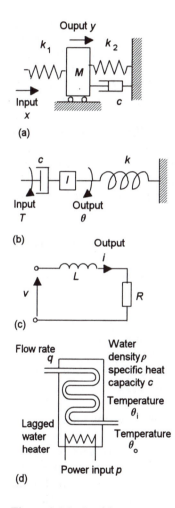

Ouput y

k_1 M k_2

Input x c

(a)

c I k

Input T Output θ

(b)

Output i

v L R

(c)

Flow rate q Water density ρ specific heat capacity c Temperature θ_i

Lagged water heater   Temperature θ_o

Power input p

(d)

Figure 8.26   *Problem 11*

9 Transfer function

9.1 Introduction

In chapter 8 we saw that the relationship between the output and the input for elements used in control systems is frequently described by a differential equation. However, in order to make life simple, what we really need is a simpler relationship than a differential equation giving the relationship between input and output for a system, even when the output varies with time. It is nice and simple to say that the output is just ten times the input and so describe the system by gain = 10. But it is not so simple when the relationship between the input and output is described by a differential equation. However, there is a way we can have such a simple form of relationship where the relationship involves time but it involves writing inputs and outputs in a different form. It is called the *Laplace transform*. In this chapter we will consider how we can carry out such transformations, but not the mathematics to justify why we can do it; the aim is to enable you to use the transform as a tool to carry out tasks. Appendix C gives an explanation of the mathematics behind the transform and the way it is used.

9.2 Transfer function

In general, when we consider inputs and outputs of systems as functions of time then the relationship between the output and input is given by a differential equation. If we have a system composed of two elements in series with each having its input–output relationships described by a differential equation, it is not easy to see how the output of the system as a whole is related to its input. There is a way we can overcome this problem and that is to transform the differential equations into a more convenient form by using the *Laplace transform* (see Appendix C for mathematical details). This form is a much more convenient way of describing the relationship than a differential equation since it can be easily manipulated by the basic rules of algebra.

To carry out the transformation we follow the following rules:

1   A variable which is a function of time, e.g. the input or output voltage v in a circuit, is often written as $v(t)$ to emphasise that v is a function of time – note that this does not mean that v is multiplied by t. When we make the transformation we make the variable a function of s and so, for example, a voltage is written as $V(s)$; note that upper case letters are used for the variables when written as functions of s and that this does not mean that V is multiplied by s.

$v(t)$ when transformed becomes $V(s)$

2   A constant k which does not vary with time remains a constant. Thus kv, where v is a function of time, becomes $kV(s)$. For example, the voltage $3v(t)$ written as an s function is $3V(s)$.

> A constant k remains a constant k when transformed

3   If the initial value of the variable v is zero at time $t = 0$, the first derivative of a function of time dv/dt becomes $sV(s)$ and kdv/dt becomes $ksV(s)$. For example, with no initial values $4dv/dt$ as an s function is $4sV(s)$.

> With no initial value at $t = 0$, dv/dt becomes $sV(s)$ when transformed

Note that if there is an initial value v_0 at $t = 0$ then the first derivative of a function of time dv/dt becomes $sV(s) - v_0$, i.e. we subtract any initial value, and kdv/dt becomes $k[sV(s) - v_0]$. For example, if we have $v_0 = 2$ at $t = 0$ then dv/dt becomes $sV(s) - 2$.

4   If the initial value of the variable v and dv/dt is zero at time $t = 0$, the second derivative of a function of time d^2v/dt^2 becomes $s^2V(s)$ and kd^2v/dt^2 becomes $ks^2V(s)$. For example, with no initial values $4d^2v/dt^2$ as an s function is $4s^2V(s)$.

> With no initial values at $t = 0$, d^2v/dt^2 becomes $s^2V(s)$ when transformed

Note that if there are initial values v_0 and $(dv/dt)_0$ then the second derivative of a function of time d^2v/dt^2 becomes $s^2V(s) - sv_0 - (dv/dt)_0$ and kd^2v/dt^2 becomes $k[s^2V(s) - sv_0 - (dv/dt)_0]$. For example, with initial values of $v_0 = 2$ and $(dv/dt)_0 = 3$ at time $t = 0$, then $4d^2v/dt^2$ as an s function is $4s^2V(s) - 2s - 3$.

5   With an integral of a function of time, when transformed we have:

> $\int_0^t v\, dt$ becomes $\frac{1}{s}V(s)$ $\int_0^t kv\, dt$ becomes $\frac{1}{s}kV(s)$

Note that, when derivatives are involved, we need to know the initial conditions of a system output prior to the input being applied before we can transform a time function into an s function.

Example

Determine the Laplace transform for the following equation where we have v and v_C as functions of time and no initial values.

$$v = RC\frac{dv_C}{dt} + v_C$$

The Laplace transform is:

$$V(s) = RCsV_C(s) + V_C(s)$$

Thus $V(s)$ is the Laplace transform of the input voltage v and $V_C(s)$ is the Laplace transform of the output voltage v_C. Rearranging gives:

$$\frac{V_C(s)}{V(s)} = \frac{1}{RCs+1}$$

The above equation thus describes the relationship between the input and output of the system when described as s functions.

Example

Determine the Laplace transform for the following equation where we have v and v_C as functions of time and no initial values.

$$v = LC\frac{d^2v_C}{dt^2} + RC\frac{dv_C}{dt} + v_C$$

The Laplace transform is:

$$V(s) = LCs^2V(s) + RCsV_C(s) + V_C(s)$$

Thus $V(s)$ is the Laplace transform of the input voltage v and $V_C(s)$ is the Laplace transform of the output voltage v_C. Rearranging gives:

$$\frac{V_C(s)}{V(s)} = \frac{1}{LCs^2 + RCs + 1}$$

The above equation thus describes the relationship between the input and output of the system when described as s functions.

9.2.1 Transfer function

In section 8.2 we used the term gain to relate the input and output of a system with gain G = output/input. When we are working with inputs and outputs described as functions of s we define the *transfer function* $G(s)$ as [output $Y(s)$/input $X(s)$] when all initial conditions before we apply the input are zero:

$$\text{transfer function} = G(s) = \frac{Y(s)}{X(s)}$$

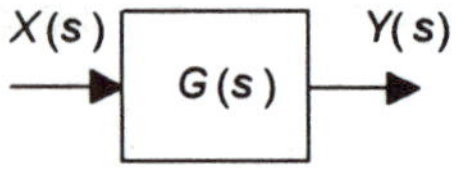

Figure 9.1 *Transfer function as the factor that multiplies the input to give the output*

A transfer function can be represented as a block diagram (Figure 9.1) with $X(s)$ the input, $Y(s)$ the output and the transfer function $G(s)$ as the operator in the box that converts the input to the output. The block represents a multiplication for the input. Thus, by using the Laplace transform of inputs and outputs, we can use the transfer function as a simple multiplication factor, like the gain discussed in Section 9.2.

Example

Determine the transfer function for an electrical system for which we have the relationship (this equation was derived in the example in the preceding section):

$$\frac{V_C(s)}{V(s)} = \frac{1}{RCs + 1}$$

The transfer function $G(s)$ is thus:

$$G(s) = \frac{V_C(s)}{V(s)} = \frac{1}{RCs + 1}$$

To get the output $V_C(s)$ we multiply the input $V(s)$ by $1/(RCs + 1)$.

Example

Determine the transfer function for the mechanical system, having mass, stiffness and damping, and input F and output y and described by the differential equation:

$$F = m\frac{d^2y}{dt^2} + c\frac{dy}{dt} + ky$$

If we now write the equation with the input and output as functions of s, with initial conditions zero:

$$F(s) = ms^2Y(s) + csY(s) + kY(s)$$

Hence the transfer function $G(s)$ of the system is:

$$G(s) = \frac{Y(s)}{F(s)} = \frac{1}{ms^2 + cs + k}$$

9.2.2 Transfer functions of common system elements

By considering the relationships between the inputs to systems and their outputs we can obtain transfer functions for them and hence describe a control system as a series of interconnected blocks, each having its input–output characteristics defined by a transfer function. The following are transfer functions which are typical of commonly encountered system elements:

1 *Gear train*
   For the relationship between the input speed and output speed with a gear train having a gear ratio N:

transfer function = N

2   *Amplifier*
For the relationship between the output voltage and the input voltage with G as the constant gain:

transfer function = G

3   *Potentiometer*
For the potentiometer acting as a simple potential divider circuit the relationship between the output voltage and the input voltage is the ratio of the resistance across which the output is tapped to the total resistance across which the supply voltage is applied and so is a constant and hence the transfer function is a constant K:

transfer function = K

4   *Armature-controlled d.c. motor*
For the relationship between the drive shaft speed and the input voltage to the armature is:

$$\text{transfer function} = \frac{1}{sL + R}$$

where L represents the inductance of the armature circuit and R its resistance.
This was derived by considering armature circuit as effectively inductance in series with resistance and hence:

$$v = L\frac{di}{dt} + Ri$$

and so, with no initial conditions:

$$(V)(s) = sLI(s) + RI(s)$$

and, since the output torque is proportional to the armature current we have a transfer function of the form $1/(sL + R)$.

5   *Valve controlled hydraulic actuator*
The output displacement of the hydraulic cylinder is related to the input displacement of the valve shaft by a transfer function of the form:

$$\text{transfer function} = \frac{K_1}{s(K_2 s + K_3)}$$

where K_1, K_2 and K_3 are constants.

6   *Heating system*
The relationship between the resulting temperature and the input to a heating element is typically of the form:

$$\text{transfer function} = \frac{1}{sC + 1/R}$$

where C is a constant representing the thermal capacity of the system and R a constant representing its thermal resistance.

7 *Tachogenerator*
The relationship between the output voltage and the input rotational speed is likely to be a constant K and so represented by:

$$\text{transfer function} = K$$

8 *Displacement and rotation*
For a system where the input is the rotation of a shaft and the output, as perhaps the result of the rotation of a screw, a displacement, since speed is the rate of displacement we have $v = dy/dt$ and so $V(s) = sY(s)$ and the transfer function is:

$$\text{transfer function} = \frac{1}{s}$$

9 *Height of liquid level in a container*
The height of liquid in a container depends on the rate at which liquid enters the container and the rate at which it is leaving. The relationship between the input of the rate of liquid entering and the height of liquid in the container is of the form:

$$\text{transfer function} = \frac{1}{sA + \rho g/R}$$

where A is the constant cross-sectional area of the container, ρ the density of the liquid, g the acceleration due to gravity and R the hydraulic resistance offered by the pipe through which the liquid leaves the container.

9.2.3 Transfer functions and systems

Consider a speed control system involving a differential amplifier to amplify the error signal and drive a motor, this then driving a shaft via a gear system. Feedback of the rotation of the shaft is via a tachogenerator.

1 The differential amplifier might be assumed to give an output directly proportional to the error signal input and so be represented by a constant transfer function K, i.e. a gain K which does not change with time.

2 The error signal is an input to the armature circuit of the motor and results in the motor giving an output torque which is proportional to the armature current. The armature circuit can be assumed to be a circuit having inductance L and resistance R and so a transfer function of $1/(sL + R)$.

3   The torque output of the motor is transformed to rotation of the drive shaft by a gear system and we might assume that the rotational speed is proportional to the input torque and so represent the transfer function of the gear system by a constant transfer function N, i.e. the gear ratio.

4   The feedback is via a tachogenerator and we might make the assumption that the output of the generator is directly proportional to its input and so represent it by a constant transfer function H.

The block diagram of the control system might thus be like that in Figure 9.2.

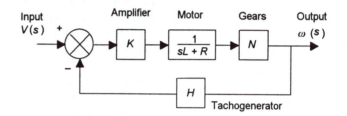

Figure 9.2   *Block diagram for the control system for speed of a shaft with the terms in the boxes being the transfer functions for the elements concerned*

9.3 System transfer functions

Consider the overall transfer functions of systems involving series-connected elements and systems with feedback loops.

9.3.1 Systems in series

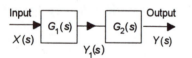

Figure 9.3 *Systems in series*

Consider a system of two subsystems in series (Figure 9.3). The first subsystem has an input of $X(s)$ and an output of $Y_1(s)$; thus, $G_1(s) = Y_1(s)/X(s)$. The second subsystem has an input of $Y_1(s)$ and an output of $Y(s)$; thus, $G_2(s) = Y(s)/Y_1(s)$. We thus have:

$$Y(s) = G_2(s)Y_1(s) = G_2(s)G_1(s)X(s)$$

The overall transfer function $G(s)$ of the system is $Y(s)/X(s)$ and so:

$$G_{\text{overall}}(s) = G_1(s)G_2(s)$$

> The overall transfer function for a system composed of elements in series is the product of the transfer functions of the individual series elements.

Example

Determine the overall transfer function for a system which consists of two elements in series, one having a transfer function of $1/(s + 1)$ and the other $1/(s + 2)$.

The overall transfer function is thus:

$$G_{\text{overall}}(s) = \frac{1}{s+1} \times \frac{1}{s+2} = \frac{1}{(s+1)(s+2)}$$

9.3.2 Systems with feedback

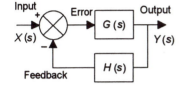

Figure 9.4   *System with negative feedback*

For systems with a negative feedback loop we can have the situation shown in Figure 9.4 where the output is fed back via a system with a transfer function $H(s)$ to subtract from the input to the system $G(s)$. The feedback system has an input of $Y(s)$ and thus an output of $H(s)Y(s)$. Thus the feedback signal is $H(s)Y(s)$. The error is the difference between the system input signal $X(s)$ and the feedback signal and is thus:

Error $(s) = X(s) - H(s)Y(s)$

This error signal is the input to the $G(s)$ system and gives an output of $Y(s)$. Thus:

$$G(s) = \frac{Y(s)}{X(s) - H(s)Y(s)}$$

and so:

$$[1 + G(s)H(s)]Y(s) = G(s)X(s)$$

which can be rearranged to give:

$$\text{overall transfer function} = \frac{Y(s)}{X(s)} = \frac{G(s)}{1 + G(s)H(s)}$$

> For a system with a negative feedback, the overall transfer function is the forward path transfer function divided by one plus the product of the forward path and feedback path transfer functions.

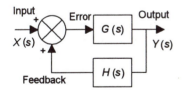

Figure 9.5   *System with positive feedback*

For a system with positive feedback (Figure 9.5), the feedback signal is $H(s)Y(s)$ and thus the input to the $G(s)$ system is $X(s) + H(s)Y(s)$. Hence:

$$G(s) = \frac{Y(s)}{X(s) + H(s)Y(s)}$$

and so:

$$[1 - G(s)H(s)]Y(s) = G(s)X(s)$$

This can be rearranged to give:

$$\text{overall transfer function} = \frac{Y(s)}{X(s)} = \frac{G(s)}{1 - G(s)H(s)}$$

For a system with a positive feedback, the overall transfer function is the forward path transfer function divided by one minus the product of the forward path and feedback path transfer functions.

Example

Determine the overall transfer function for a control system (Figure 9.6) which has a negative feedback loop with a transfer function 4 and a forward path transfer function of $2/(s + 2)$.

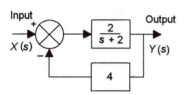

Figure 9.6 *Example*

The overall transfer function of the system is:

$$G_{overall}(s) = \frac{\frac{2}{s+2}}{1 + 4 \times \frac{2}{s+2}} = \frac{2}{s+10}$$

Example

Determine the overall transfer function for a system (Figure 9.7) which has a positive feedback loop with a transfer function 4 and a forward path transfer function of $2/(s + 2)$.

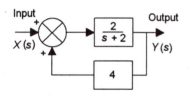

Figure 9.7 *Example*

The overall transfer function is:

$$G_{overall}(s) = \frac{\frac{2}{s+2}}{1 - 4 \times \frac{2}{s+2}} = \frac{2}{s-6}$$

9.4 Block manipulation

Very often, systems may have many elements and sometimes more than one input. A single input–single output system is often termed a SISO system while a multiple input–multiple output system is a MISO system. The following are some of the ways we can reorganise the blocks in a block diagram of a system in order to produce simplification and still give the same overall transfer function for the system. To simplify the diagrams, the (s) has been omitted; it should, however, be assumed for all dynamic situations.

9.4.1 Blocks in series

As indicated in Section 9.3.1, Figure 9.8 shows the basic rule for simplifying blocks in series.

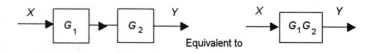

Figure 9.8 *Blocks in series*

9.4.2 Moving takeoff points

As a means of simplifying block diagrams it is often necessary to move takeoff points. The following figures (Figures 9.9 and 9.10) give the basic rules for such movements.

Figure 9.9 *Moving a takeoff point to beyond a block*

Figure 9.10 *Moving a takeoff point to ahead of a block*

9.4.3 Moving a summing point

As a means of simplifying block diagrams it is often necessary to move summing points. The following figures (Figures 9.11–9.14) give the basic rules for such movements.

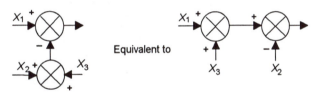

Figure 9.11 *Rearrangement of summing points*

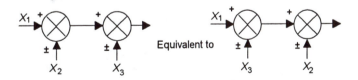

Figure 9.12 *Interchange of summing points*

Figure 9.13 *Moving a summing point ahead of a block*

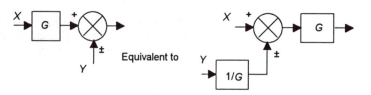

Figure 9.14 *Moving a summing point beyond a block*

9.4.4 Changing feedback and forward paths

Figures 9.15 and 9.16 show block simplification techniques when changing feedforward and feedback paths.

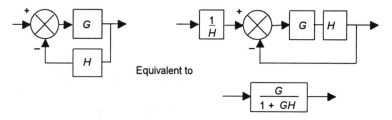

Figure 9.15 *Removing a block from a feedback path*

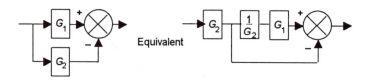

Figure 9.16 *Removing a block from a forward path*

Example

Use block simplification techniques to simplify the system shown in Figure 9.17.

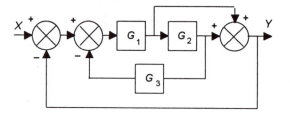

Figure 9.17 *The circuit to be simplified*

Figures 9.18–9.23 show the various stages in the simplification.

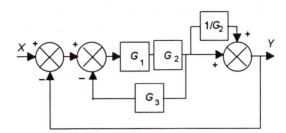

Figure 9.18   *Moving a takeoff point*

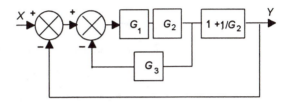

Figure 9.19   *Eliminating a feedforward loop*

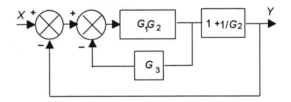

Figure 9.20   *Simplifying series elements*

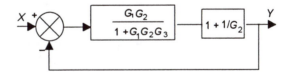

Figure 9.21   *Simplifying a feedback element*

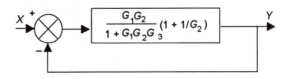

Figure 9.22   *Simplifying series elements*

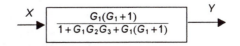

Figure 9.23   *Simplifying negative feedback*

9.5 Multiple inputs

When there is more than one input to a system, the *superposition principle* can be used. This is that:

> The response to several inputs simultaneously applied is the sum of the individual responses to each input when applied separately.

Thus, the procedure to be adopted for a multiple input–single output (MISO) system is:

1 Set all but one of the inputs to zero.

2 Determine the output signal due to this one non-zero input.

3 Repeat the above steps for each input in turn.

4 The total output of the system is the algebraic sum of the outputs due to each of the inputs.

Example

Determine the output $Y(s)$ of the system shown in Figure 9.24 when there is an input $X(s)$ to the system as a whole and a disturbance signal $D(s)$ at the point indicated.

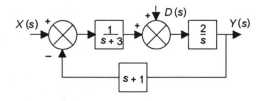

Figure 9.24   *System with a disturbance input*

If we set $D(s)$ to zero we have the system shown in Figure 9.25 and the output is given by:

$$\frac{Y(s)}{X(s)} = \frac{2}{s(s+3)+2(s+1)}$$

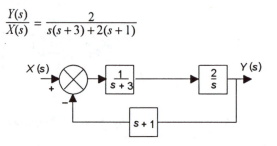

Figure 9.25   *System with disturbance put equal to zero*

If we now set $X(s)$ to zero we have the system shown in Figure 9.26. This is a system with a forward path transfer function of $2/s$

and a positive feedback of $(1/s + 3)[-(s + 1)]$. This gives an output of:

$$\frac{Y(s)}{D(s)} = \frac{2(s+3)}{s(s+3)+2(s+1)}$$

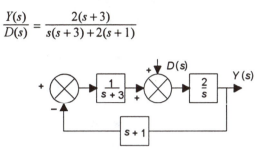

Figure 9.26 *System with input equal to zero*

The total input is the sum of the outputs due to each of the inputs and so:

$$Y(s) = \frac{2}{s(s+3)+2(s+1)}X(s) + \frac{2(s+3)}{s(s+3)+2(s+1)}D(s)$$

9.6 Sensitivity

The *sensitivity* of a system is the measure of the amount by which the overall gain of the system is affected by changes in the gain of system elements or particular inputs. In the following, we consider the effects of changing the gain of elements and also the effect of disturbances.

9.6.1 Sensitivity to changes in parameters

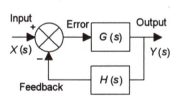

Figure 9.27 *System with negative feedback*

With a control system, the transfer functions of elements may drift with time and thus we need to know how such drift will affect the overall performance of the system.

For a closed-loop system with negative feedback (Figure 9.27):

$$\text{overall transfer function} = \frac{G(s)}{1+G(s)H(s)}$$

If $G(s)H(s)$ is large then the above equation reduces to:

$$\text{overall transfer function} \simeq \frac{G(s)}{G(s)H(s)} \simeq \frac{1}{H(s)}$$

Thus, in such a situation, the system is relatively insensitive to variations in the forward path transfer function but is sensitive to variations in the feedback path transfer function. For example, a change in the feedback path transfer function of, say, 10%, i.e. from $H(s)$ to $1.1H(s)$, will result in a change in the overall transfer function from $1/H(s)$ to $1/1.1H(s)$ or about $0.9/H(s)$ and so a change of about 10%.

This sensitivity is because the feedback transfer function is for the measurement system supplying the signal which is compared with the set value signal to determine the error and so variations in the feedback transfer function directly affect the computation of the error.

If the forward path transfer function $G(s)$ changes then the overall transfer function $G_{overall}(s)$ will change. We can define the sensitivity of the system to changes in the transfer function of the forward element as the fractional change in the overall system transfer function $G_{overall}(s)$ divided by the fractional change in the forward element transfer function $G(s)$, i.e. $(\Delta G_{overall}/G_{overall})/(\Delta G/G)$ where $\Delta G_{overall}$ is the change in overall gain producing a change of ΔG in the forward element transfer function. Thus, the sensitivity can be written as:

$$\text{sensitivity} = \frac{\Delta G_{overall}(s)}{\Delta G(s)} \frac{G(s)}{G_{overall}(s)}$$

If we differentiate the equation given above for the overall transfer function we obtain:

$$\frac{dG_{overall}(s)}{dG(s)} = \frac{1}{[1 + G(s)H(s)]^2}$$

and since $G_{overall}(s)/G(s) = 1/[1 + G(s)H(s)]$, the sensitivity is:

$$\text{sensitivity} = \frac{1}{1 + G(s)H(s)}$$

Application
For the feedback op-amp amplifier discussed in Section 8.2.2, the forward path transfer function for the op amp is very large and so gives a system with low sensitivity to changes in the op amp gain and hence a stable system which can have its gain determined by purely changing the feedback loop gain, i.e. the resistors in a potential divider.

Thus the bigger the value of $G(s)H(s)$ the lower the sensitivity of the system to changes in the forward path transfer function.

Example

A closed-loop control system with negative feedback has a feedback transfer function of 0.1 and a forward path transfer function of (a) 50, (b) 5. What will be the effect of a change in the forward path transfer function of an increase by 10%?

(a) We have, before the change:

$$\text{overall transfer function} = \frac{G(s)}{1 + G(s)H(s)} = \frac{50}{1 + 50 \times 0.1} = 8.3$$

After the change we have:

$$\text{overall transfer function} = \frac{G(s)}{1 + G(s)H(s)} = \frac{55}{1 + 55 \times 0.1} = 8.5$$

The change is thus about 2%.
(b) We have, before the change:

$$\text{overall transfer function} = \frac{G(s)}{1 + G(s)H(s)} = \frac{5}{1 + 5 \times 0.1} = 3.3$$

After the change we have:

$$\text{overall transfer function} = \frac{G(s)}{1+G(s)H(s)} = \frac{5.5}{1+5.5\times0.1} = 3.5$$

The change is thus about 6%.

Thus the sensitivity of the system to changes in the forward path transfer function is reduced as the gain of the forward path is increased.

9.6.2 Sensitivity to disturbances

An important effect of having feedback in a system is the reduction of the effects of disturbance signals on the system. A disturbance signal is an unwanted signal which affects the output signal of a system, e.g. noise in electronic amplifiers or a door being opened in a room with temperature controlled by a central heating system.

As an illustration, consider the effect of external disturbances on the overall gain of a system. Firstly we consider the effect on an open-loop system and then on a closed-loop system.

For the two-element open-loop system shown in Figure 9.28, there is a disturbance which gives an input between the two elements. For an input $X(s)$ to the system, the first element gives an output of $G_1(s)X(s)$. To this is added the disturbance $D(s)$ to give an input of $G_1(s)X(s) + D(s)$. The overall system output will then be

$$Y(s) = G_2(s)[G_1(s)X(s) + D(s)] = G_1(s)G_2(s)X(s) + G_2(s)D(s)$$

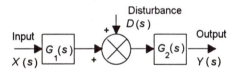

Figure 9.28 *Disturbance with an open-loop system*

For the system with feedback (Figure 9.29), the input to the first forward element $G_1(s)$ is $X(s) - H(s)Y(s)$ and so its output is $G_1(s)[X(s) - H(s)Y(s)]$. The input to $G_2(s)$ is $G_1(s)[X(s) - H(s)Y(s)] + D(s)$ and so its output is $X(s) = G_2(s)\{G_1(s)[X(s) - H(s)Y(s)] + D(s)\}$. Thus:

$$Y(s) = \frac{G_1(s)G_2(s)}{1+G_1(s)G_2(s)H(s)}X(s) + \frac{G_2(s)}{1+G_1(s)G_2(s)H(s)}D(s)$$

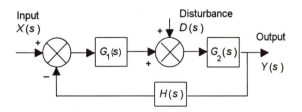

Figure 9.29 *Disturbance with closed-loop system*

Comparing this with the equation for the open-loop system of $Y(s) = G_2(s)[G_1(s)X(s) + D(s)] = G_1(s)G_2(s)X(s) + G_2(s)D(s)$ indicates that the effect of the disturbance on the output of the system has been reduced by a factor of $[1 + G_1(s)G_2(s)H(s)]$. This factor is thus a measure of how much the effects of a disturbance are reduced by feedback.

Problems *Questions 1 to 4 have four answer options: A, B, C or D. Choose the correct answer from the answer options.*

1 Decide whether each of these statements is True (T) or False (F).

A system has an input x which gives rise to an output y with the following differential equation describing the relationship:

$$5\frac{dy}{dt} + y = 2x$$

With no initial conditions:
(i) The Laplace transform of the equation is $5Y(s) = 2X(s)$.
(ii) The system transfer function is 2/5.

A (i) T (ii) T
B (i) T (ii) F
C (i) F (ii) T
D (i) F (ii) F

2 An open-loop control system consists of two elements in series, the first having a transfer function of $2s$ and the second a transfer function of $1/(s + 1)$. The overall transfer function of the system is:

A $2s + 1/(s + 1)$
B $2s/(s + 1)$
C $2s(s + 1)$
D $(s + 1)/2s$

3 A closed-loop control system has a forward loop with a transfer function of $3/(s + 2)$ and a negative feedback loop with a transfer function of 5. The overall transfer function of the system is:

A $5 + 3/(s + 2)$
B $5(s + 2)/3$
C $3/(s + 17)$
D $15/(s + 2)$

4 An open-loop control system consists of a d.c. motor with a transfer function of $2/(0.5s + 2)$ and a process, its shaft and load, with a transfer function of $1/(0.1s + 0.5)$. The overall transfer function of the system is:

A $2/[(0.5s + 20)(0.1s + 0.5)]$
B $[2/(0.5s + 2)] + [1/(0.1s + 0.5)]$
C $2(0.1s + 0.5)/0.5s + 2)$
D $(0.5s + 2)(0.1s + 0.5)/2$

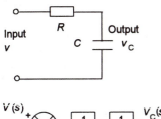

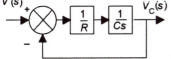

Figure 9.30   *Problem 8*

5 An open-loop system consists of three elements in series, the elements having transfer functions of 5, $1/s$ and $1/(s + 1)$. What is the overall transfer function of the system?

6 What is the overall gain of a closed-loop negative feedback system having a forward path gain of 2 and a feedback path gain of 0.1?

7 What is the overall transfer function of a closed-loop negative feedback system having a forward path transfer function of $2/(s + 1)$ and a feedback path transfer function of 0.1?

8 Figure 9.30 shows an electrical circuit and its block diagram representation. What is the overall transfer function of the system?

9 Use block simplification to arrive at the overall transfer function of the systems shown in Figure 9.31.

10 What is the overall transfer function for the systems shown in Figure 9.32?

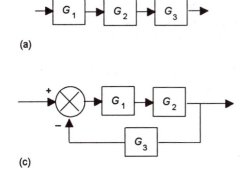

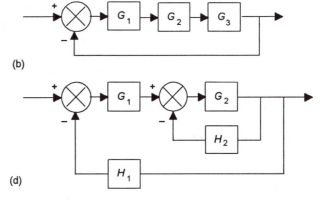

Figure 9.31   *Problem 9*

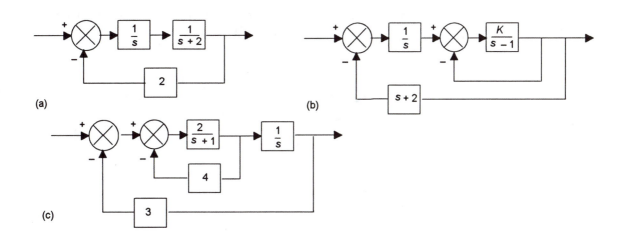

Figure 9.32   *Problem 10*

11 A closed-loop negative feedback system to be used for controlling the position of a load has a differential amplifier with transfer function K_1 operating a motor with transfer function $1/(sL + R)$. The output of the motor operates a gear system with gear ratio N and this, in turn, operates a screw with transfer function $1/s$ to give the resulting displacement. The position sensor is a potentiometer and this gives a feedback voltage related to the position of the load by the transfer function K_2. Derive the transfer function for the system as a whole, relating the input voltage to the system to the displacement output.

12 A closed-loop negative feedback system for the control of the height of liquid in a tank by pumping liquid from a reservoir tank can be considered to be a system with a differential amplifier having a transfer function of 5, its output operating a pump with a transfer function $5/(s + 1)$. The coupled system of tanks has a transfer function, relating height in the tank to the output from the pump, of $3/(s + 1)(s + 2)$. The feedback sensor of the height level in the tank has a transfer function of 0.1. Determine the overall transfer function of the system, relating the input voltage signal to the system to the height of liquid in the tank.

13 For the control system shown in Figure 9.33, determine the output $Y(s)$ in terms of the inputs $X_1(s)$ and $X_2(s)$.

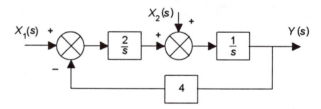

Figure 9.33 *Problem 13*

14 For the control system shown in Figure 9.34, determine the output $Y(s)$ in terms of the inputs $X_1(s)$ and $X_2(s)$.

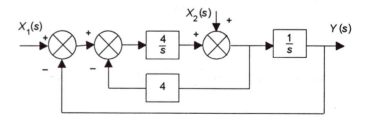

Figure 9.34 *Problem 14*

10 System response

10.1 Introduction

In Chapter 9 the method of describing a system by means of a transfer function was introduced; the transfer function is the output divided by the input when both are written as Laplace transforms. In this chapter we consider how we can use such transfer functions to determine how the output of a system will change with time for particular inputs, namely step, impulse and ramp.

10.2 Inputs

Inputs to systems commonly take a number of standard forms (Figure 10.1). With the *step input* we have the input suddenly being switched to a constant value at some particular time. With the *impulse input* we have the input existing for just a very brief time before dropping back to zero. With the *ramp input*, we have the input starting at some time and then increasing at a constant rate. The Laplace transforms of particular forms of such signals are:

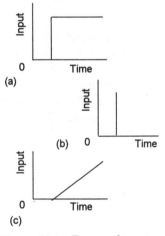

Figure 10.1 *Forms of input: (a) step, (b) impulse, (c) ramp*

A *unit step input* which starts at a time $t = 0$ and rises to the constant value 1 has a Laplace transform of $1/s$.
A *unit impulse input* which starts at a time $t = 0$ and rises to the value 1 has a Laplace transform of 1.
A *unit ramp input* which starts at time $t = 0$ and rises by 1 each second has a Laplace transform of $1/s^2$.

In general, if a function of time is multiplied by some constant, then the Laplace transform of that function is multiplied by the same constant. Thus, if we have a step input of size 5 at time $t = 0$ then the Laplace transform is 5 times the transform of a unit step and so is $5/s$. If we have an impulse of size 5 at time $t = 0$ then its transform is 5.

Example

An electrical system has an input of a voltage of 2 V which is suddenly applied by a switch being closed. What is the Laplace transform of the input?

Assume the input occurs at time $t = 0$. The input is a step voltage of size 2 V. An input of a unit step voltage has a Laplace transform of $(1/s)$ and thus for a 2 V step is $(2/s)$ V.

Example

A controlled speed motor has a voltage input which is increased at the rate of 3 V per second. What is the Laplace transform of the input?

Assume the input starts at time t = 0. The input is a ramp voltage of 3 V/s. An input of a unit ramp voltage as an s function is $(1/s^2)$ and thus for a 3 V/s ramp is $(3/s^2)$.

10.3 Determining outputs

The procedure we can use to determine how the output of a system will change with time when there is some input to the system is:

1   *Determine the output in the form of a Laplace transform*
    In terms of the transfer function $G(s)$ we have:

$$\text{Output } (s) = G(s) \times \text{Input } (s)$$

We can thus obtain the output of a system as a Laplace transform by multiplying its transfer function by the Laplace transform of the input.

2   *Determine the time function corresponding to the output transform*
    To obtain the output as a function of time we need to find the time function that will give the particular output transform that we have obtained; this is said to be determining the *inverse transform*. Tables of Laplace transforms and their corresponding time functions can be used; Table 10.1 is a table of commonly encountered functions. Often, however, the s function output has to be rearranged to put it into a form given in the table.

In obtaining the time function corresponding to a particular s function we can utilise the following properties of s functions:

1   If a Laplace transform is multiplied by some constant then the corresponding time function is multiplied by the same constant. For example, if we have $3/(s + 1)$ then the corresponding time function is that of 3 times the time function for $1/(s + 1)$.

2   If we have two separate s terms then the corresponding time function is the sum of the time functions corresponding to each of the separate s terms. For example, if we have $[1/(s + 2)] + [1/(s + 1)]$ then the time function is the time function for $[1/(s + 2)]$ plus the time function for $[1/(s + 1)]$.

Example

A system gives an output of $1/(s + 5)$. What is the output as a function of time?

Table 10.1 *Laplace functions and their corresponding time functions*

	Time function $f(t)$	Laplace transform $F(s)$
1	A unit impulse	1
2	A unit step	$\dfrac{1}{s}$
3	t, a unit ramp	$\dfrac{1}{s^2}$
4	e^{-at}, exponential decay	$\dfrac{1}{s+a}$
5	$1 - e^{-at}$, exponential growth	$\dfrac{a}{s(s+a)}$
6	te^{-at}	$\dfrac{1}{(s+a)^2}$
7	$t - \dfrac{1-e^{-at}}{a}$	$\dfrac{a}{s^2(s+a)}$
8	$e^{-at} - e^{-bt}$	$\dfrac{b-a}{(s+a)(s+b)}$
9	$(1-at)e^{-at}$	$\dfrac{s}{(s+a)^2}$
10	$1 - \dfrac{b}{b-a}e^{-at} + \dfrac{a}{b-a}e^{-bt}$	$\dfrac{ab}{s(s+a)(s+b)}$
11	$\dfrac{e^{-at}}{(b-a)(c-a)} + \dfrac{e^{-bt}}{(c-a)(a-b)} + \dfrac{e^{-ct}}{(a-c)(b-c)}$	$\dfrac{1}{(s+a)(s+b)(s+c)}$
12	$\sin \omega t$, a sine wave	$\dfrac{\omega}{s^2+\omega^2}$
13	$\cos \omega t$, a cosine wave	$\dfrac{s}{s^2+\omega^2}$
14	$e^{-at}\sin \omega t$, a damped sine wave	$\dfrac{\omega}{(s+a)^2+\omega^2}$
15	$e^{-at}\cos \omega t$, a damped cosine wave	$\dfrac{s+a}{(s+a)^2+\omega^2}$
16	$\dfrac{\omega}{\sqrt{1-\zeta^2}}e^{-\zeta\omega t}\sin \omega\sqrt{1-\zeta^2}\,t$	$\dfrac{\omega^2}{s^2+2\zeta\omega s+\omega^2}$
17	$1 - \dfrac{1}{\sqrt{1-\zeta^2}}e^{-\zeta\omega t}\sin\left(\omega\sqrt{1-\zeta^2}\,t+\phi\right)$, $\cos\phi = \zeta$	$\dfrac{\omega^2}{s(s^2+2\zeta\omega s+\omega^2)}$

The output is of the form given in Table 10.1 as item 4 with $a = 5$. Hence the time function is e^{-5t} and thus describes an output which decays exponentially with time.

Example

A system gives an output of $10/[s(s + 5)]$. What is the output as a function of time?

The nearest form we have in Table 10.1 to the output is item 5 as $2 \times a/[s(s + a)]$ with $a = 5$. Thus the output, as a function of time, is $2(1 - e^{-5t})$.

Example

A system has a transfer function of $1/(s + 2)$. What will be its output as a function of time when it is subject to a step input of 1 V?

The step input has a Laplace transform of $(1/s)$. Thus:

Output $(s) = G(s) \times$ Input (s)

$$= \frac{1}{s+2} \times \frac{1}{s} = \frac{1}{s(s+2)}$$

The nearest form we have in Table 10.1 to the output of $1/[s(s + 2)]$ is item 5 as $\frac{1}{2} \times 2/[s(s + 2)]$. Thus the output, as a function of time, is $\frac{1}{2}(1 - e^{-5t})$ V.

Example

A system has a transfer function of $4/(s + 2)$. What will be its output as a function of time when subject to a ramp input of 2 V/s?

The ramp input has a Laplace transform of $(2/s^2)$. Thus:

Output $(s) = G(s) \times$ Input (s)

$$= \frac{4}{s+2} \times \frac{2}{s^2} = \frac{8}{s^2(s+2)}$$

The nearest form we have in Table 10.1 to the output is item 7 when written as $4 \times 2/[s^2(s + 2)]$. Thus the output, as a function of time, is $4[t - (1 - e^{-2t})/2] = 4t - 2(1 - e^{-2t})$ V.

10.3.1 Partial fractions

A technique that is often required to put a Laplace transform in terms which identify with standard forms, so enabling the corresponding time function to be obtained in Table 10.1, is *partial fractions*. The term partial fractions is used for the process of converting an expression involving a complex fraction into a number of simpler fraction terms. This technique can be used provided the highest power of s in the numerator of the expression is less than that in the denominator. When the highest power in the numerator is equal to or higher than that of the denominator, the denominator must be divided into the numerator until the result is the sum of terms with the remainder fractional term having a numerator with a lower power than the denominator.

There are basically three types of partial fractions:

1   The numerator is some function of s and the denominator contains factors which are only of the form $(s + a)$, $(s + b)$, $(s + c)$, etc. and so is of the form:

$$\frac{F(s)}{(s+a)(s+b)(s+c)}$$

and has the partial fractions of

$$\frac{A}{(s+a)} + \frac{B}{(s+b)} + \frac{C}{(s+c)}$$

There is a partial fraction term for each bracketed term in the denominator. Thus, if we have $1/(s + a)(s + b)$ there will be two partial fraction terms.

2   There are repeated $(s + a)$ factors in the denominator and the expression is of the form:

$$\frac{F(s)}{(s+a)^n}$$

and has the partial fractions of:

$$\frac{A}{(s+a)^1} + \frac{B}{(s+a)^2} + \frac{C}{(s+a)^3} + \cdots + \frac{N}{(s+a)^n}$$

A multiple root expression has thus a partial fraction term for each power of the factor. Thus, if we have $1/(s + a)^2$ there will be two partial fraction terms; if we have $1/(s + a)^3$ there will be three partial fraction terms.

3   The denominator contains quadratic factors and the quadratic does not factorise, being of the form:

$$\frac{F(s)}{(as^2 + bs + c)(s+d)}$$

and has the partial fractions of:

$$\frac{As+B}{as^2 + bs + c} + \frac{C}{s+d}$$

The values of the constants A, B, C, etc. can be found by either making use of the fact that the equality between the expression and the partial fractions must be true for all values of s and so considering particular values of s that make calculations easy or that the coefficients of s^n in the expression must equal those of s^n in the partial fraction expansion.

Example

Determine the partial fractions of:

$$\frac{s+4}{(s+1)(s+2)}$$

The partial fractions are of the form:

$$\frac{A}{s+1} + \frac{B}{s+2}$$

Then, for the partial fraction expression to equal the original fraction, we must have:

$$\frac{s+4}{(s+1)(s+2)} = \frac{A(s+2)+B(s+1)}{(s+1)(s+2)}$$

and consequently:

$$s+4 = A(s+2) + B(s+1)$$

This must be true for all values of s. The procedure is then to pick values of s that will enable some of the terms involving constants to become zero and so enable other constants to be determined. Thus if we let $s = -2$ then we have

$$(-2)+4 = A(-2+2) + B(-2+1)$$

and so $B = -2$. If we now let $s = -1$ then

$$(-1)+4 = A(-1+2) + B(-1+1)$$

and so $A = 3$. Thus

$$\frac{s+4}{(s+1)(s+2)} = \frac{3}{s+1} - \frac{2}{s+2}$$

Example

Determine the partial fractions of:

$$\frac{3s+1}{(s+2)^3}$$

This has partial fractions of:

$$\frac{A}{(s+2)} + \frac{B}{(s+2)^2} + \frac{C}{(s+2)^3}$$

Then, for the partial fraction expression to equal the original fraction, we must have:

$$\frac{3s+1}{(s+2)^3} = \frac{A}{(s+2)} + \frac{B}{(s+2)^2} + \frac{C}{(s+2)^3}$$

and so consequently have:

$$3s + 1 = A(s + 2)^2 + B(s + 2) + C$$

$$= A(s^2 + 2s + 1) + B(s + 2) + C$$

Equating s^2 terms gives $0 = A$. Equating s terms gives $3 = 2A + B$. and so $B = 3$. Equating the numeric terms gives $1 = A + 2B + C$ and so $C = -5$. Thus:

$$\frac{3s + 1}{(s + 2)^3} = \frac{3}{(s + 2)^2} - \frac{5}{(s + 2)^3}$$

Example

Determine the partial fractions of:

$$\frac{2s + 1}{(s^2 + s + 1)(s + 2)}$$

This will have partial fractions of:

$$\frac{As + B}{s^2 + s + 1} + \frac{C}{s + 2}$$

Thus we must have:

$$\frac{2s + 1}{(s^2 + s + 1)(s + 2)} = \frac{As + B}{s^2 + s + 1} + \frac{C}{s + 2}$$

and so:

$$2s + 1 = (As + B)(s + 2) + C(s^2 + s + 1)$$

With $s = -2$ then $-3 = 3C$ and so $C = -1$. Equating s^2 terms gives $0 = A + C$ and so $A = 1$. Equating s terms gives $2 = 2A + B + C$ and so $B = 1$. As a check, equating numeric terms gives $1 = 2B + C$. Thus:

$$\frac{2s + 1}{(s^2 + s + 1)(s + 2)} = \frac{s + 1}{s^2 + s + 1} - \frac{1}{s + 2}$$

Example

Determine the partial fractions of:

$$\frac{2s^2 + 2}{(s + 4)(s - 2)}$$

This expression has a numerator with the highest power of s the same as that in the denominator. Thus we have to use division to put the expression in a form where we can use partial fractions.

$$\frac{2}{s^2 + 2s - 8 \overline{\smash{\big)}\ 2s^2 + 2}}$$
$$\underline{2s^2 + 4s - 16}$$
$$- 4s + 18$$

Thus we can write the expression as:

$$2 + \frac{-4s + 18}{(s + 4)(s - 2)}$$

Normal partial fraction methods can now be used for the fraction.

$$\frac{-4s + 18}{(s + 4)(s - 2)} = \frac{A}{s + 4} + \frac{B}{s - 2}$$

$$-4s + 18 = A(s - 2) + B(s + 4)$$

With $s = 2$, then $B = 5/3$. With $s = -4$, then $A = -17/3$. Thus the expression can be written as:

$$2 - \frac{17}{3(s + 4)} + \frac{5}{3(s - 2)}$$

10.4 First-order systems

A first-order system has a differential equation of the form (see Section 8.4.1):

$$\tau \frac{dy}{dt} + y = kx$$

Application
A temperature sensor is typically a first-order system. First-order systems are generally systems which have just capacitive and resistive elements.

where τ is the time constant and k the steady-state gain. The Laplace transform of this equation can be written as:

$$\tau Y(s) + Y(s) = kX(s)$$

and so a transfer function of the form:

$$G(s) = \frac{Y(s)}{X(s)} = \frac{k}{\tau s + 1}$$

When a first-order system is subject to a *unit impulse input* then $X(s)$ = 1 and the output transform $Y(s)$ is:

$$Y(s) = G(s)X(s) = \frac{k}{\tau s + 1} \times 1 = k\frac{(1/\tau)}{s + 1/\tau}$$

Hence, since we have the transform in the form $1/(s + a)$, using item 4 in Table 10.1 gives:

$$y = k(1/\tau)\, e^{-t/\tau}$$

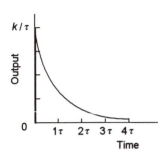

Figure 10.2 *Output with a unit impulse input to a first-order system*

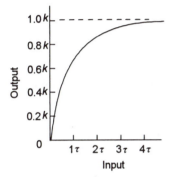

Figure 10.3 *Behaviour of a first-order system when subject to a unit step input*

Application
As a general guide, a first-order system is considered to have settled down to its steady-state value after a time equal to four time constants.

Figure 10.2 shows how the output varies with time; it is an exponential decay. The output rises to its maximum value at time $t = 0$ and then after 1τ it drops to 0.37 of the initial value, after 2τ it is 0.14 of the initial value and after 3τ it is 0.05 of the initial value. Thus by about a time equal to four times the time constant the output is effectively zero. The exponential term tends to a zero value as the time t tends to an infinite value.

When a first-order system is subject to a *unit step input* then $X(s) = 1/s$ and the output transform $Y(s)$ is:

$$Y(s) = G(s)X(s) = \frac{k}{s(\tau s + 1)} = k\frac{(1/\tau)}{s(s + 1/\tau)}$$

Hence, since we have the transform in the form $a/s(s + a)$, using item 5 in Table 10.1 gives:

$$y = k(1 - e^{-t/\tau})$$

Figure 10.3 shows how the output varies with time. Initially, at time $t = 0$, the output is zero. It then rises to 0.63 of the steady-state value after 1τ, then 0.86 of the steady-state value after 2τ and 0.95 of the steady-state value after 3τ. After 4τ the output is effectively at the steady-state value of k, the exponential term in the above equation becoming zero as time t tends to infinity.

Example

A circuit has a resistance R in series with a capacitance C. The differential equation relating the input v and output v_C, i.e. the voltage across the capacitor, is:

$$v = RC\frac{dv_C}{dt} + v_C$$

Determine the output of the system when there is a 3 V impulse input.

As a function of s the differential equation becomes:

$$V(s) = RCsV_C(s) + V_C(s)$$

Hence the transfer function is

$$G(s) = \frac{V_C(s)}{V(s)} = \frac{1}{RCs + 1}$$

The output when there is 3 V impulse input is:

$$V_C(s) = G(s)V(s) = \frac{1}{RCs + 1} \times 3 = \frac{3/RC}{s + 1/RC}$$

Hence, since we have the transform in the form $1/(s + a)$, using item 4 in Table 10.1 gives:

$$v_C = (3/RC)\, e^{-t/RC}$$

Example

A thermocouple which has a transfer function linking its voltage output V and temperature input of:

$$G(s) = \frac{30 \times 10^{-6}}{10s + 1}\ \text{V/°C}$$

Determine the response of the system when it is suddenly immersed in a water bath at 100°C.

The output as an s function is:

$$V(s) = G(s) \times \text{input }(s)$$

The sudden immersion of the thermometer gives a step input of size 100°C and so the input as an s function is $100/s$. Thus:

$$V(s) = \frac{30 \times 10^{-6}}{10s + 1} \times \frac{100}{s} = \frac{30 \times 10^{-4}}{10s(s + 0.1)} = 30 \times 10^{-4}\frac{0.1}{s(s + 0.1)}$$

The fraction element is of the form $a/s(s + a)$ and so the output as a function of time is:

$$V = 30 \times 10^{-4}\,(1 - e^{-0.1t})\ \text{V}$$

10.4.1 First-order system parameters

The following are parameters used to describe the transient performance of a first-order system while it is moving towards the steady-state value:

1 *Time constant*
   When the time $t = \tau$ then we have $y = k(1 - e^{-1}) = 0.63k$. Thus, the *time constant* τ for a first-order system when subject to a step input is the time taken for the output to reach 0.63, of the steady-state value.

2 *Delay time*
   The delay time t_d is the time required for the output response to reach 50% of its steady-state value. Thus, since k is the final value, the time taken to reach 50% of this value is given by:

$$\tfrac{1}{2}k = k(1 - e^{-t_d/\tau})$$

$$\tfrac{1}{2} = e^{-t_d/\tau}$$

$$\ln 2 = \frac{t_d}{\tau}$$

$$t_d = \tau \ln 2$$

3   *Rise time*

The rise time t_r is the time required for the output to rise from 10% to 90% of its steady-state value. Note that the specification is not always in terms of 10% to 90%, sometimes it is 0% to 100%. Since k is the steady-state value then the time taken to reach 10% of that value is:

$$\frac{10}{100}k = k(1 - e^{-t_{10}/\tau})$$

$$\frac{1}{10} = e^{-t_{10}/\tau}$$

$$\ln 10 = \frac{t_{10}}{\tau}$$

$$t_{10} = \tau \ln 10$$

The time taken to reach 90% of the steady-state value is given by:

$$\frac{90}{100}k = k(1 - e^{-t_{90}/\tau})$$

$$\frac{9}{10} = e^{-t_{90}/\tau}$$

$$\ln 10 - \ln 9 = \frac{t_{90}}{\tau}$$

$$t_{90} = \tau \ln 10 - \tau \ln 9$$

Hence the rise time is:

$$t_r = t_{90} - t_{10} = \tau \ln 9$$

Example

Determine the delay time and the rise time for a first-order system with the transfer function:

$$G(s) = \frac{3}{2s + 1}$$

Comparing the transfer function with $k/(\tau s + 1)$ indicates that the steady-state gain is 3 and the time constant is 2 s. Thus, the delay time is:

$$t_{10} = \tau \ln 10 = 2 \ln 10 = 4.6 \text{ s}$$

The rise time is:

$$t_r = t_{90} - t_{10} = \tau \ln 9 = 2 \ln 9 = 4.4 \text{ s}$$

Example

A mercury-in-glass thermometer acts as a first-order system with an input of temperature and an output of the mercury position against a scale. The thermometer is initially at 0°C and is then suddenly placed in water at 100°C. After 80 s the thermometer reads 98°C. Determine (a) the time constant, (b) the delay time, (c) the rise time.

(a) For such a system the output θ is related to the input by the equation:

$$\theta = 100(1 - e^{-t/\tau})$$

Hence:

$$98 = 100(1 - e^{-80/\tau})$$

$$0.2 = e^{-80/\tau}$$

$$\ln 0.2 = -\frac{80}{\tau}$$

Hence, the time constant τ is 49.7 s.

(b) The delay time is:

$$t_{10} = \tau \ln 10 = 49.7 \ln 10 = 114.4 \text{ s}$$

(c) The rise time is:

$$t_r = t_{90} - t_{10} = \tau \ln 9 = 49.7 \ln 9 = 109.2 \text{ s}$$

10.5 Second-order systems

The differential equation for a second-order system is of the form (see Section 8.4.2):

$$\frac{d^2 y}{dt^2} + 2\zeta\omega_n \frac{dy}{dt} + \omega_n^2 y = k\omega_n^2 x$$

where ω_n is the natural angular frequency with which the system oscillates and ζ is the damping ratio. Hence, for the Laplace transform we have:

$$s^2 Y(s) + 2\zeta\omega_n s Y(s) + \omega_n^2 Y(s) = k\omega_n^2 X(s)$$

and so a transfer function of:

$$G(s) = \frac{Y(s)}{X(s)} = \frac{k\omega_n^2}{s^2 + 2\zeta\omega_n s + \omega_n^2}$$

Application
A mechanical second-order system will have a spring element and an inertia element, an electrical system capacitance and inductance.

An example of a control system which behaves as a second-order system is the idling control system used with a car engine. The important issue is how the engine runs when there is a sudden change in engine load. The engine speed must not drop and cause the engine to stall, thus any transient drop must not be excessive and the speed should recover to the required idling speed as fast as possible. We thus need to know how the second-order system behaves when the input changes.

When a second-order system is subject to a *unit step input*, i.e. $X(s) = 1/s$, then the output transform $Y(s)$ is:

$$Y(s) = G(s)X(s) = \frac{k\omega_n^2}{s(s^2 + 2\zeta\omega_n s + \omega_n^2)}$$

There are three different forms of answer to this equation for the way the output varies with time; these depending on the value of the damping constant and whether it gives an overdamped, critically damped or underdamped system (see Figure 8.24). We can determine the condition for these three forms of output by putting the equation in the form:

$$Y(s) = \frac{k\omega_n^2}{s(s+p_1)(s+p_2)}$$

where p_1 and p_2 are the roots of the denominator quadratic term, the so termed *characteristic equation*:

$$s^2 + 2\zeta\omega_n s + \omega_n^2 = 0$$

Hence, if we use the standard equation to determine the roots of a quadratic equation, we obtain:

$$p = \frac{-2\zeta\omega_n \pm \sqrt{4\zeta^2\omega_n^2 - 4\omega_n^2}}{2}$$

and so the two roots are given by:

$$p_1 = -\zeta\omega_n + \omega_n\sqrt{\zeta^2 - 1} \quad \text{and} \quad p_2 = -\zeta\omega_n - \omega_n\sqrt{\zeta^2 - 1}$$

The important issue in determining the form of the roots is the value of the square root term and this is determined by the value of the damping factor ζ.

Application
A stepper motor is essentially a second-order system. At very low stepping rates, the motor has time to move from one step to another and settle down to its steady-state value. However, if the stepping rate is increased, at some rate of stepping the rotor does not have time to reach its steady-state position before the next step occurs. This constitutes the limiting stepping value for the motor.

1   $\zeta > 1$
With the damping factor ζ greater than 1 the square root term is real and will factorise. To find the inverse transform we can either use partial fractions to break the expression down into a number of simple fractions or use item 10 in Table 10.1. The output is thus:

$$y = \frac{k\omega_n^2}{p_1 p_2}\left[1 - \frac{p_2}{p_2 - p_1}e^{-p_1 t} + \frac{p_1}{p_2 - p_1}e^{-p_2 t}\right]$$

This describes an output which does not oscillate but dies away with time and thus the system is *overdamped*. As the time t tends to infinity then the exponential terms tend to zero and the output becomes the steady value of $k\omega_n^2/(p_1 p_2)$. Since $p_1 p_2 = \omega_n^2$, the steady-state value is k.

2   $\zeta = 1$

With $\zeta = 1$ the square root term is zero and so $p_1 = p_2 = -\omega_n$; both roots are real and both the same. The output equation then becomes:

$$Y(s) = \frac{k\omega_n^2}{s(s + \omega_n)^2}$$

This equation can be expanded by means of partial fractions to give:

$$Y(s) = k\left[\frac{1}{s} - \frac{1}{s + \omega_n} - \frac{\omega_n}{(s + \omega_n)^2}\right]$$

Hence:

$$y = k[1 - e^{-\omega_n t} - \omega_n t e^{-\omega_n t}]$$

This is the critically damped condition and describes an output which does not oscillate but dies away with time. As the time t tends to infinity then the exponential terms tend to zero and the output tends to the steady-state value of k.

3   $\zeta < 1$

With $\zeta < 1$ the square root term does not have a real value. Using item 17 in Table 10.1 then gives:

$$y = k\left[1 - \frac{e^{-\zeta\omega_n t}}{\sqrt{1 - \zeta^2}} \sin\left(\omega_n \sqrt{(1 - \zeta^2)} \, t + \phi\right)\right]$$

where $\cos \phi = \zeta$. This is an under damped oscillation. The angular frequency of the damped oscillation is:

$$\omega = \omega_n \sqrt{1 - \zeta^2}$$

Only when the damping is very small does the angular frequency of the oscillation become nearly the natural angular frequency ω_n. As the time t tends to infinity then the exponential term tends to zero and so the output tends to the value k.

Example

What will be the state of damping of a system having the following transfer function and subject to a unit step input?

$$G(s) = \frac{1}{s^2 + 8s + 16}$$

The output $Y(s)$ from such a system is given by:

$$Y(s) = G(s)X(s)$$

For a unit step input $X(s) = 1/s$ and so the output is given by:

$$Y(s) = \frac{1}{s(s^2 + 8s + 16)} = \frac{1}{s(s+4)(s+4)}$$

The roots of $s^2 + 8s + 16$ are $p_1 = p_2 = -4$. Both the roots are real and the same and so the system is critically damped.

Example

A system has an output y related to the input x by the differential equation:

$$\frac{d^2y}{dt^2} + 5\frac{dy}{dt} + 6y = x$$

What will be the output from the system when it is subject to a unit step input? Initially both the output and input are zero.

We can write the Laplace transform of the equation as:

$$s^2Y(s) + 5sY(s) + 6Y(s) = X(s)$$

The transfer function is thus:

$$G(s) = \frac{Y(s)}{X(s)} = \frac{1}{s^2 + 5s + 6}$$

For a unit step input the output is given by:

$$Y(s) = \frac{1}{s(s^2 + 5s + 6)} = \frac{1}{s(s+3)(s+2)}$$

Because the quadratic term has two real roots, the system is overdamped. We can directly use one of the standard forms given in Table 10.1 or partial fractions to first simplify the expression before using Table 10.1. Using partial fractions:

$$\frac{1}{s(s+3)(s+2)} = \frac{A}{s} + \frac{B}{s+3} + \frac{C}{s+2}$$

Thus, we have $1 = A(s + 3)(s + 2) + Bs(s + 2) + Cs(s + 3)$. When $s = 0$ then $1 = 6A$ and so $A = 1/6$. When $s = -3$ then $1 = 3B$ and so $B = 1/3$. When $s = -2$ then $1 = -2C$ and so $C = -1/2$. Hence we can write the output in the form:

$$Y(s) = \frac{1}{6s} + \frac{1}{3(s+3)} - \frac{1}{2(s+2)}$$

Hence, using Table 10.1 gives:

$$y = 0.17 + 0.33\ e^{-3t} - 0.5\ e^{-2t}$$

10.5.1 Second-order system parameters

For the underdamped oscillations of a system we have the output y given by:

$$y = k\left[1 - \frac{e^{-\zeta\omega_n t}}{\sqrt{1-\zeta^2}}\ \sin\!\left(\omega_n\sqrt{(1-\zeta^2)}\ t + \phi\right)\right]$$

with the damped natural frequency ω given by:

$$\omega = \omega_n\sqrt{(1-\zeta^2)}$$

We can write the above equation for the output in what is often a more convenient form. Since $\sin(A + B) = \sin A \cos B + \cos A \sin B$, the sine term can be written as:

$$\sin(\omega t + \phi) = \sin \omega t \cos \phi + \cos \omega t \sin \phi$$

and since ϕ is a constant:

$$\sin(\omega t + \phi) = P \sin \omega t + Q \cos \omega t$$

where P and Q are constants. Thus the output can be written as:

$$y = k\left[1 - \frac{e^{-\zeta\omega_n t}}{\sqrt{1-\zeta^2}}\,(P \sin \omega t + Q \cos \omega t)\right]$$

The performance of an underdamped second-order system to a unit step input (Figure 10.4) can be specified by:

1   *Rise time*
    The rise time t_r is the time taken for the response x to rise from 0 to the steady-state value y_{SS}. This is the time for the oscillating response to complete a quarter of a cycle, i.e. $\tfrac{1}{2}\pi$. Thus:

$$\omega t_r = \tfrac{1}{2}\pi$$

We can thus reduce the rise time by increasing the damped natural frequency, this value being determined by the undamped natural angular frequency and the damping factor, i.e.

$$\omega = \omega_n\sqrt{(1-\zeta^2)}$$

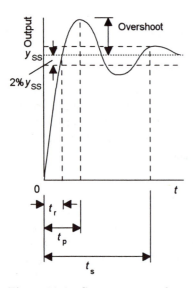

Figure 10.4   *Step response of an under damped system*

and so:

$$t_r = \frac{\pi}{2\omega_n \sqrt{1 - \zeta^2}}$$

The rise time is sometimes specified as the time taken for the response to rise from 10% to 90% of the steady-state value.

2   *Peak time*
The peak time t_p is the time taken for the response to rise from 0 to the first peak value. This is the time for the oscillating response to complete one half-cycle, i.e. π. Thus:

$$\omega t_p = \pi$$

and so using:

$$\omega = \omega_n \sqrt{(1 - \zeta^2)}$$

we can write:

$$t_p = \frac{\pi}{\omega_n \sqrt{1 - \zeta^2}}$$

When ζ is 1 then the peak time becomes infinite; this indicates that at critical damping the steady-state value is never reached but only approached asymptotically.

3   *Overshoot*
The overshoot is the maximum amount by which the response overshoots the steady-state value and is thus the amplitude of the first peak. The overshoot is often written as a percentage of the steady-state value.

The steady-state value is when t tends to infinity and thus $y_{ss} = k$. Since $y = 0$ when $t = 0$ then, since $e^0 = 1$, then using:

$$y = k\left[1 - \frac{e^{-\zeta\omega_n t}}{\sqrt{1 - \zeta^2}}(P \sin \omega t + Q \cos \omega t)\right]$$

we have:

$$0 = k\left[1 - \frac{1}{\sqrt{1 - \zeta^2}}(0 + Q)\right]$$

and so $Q = \sqrt{(1 - \zeta^2)}$.
The overshoot occurs at $\omega t = \pi$ and thus:

$$y = k\left[1 - \frac{e^{-\zeta\omega_n t}}{\sqrt{1-\zeta^2}}(P\sin\omega t + Q\cos\omega t)\right]$$

becomes:

$$y = y_{SS}\left[1 - \frac{e^{-\zeta\omega_n \pi/\omega}}{\sqrt{1-\zeta^2}}(0 - Q)\right]$$

The overshoot is the difference between the output at that time and the steady-state value. Hence:

$$\text{overshoot} = y_{SS}\frac{e^{-\zeta\omega_n \pi/\omega}}{\sqrt{1-\zeta^2}}Q = y_{SS}\,e^{-\zeta\omega_n \pi/\omega}$$

Since $\omega = \omega_n\sqrt{(1-\zeta^2)}$ then we can write:

$$\text{Overshoot} = y_{SS}\exp\left(\frac{-\zeta\omega_n \pi}{\omega_n\sqrt{1-\zeta^2}}\right)$$

$$= y_{SS}\exp\left(\frac{-\zeta\pi}{\sqrt{1-\zeta^2}}\right)$$

Expressed as a percentage of y_{SS}:

$$\text{percentage overshoot} = \exp\left(\frac{-\zeta\pi}{\sqrt{1-\zeta^2}}\right)\times 100\%$$

Table 10.2 *Percentage peak overshoot*

Damping ratio	Percentage overshoot
0.2	52.7
0.3	37.2
0.4	25.4
0.5	16.3
0.6	9.5
0.7	4.6
0.8	1.5
0.9	0.2

Note that the overshoot does not depend on the natural frequency of the system but only on the damping factor. As the damping factor approaches 1 so the percentage overshoot approaches zero. Table 10.2 gives values of the percentage overshoot for particular damping ratios.

4   *Subsidence ratio*

An indication of how fast oscillations decay is provided by the *subsidence ratio* or *decrement*. This is the amplitude of the second overshoot divided by the amplitude of the first overshoot. The first overshoot occurs when we have $\omega t = \pi$ and so:

$$\text{first overshoot} = y_{SS}\exp\left(\frac{-\zeta\pi}{\sqrt{1-\zeta^2}}\right)$$

The second overshoot occurs when $\omega t = 3\pi$ and so:

$$\text{second overshoot} = y_{ss} \exp\left(\frac{-3\zeta\pi}{\sqrt{1-\zeta^2}}\right)$$

Thus the subsidence ratio is given by:

$$\text{subsidence ratio} = \frac{\text{second overshoot}}{\text{first overshoot}} = \exp\left(\frac{-2\zeta\pi}{\sqrt{1-\zeta^2}}\right)$$

5   *Settling time*
The settling time t_s is used as a measure of the time taken for the oscillations to die away. It is the time taken for the response to fall within and remain within some specified percentage of the steady-state value (see Table 10.2). Thus for the 2% settling time, the amplitude of the oscillation should fall to be less than 2% of y_{ss}. We have:

$$y = k\left[1 - \frac{e^{-\zeta\omega_n t}}{\sqrt{1-\zeta^2}}(P\sin\omega t + Q\cos\omega t)\right]$$

with $y_{ss} = k$, $\omega = \omega_n\sqrt{(1-\zeta^2)}$ and, as derived earlier in item 3, $Q = \sqrt{(1-\zeta^2)}$. The amplitude of the oscillation is $(y - y_{ss})$ when y is a maximum value. The maximum values occur when ωt is some multiple of π and thus we have $\cos\omega t = 1$ and $\sin\omega t = 0$. For the 2% settling time, the settling time t_s is when the maximum amplitude is 2% of y_{ss}, i.e. $0.02y_{ss}$. Thus:

$$0.02y_{ss} = y_{ss}\, e^{-\zeta\omega_n t_s}$$

Taking logarithms gives $\ln 0.02 = -\zeta\omega_n t_s$ and since $\ln 0.02 = -3.9$ or approximately 4 then:

$$t_s = \frac{4}{\zeta\omega_n}$$

The above is the value of the settling time if the specified percentage is 2%. If the specified percentage is 5% the equation becomes

$$t_s = \frac{3}{\zeta\omega_n}$$

6   *Number of oscillations to settling time*
The time taken to complete one cycle, i.e. the periodic time, is $1/f$, where f is the frequency, and since $\omega = 2\pi f$ then the time to

complete one cycle is $2\pi/f$. In a settling time of t_s the number of oscillations that occur is:

$$\text{number of oscillations} = \frac{\text{settling time}}{\text{periodic time}}$$

and thus for a settling time defined for 2% of the steady-state value:

$$\text{number of oscillations} = \frac{4/\zeta\omega_n}{2\pi/\omega}$$

Since $\omega = \omega_n\sqrt{(1 - \zeta^2)}$, then:

$$\text{number of oscillations} = \frac{2\omega_n\sqrt{1-\zeta^2}}{\pi\zeta\omega_n} = \frac{2}{\pi}\sqrt{\frac{1}{\zeta^2} - 1}$$

In designing a system the following are the typical points that are considered:

1   For a rapid response, i.e. small rise time, the natural frequency must be large. Figure 10.5 shows the types of response obtained to a unit step input to systems having the same damping factor of 0.2 but different natural angular frequencies. The response time with the natural angular frequency of 10 rad/s, damped frequency 9.7 rad/s, is much higher than that with a natural angular frequency of 1 rad/s, damped frequency 0.97 rad/s.

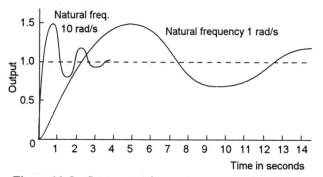

Figure 10.5   *Response of a unit gain second-order system to a unit step input, the damping factor being the same for both responses*

2   The damping factor is typically in the range 0.4 to 0.8 since smaller values give an excessive overshoot and a large number of oscillations before the system settles down. Larger values render the system sluggish since they increase the response time. Though, in some systems where no overshoot can be tolerated, a high value of damping factor may have to be used. Figure 10.6 shows the effect on the response of a second-order system of a change of damping factor when the natural angular frequency remains unchanged.

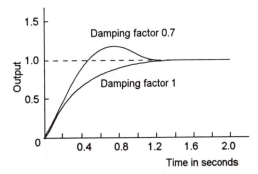

Figure 10.6   *Response of a unit gain second-order system to a unit step input, the natural angular frequency being the same for both responses*

Example

A second-order system has a natural angular frequency of 2.0 rad/s and a damped frequency of 1.8 rad/s. What are its (a) damping factor, (b) 100% rise time, (c) percentage overshoot, (c) 2% settling time, and (d) the number of oscillations within the 2% settling time?

(a) Since $\omega = \omega_n \sqrt{(1 - \zeta^2)}$, then the damping factor is given by:

$$1.8 = 2.0 \sqrt{1 - \zeta^2}$$

and $\zeta = 0.44$.
(b) Since $\omega t_r = \frac{1}{2}\pi$, then the 100% rise time is given by

$$t_r = \frac{\pi}{2 \times 1.8} = 0.87 \text{ s}$$

(c) The percentage overshoot is given by:

$$\% \text{ overshoot} = \exp\left(\frac{-\zeta\pi}{\sqrt{1-\zeta^2}}\right) \times 100\%$$

$$= \exp\left(\frac{-0.44\pi}{\sqrt{1-0.44^2}}\right) \times 100\% = 21\%$$

(d) The 2% settling time is given by:

$$t_s = \frac{4}{\zeta\omega_n} = \frac{4}{0.44 \times 2.0} = 4.5 \text{ s}$$

(e) The number of oscillations occurring within the 2% settling time is given by:

$$\text{number of oscillations} = \frac{2}{\pi}\sqrt{\frac{1}{\zeta^2} - 1} = \frac{2}{\pi}\sqrt{\frac{1}{0.44^2} - 1} = 1.3$$

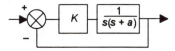

Figure 10.7 *Example*

Example

The feedback system shown in Figure 10.7 has the transfer function of the forward path as $K/[s(s + a)]$ and the transfer function of the feedback path as 1. What will be the effect on the system response of changing the gain K?

The transfer function of the closed-loop system is:

$$G(s) = \frac{K/[s(s+a)]}{1 + K/[s(s+a)]} = \frac{K}{s(s+a)+K} = \frac{K}{s^2 + as + K}$$

By comparison with the standard form of the transfer function for a second-order system, we have $\omega_n = \sqrt{K}$ and $\zeta = a/2\sqrt{K}$. Thus the rise time is given by $\omega t_r = \frac{1}{2}\pi$, with $\omega = \omega_n\sqrt{(1 - \zeta^2)}$, and so is:

$$t_r = \frac{\pi}{2\sqrt{K}\sqrt{1 - (a^2/4K)}}$$

The rise time thus decreases as K increases. Thus, increasing the gain decreases the rise time and so increases the speed of response of the system.

The percentage overshoot is given by:

$$\% \text{ overshoot} = \exp\left(\frac{-\zeta\pi}{\sqrt{1-\zeta^2}}\right) \times 100\%$$

$$= \exp\left(\frac{-(a/2\sqrt{K})\pi}{\sqrt{1-(a^2/4K)}}\right) \times 100\%$$

$$= \exp\left(\frac{a\pi}{\sqrt{4K - a^2}}\right) \times 100\%$$

Thus, increasing K results in an increase in the percentage overshoot.

The 2% settling time is given by:

$$t_s = \frac{4}{\zeta\omega_n} = \frac{4}{a/2} = \frac{8}{a}$$

Thus, the settling time is independent of the gain K.

10.6 Stability

We can define a system as being a *stable system* if, when given an input or a change in input, it has transients which die away with time and leave the system in a steady-state condition. The system would be *unstable* if the transients did not die away with time but grew with time and so steady-state conditions were never reached.

Consider a second-order system with the transfer function:

$$G(s) = \frac{1}{(s+1)(s+2)}$$

The values of s which make the transfer function infinite are termed the *poles* of the system, i.e. they are the roots of the denominator polynomial (the characteristic equation). Thus, the above system has the poles $s = -1$ and $s = -2$. A unit step input to such a system gives an output $Y(s)$:

$$Y(s) = \frac{1}{s(s+1)(s+2)} = \frac{1}{2s} - \frac{1}{s+1} + \frac{1}{2(s+2)}$$

This varies with time as:

$$y = \tfrac{1}{2} - e^{-t} + \tfrac{1}{2}e^{-2t}$$

Each of the poles gives rise to a transient term. Both the resulting exponential terms die away with time to give a steady-state value of 0.5, the more negative the value of s for a pole the more rapidly the corresponding term dies away. Thus, the system is stable.

Now consider a second-order system with the transfer function:

$$G(s) = \frac{1}{(s-1)(s-2)}$$

This system has the poles $s = +1$ and $s = +2$. A unit step input to such a system gives an output $Y(s)$:

$$Y(s) = \frac{1}{s(s-1)(s-2)} = \frac{1}{2s} - \frac{1}{s-1} + \frac{1}{2(s-2)}$$

This varies with time as:

$$y = \tfrac{1}{2} - e^{+t} + \tfrac{1}{2}e^{+2t}$$

Each of the positive poles gives rise to exponential terms which grow with time, the larger the value of s for a pole the more rapidly the corresponding term grows. Thus, the transients do not die away but increase and so the system is unstable.

In general, if a system has a transfer function with poles which are negative then it gives rise to a transient which dies away with time, whereas if it has a pole which is positive then the transient grows with time. Thus, if a transfer function has a pole which is positive then it is unstable.

In general, for a second-order system we can write the transfer function in the form:

$$G(s) = \frac{k\omega_n^2}{s^2 + 2\zeta\omega_n s + \omega_n^2} = \frac{k\omega_n^2}{(s+p_1)(s+p_2)}$$

The roots of the quadratic denominator, i.e. the poles, are given by:

$$p_1 = -\zeta\omega_n + \omega_n\sqrt{\zeta^2 - 1} \qquad p_2 = -\zeta\omega_n - \omega_n\sqrt{\zeta^2 - 1}$$

With $\zeta > 1$ we have real roots and the square root is of a positive quantity and thus the overall root can be written as a real number, as in the examples given above when we had $s = +1$, $s = +2$, $s = -1$ and $s = -2$. The output will have exponential terms and depending on whether the terms have negative or positive power terms then so the system will be stable or unstable.

With $\zeta < 1$ then the square root is of a negative quantity. If we write j for the square root of minus 1 then the roots can be written as:

$$p_1 = -\zeta\omega_n + j\omega_n\sqrt{1 - \zeta^2} \qquad p_2 = -\zeta\omega_n - j\omega_n\sqrt{1 - \zeta^2}$$

We can thus write the values of the roots, and so poles, in the form $a + jb$; the jb part of the value is known as an imaginary number. The output can be written as exponential terms with the terms having powers of the form $a + jb$. Such values give rise to oscillatory transients. However, the same rule applies for stability, namely that if the a term is negative then the system is stable and if it is positive it is unstable.

> The poles of a system, i.e. the roots of the characteristic equation, determine how the system behaves during its free response to an input. The poles can be real or complex, i.e. having an imaginary component. When the poles are purely imaginary or have imaginary parts, the output will have oscillatory components. For a system to be stable, *all* the poles must be negative or, if complex, the real part must be negative.

10.6.1 The *s* plane

We can plot the positions of the poles on a graph with the real part of the pole value as the x-axis and the imaginary part as the y-axis. The resulting graph describes what is termed the *s-plane*.

As an illustration, Figure 10.8 shows the *s*-plane for the transfer function:

$$G(s) = \frac{1}{(s + 1)(s + 2)}$$

with $s = -1$ and $s = -2$, there being no imaginary terms. This describes a stable system since the roots are negative and give rise to exponential terms which decrease with time.

Figure 10.9 shows the *s*-plane for the transfer function:

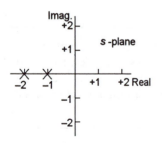

Figure 10.8 *Poles at –1 and –2 for a stable system*

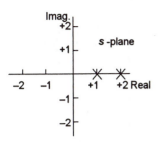

Figure 10.9 *Poles at +1 and +2 for an unstable system*

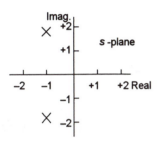

Figure 10.10 *Poles at −1 ± j1.73 for a stable system*

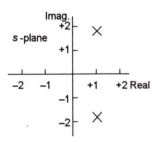

Figure 10.11 *Poles at +1 ± j1.73 for an unstable system*

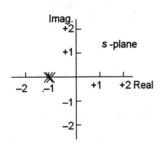

Figure 10.12 *Double pole at −1 for critical damping*

$$G(s) = \frac{1}{(s-1)(s-2)}$$

with $s = +1$ and $s = +2$, there being no imaginary terms. This shows an unstable system since the roots are positive and give rise to exponential terms which increase with time.

For the transfer function:

$$G(s) = \frac{1}{s^2 + 2s + 4}$$

we have the roots of the quadratic given by:

$$s = \frac{-2 \pm \sqrt{4 - 16}}{2} = -1 \pm j1.73$$

Figure 10.10 shows the s-plane for this transfer function, this describing an underdamped but stable system. The system is stable because the real parts of the roots are negative.

For the transfer function:

$$G(s) = \frac{1}{s^2 - 2s + 4}$$

we have the roots of the quadratic given by:

$$s = \frac{2 \pm \sqrt{4 - 16}}{2} = +1 \pm j1.73$$

Figure 10.11 shows the s-plane for this transfer function. For the transfer function:

$$G(s) = \frac{1}{(s+1)^2}$$

we have the roots $s = -1$ and $s = -1$. This is critical damping. Figure 10.12 shows the s-plane for this transfer function.

In general we can state (Figure 10.13):

> A system is stable if all the system poles lie in the left half of the s-plane.

The relationship between the location of a pole and the form of transient is shown in Figure 10.14. The more negative the real part of the pole the more rapidly the transient dies away. The larger the imaginary part of the pole the higher the frequency of the oscillation. A system having a pole which has a positive real part is unstable.

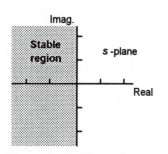

Figure 10.13 *The s-plane: stability when poles are in the left half*

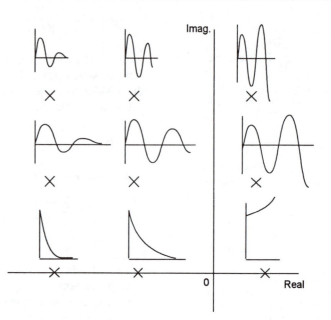

Figure 10.14   *Relationship between pole location and the resulting transient: each oscillatory transient arises from a pair of roots a ± jb with only one of them shown in the figure*

Example

Which of the following systems are stable:
(a) $G(s) = 1/(s^2 + s + 1)$, (b) $G(s) = 1/(s^2 - 5s + 4)$,
(c) $G(s) = 1/(s^2 - 2s + 3)$, (d) $G(s) = 1/(s^2 + s + 6)$?

(a) This has poles of :

$$s = \frac{-1 \pm \sqrt{1-4}}{2} = -0.5 \pm j1.73$$

The poles all lie in the left half of the s-plane and so the system is stable.
(b) This has poles of $s = +4$ and $s = +1$ and so both the poles lie in the right half of the s-plane and the system is unstable.
(c) This has poles of:

$$s = \frac{2 \pm \sqrt{4-12}}{2} = +1 \pm j2.8$$

The poles will both lie in the right half of the s-plane and so the system is unstable.
(d) The denominator can be written as $(s-2)(s+3)$ and so has poles at $+2$ and -3. Because one of the poles is in the positive part of the s-plane, the system is unstable.

10.7 Software   A very widely used software package used in the design and analysis of control systems is MATLAB. The following is a brief indication of how it can be used in the context of this chapter.

Commands can be entered by typing them after the prompt and then pressing the return key to execute the command. Numerical values can be assigned to a variable by using the equal (=) sign, e.g. typing the variable a, then the equal sign and then the value 2, so entering a = 2, and then pressing the return key.

Each element of a row vector (the basic format used for data in MATLAB is the matrix, this being a rectangular table with the elements ordered in rows and columns, when a matrix consists of only one row it is termed a row vector) is entered between square brackets, with a space between each term, and set equal to a variable. We can enter the coefficients of a polynomial as a row vector, the coefficients being entered in descending order of the powers of the variable; if a coefficient is not present we must enter a 0. For example, for the polynomial $p(x) = x^2 + 5x + 2$ we type in:

p = [1 5 2]

and then press the return key. To determine the roots of a polynomial we use the following command:

roots ([1 5 2]) or roots (p)

followed by pressing the return key.

One way of entering a transfer function into MATLAB is to use the command tf (num,den), where num and den are row vectors of the coefficients of the numerator and denominator polynomials. For example, to enter the transfer function $(s + 3)/(s^2 + 5s + 1)$ the sequence can be:

num = [1 3] followed by pressing return key
den = [1 5 1] followed by pressing return key
g1 = tf (num, den) followed by pressing return key

An alternative way of entering a transfer function is to use the command zpk (zeros,poles,gain), where zeros, poles and gain are row vectors of the zeros, poles and gain of the transfer function (note that zeros are the values of s that make the numerator 0). Thus, the transfer function $6(s + 3)/(s + 2)(s + 5)$ would be entered as:

zeros = [–3] followed by pressing return key
poles = [–2 –5] followed by pressing return key
gain = 6 followed by pressing return key
g = zpk (zeros,poles,gain) followed by pressing return key

Once transfer functions have been entered, we can combine them. Thus for two series components g1 and g2 we can enter:

$$g = g1 * g2$$

* is used for multiplication.

We can use the transfer function g to find the time response of a system. For the response to a unit impulse we use the command:

impulse (g) followed by pressing return key

For a unit step we use the command:

step (g) followed by pressing return key

Where the input is not unit but some value k we enter the command as:

impulse (k*g) followed by pressing return
step (k*g) followed by pressing return key

As an illustration, for the second-order system:

$$G(s) = \frac{\omega_n^2}{s^2 + 2\zeta\omega_n s + \omega_n^2}$$

with the natural angular frequency as 3 rad/s and the damping factor as 0.3, the output when it is subject to a unit impulse input can be obtained by the MATLAB program (the symbol ^ is used to indicate the term following is a power and the semicolon to separate rows in the program):

```
Wn = 3;
Ze = 0.3;
num = Wn^2;
den = [1 2*Ze*Wn Wn^2];
g = tf(num,den);
impulse(g)
```

10.7.1 Simulink

Simulink is a software program that uses a graphical interface for modelling systems so that systems can be specified on screen by connecting boxes rather than writing a series of commands. Simulink is started from within MATLAB by typing the command Simulink. This then opens Simulink and gives access to a number of Simulink libraries from which selections can be made. Thus, if the Continuous library is opened, the Transfer Fcn entry in the library can be dragged by the mouse into a modelling window. Other components can similarly be dragged from libraries and a graphical model of the system built up on screen. Figure 10.15 shows what the screen might look like for a first-order system. By double clicking on boxes, data values can be entered for a particular system. The model can then be run with the appropriate input by selecting parameters from the simulation menu. The output response is plotted by using the command plot(t,y).

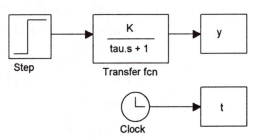

Figure 10.15 *Simulink screen*

Problems   *Questions 1 to 10 have four answer options: A, B, C or D. Choose the correct answer from the answer options.*

1   The Laplace transform of a step input of size 5 is:

A   5
B $5s$
C $5/s$
D $s/5$

2   The Laplace transform of the time function $2e^{-3t}$ is:

A $2/(s + 3)$
B $2/(s - 3)$
C $3/(s + 2)$
D $3/(s - 2)$

3   The Laplace transform of the time function $t\,e^{-3t}$ is:

A $3/[s(s - 3)]$
B $3/[s(s + 3)]$
C $1/(s - 3)^2$
D $1/(s + 3)^2$

4   The time function corresponding to the Laplace transform $5/(s + 3)$ is:

A $5\,e^{3t}$
B $5\,e^{-3t}$
C $3\,e^{5t}$
D $3\,e^{-5t}$

5   A system has a transfer function of $1/(s + 3)$. When subject to a unit impulse input, the output of the system will have the Laplace transform:

A $1/(s + 3)$
B $s/(s + 3)$
C $1/[s(s + 3)]$
D $(s + 3)/1$

6 Decide whether each of these statements is True (T) or False (F).

The differential equation $2(dy/dt) + y = 5x$ relating the input x and the output y has:
(i)   A time constant value of 2.
(ii) A steady-state gain of 5

A   (i) T (ii) T
B   (i) T (ii) F
C   (i) F (ii) T
D   (i) F (ii) F

7 Decide whether each of these statements is True (T) or False (F).

A system with the transfer function $8/(s^2 + 4s + 4)$ is subject to a unit step input. The output:
(i)   Has a steady-state gain of 8
(ii) Is critically damped

A   (i) T (ii) T
B   (i) T (ii) F
C   (i) F (ii) T
D   (i) F (ii) F

8 The Laplace transform $6/[s(s + 3)]$ can be written as:

A $(6/s) + 1/(s + 3)$
B $(2/s) - 3/(s + 3)$
C $(2/s) - 2/(s + 3)$
D $(2/s) + 2/(s + 3)$

9 Decide whether each of these statements is True (T) or False (F).

A second-order system has a natural angular frequency of 1 rad/s and a damped angular frequency of 2 rad/s. It thus has:
(i)   A damping factor of 0.87
(ii) A 100% rise time of 1.57 s.

A   (i) T (ii) T
B   (i) T (ii) F
C   (i) F (ii) T
D   (i) F (ii) F

10 Decide whether each of these statements is True (T) or False (F).

A system with the transfer function $12/(s^2 + 5s + 6)$ is subject to a unit step input. The output:
(i)   Is stable.
(ii) Oscillates about the steady-state value.

A   (i) T (ii) T
B   (i) T (ii) F
C   (i) F (ii) T
D   (i) F (ii) F

11 A system has an input of a voltage of 3 V which is suddenly applied by a switch being closed. What is the input as an *s* function?

12 A system has an input of a voltage impulse of 2 V. What is the input as an *s* function?

13 A system has an input of a voltage of a ramp voltage which increases at 5 V per second. What is the input as an *s* function?

14 A system gives an output of 1/(s + 5) $V(s)$. What is the output as a function of time?

15 A system has a transfer function of 5/(s + 3). What will be its output as a function of time when subject to (a) a unit step input of 1 V, (b) a unit impulse input of 1 V?

16 A system has a transfer function of 2/(s + 1). What will be its output as a function of time when subject to (a) a step input of 3 V, (b) an impulse input of 3 V?

17 A system has a transfer function of 1/(s + 2). What will be its output as a function of time when subject to (a) a step input of 4 V, (b) a ramp input unit impulse of 1 V/s?

18 Use partial fractions to simplify the following expressions:

(a) $\dfrac{s-6}{(s-1)(s-2)}$, (b) $\dfrac{s+5}{s^2+3s+2}$, (c) $\dfrac{2s-1}{(s+1)^2}$

19 A system has a transfer function of:

$$\frac{8(s+3)(s+8)}{(s+2)(s+4)}$$

What will be the output as a time function when it is subject to a unit step input? Hint: use partial fractions.

20 A system has a transfer function of:

$$G(s) = \frac{8(s+1)}{(s+2)^2}$$

What will be the output from the system when it is subject to a unit impulse input? Hint: use partial fractions.

21 What will be the state of damping of systems having the following transfer functions and subject to a unit step input?

(a) $\dfrac{1}{s^2+2s+1}$, (b) $\dfrac{1}{s^2+7s+12}$, (c) $\dfrac{1}{s^2+s+1}$

22 The input *x* and output *y* of a system are described by the differential equation:

$$\frac{dy}{dt} + 2y = x$$

Determine how the output will vary with time when there is an input which starts at zero time and then increases at the constant rate of 6 units/s. The initial output is zero.

23   The input x and output y of a system are described by the differential equation:

$$\frac{d^2y}{dt^2} + 3\frac{dy}{dt} + 2y = x$$

If initially the input and output are zero, what will be the output when there is a unit step input?

24   The input x and output y of a system are described by the differential equation:

$$\frac{d^2y}{dt^2} + 4\frac{dy}{dt} + 3y = x$$

If initially the input and output are zero, what will be the output when there is a unit impulse input?

25   A control system has a forward path transfer function of $2/(s + 2)$ and a negative feedback loop with transfer function 4. What will be the response of the system to a unit step input?

26   A system has a transfer function of $100/(s^2 + s + 100)$. What will be its natural frequency ω_n and its damping ratio ζ?

27   A system has a transfer function of $10/(s^2 + 4s + 9)$. Is the system under-damped, critically damped or over-damped?

28   A system has a transfer function of $3/(s^2 + 6s + 9)$. Is the system under-damped, critically damped or over-damped?

29   A system has a forward path transfer function of $10/(s + 3)$ and a negative feedback loop with transfer function 5. What is the time constant of the resulting first-order system?

30   Determine the delay time and the rise time for the following first-order systems: (a) $G(s) = 1/(4s + 1)$, (b) $G(s) = 5/(s + 1)$, (c) $G(s) = 2/(s + 3)$.

31   A first-order system has a time constant of 30 s. What will be its delay time and rise time when subject to a unit step input?

32   A first-order system when subject to a unit step input rises to 90% of its steady-state value in 20 s. Determine its time constant, delay time and rise time?

33   Determine the natural angular frequency, the damping factor, the rise time, percentage overshoot and 2% settling time for systems with the following transfer functions: (a) $100/(s^2 + 4s + 100)$, (b) $49/(s^2 + 4s + 49)$.

34   Determine the natural angular frequency, the damping factor, the rise time, percentage overshoot and 2% settling time for a system where the output y is related to the input x by the differential equation:

$$\frac{d^2y}{dt^2} + 5\frac{dy}{dt} + 16y = 16x$$

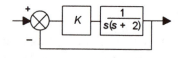

Figure 10.16 *Problem 35*

35   For the feedback system shown in Figure 10.16, what gain K should be used to give a rise time of 2 s?

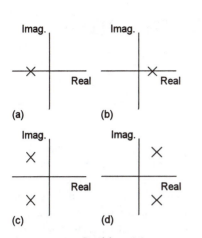

(a)   (b)

(c)   (d)

Figure 10.17 *Problem 36*

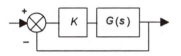

Figure 10.18 *Problem 38*

36 For the poles shown on the *s*-planes in Figure 10.17, which will give rise to stable transients and which to oscillating transients?

37 Are the systems with the following transfer functions stable?

$$(a) \; \frac{1}{s^2 + 2s + 1}, \; (b) \; \frac{1}{s^2 - 2s + 10}, \; (c) \; \frac{1}{(s+1)(s-3)}$$

38 Figure 10.18 shows a feedback control system with unity feedback. Will the system be stable when (a) $K = 1$, $G(s) = 1/[s(s + 1)]$, (b) $K = 3$, $G(s) = 1/[(s + 4)(s - 1)]$, (c) $K = 5$, $G(s) = 1/[(s + 4)(s - 1)]$?

39 State if the following systems are stable, the relationship between input *x* and output *y* being described by the differential equations :

$$(a) \; \frac{d^2y}{dt^2} + 3\frac{dy}{dt} + 2y = x, \; (b) \; \frac{d^2y}{dt^2} + \frac{dy}{dt} - 6y = x$$

11 Frequency response

11.1 Introduction

In earlier chapters we have considered the outputs that arise from systems when subject to step, impulse and ramp inputs. In this chapter we consider the steady-state output responses of systems when the inputs are sinusoidal signals. This leads to powerful methods of analysing systems in considering how the amplitude and phase of the output signal varies as the frequency of the input sinusoidal signal is changed. This variation is termed the *frequency response* of the system and can be described by, what are termed, *Bode diagrams*.

The derivation of a transfer function for system involves making assumptions about the physical model, e.g. springs, masses and damping, that can be used to represent the system. We can derive the frequency response of systems from a knowledge of their transfer functions. Thus the validity of the resulting transfer function can be tested by experiment using sinusoidal inputs and comparing the experimental frequency response with that which is predicted by the transfer function. Conversely, we can determine the frequency response for a system and then use it to predict the form of the transfer function.

This chapter shows how frequency response information can be obtained from the transfer function and how it can be presented graphically by means of a Bode diagram. The use of experimentally determined Bode plots to estimate the transfer functions of systems is then discussed. The chapter concludes with a discussion of the parameters used to describe the stability of systems and their determination from Bode plots, also compensation techniques which can be used to enhance the stability of systems.

11.1.1 Sinusoidal signals

When we have a sinusoidal signal input to a system, the steady-state output will also be a sinusoidal signal with the same frequency. The things that can differ between input and output are the *magnitude*, or amplitude, and the *phase* ϕ. Figure 11.1 illustrates what is meant by these terms. A convenient way of representing sinusoidal signals is by *phasors*. We imagine a sinusoidal signal $y = Y \sin \omega t$, i.e. amplitude Y and angular frequency ω, as being produced by a radial line of length Y rotating with a constant angular velocity ω (Figure 11.2), taking the vertical projection y of the line at any instant of time to represent the value of the sinusoidal signal. If we have another sinusoidal signal of different amplitude then the radial line will be of a different length. If we have a sinusoidal signal with a different phase then it will start with a different value at time $t = 0$ and so the radial line will start at $t = 0$ at

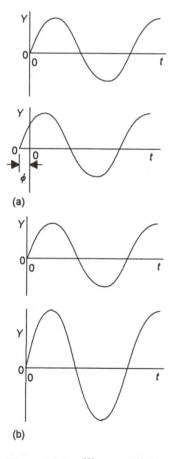

(a)

(b)

Figure 11.1 *Waves with (a) same magnitude but different phase, (b) different magnitude but same phase*

some angle, termed the *phase angle*, to the reference axis. The reference axis is usually taken as the horizontal axis. Such lines are termed *phasors*. In order to clearly indicate when we are talking of the magnitude of a sinusoidal signal we often write $|Y|$ for the magnitude of the sinusoidal signal represented by the phasor and bold, non-italic, letters for the symbols for phasors, e.g. **Y**. Thus, **Y** implies a phasor with both magnitude and phase.

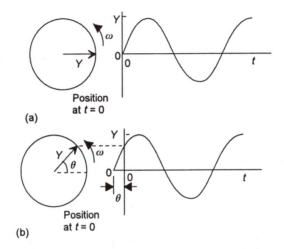

Figure 11.2   *Sinusoidal signals represented by rotating lines, i.e. phasors: (a) $y = Y \sin \omega t$, (b) $y = Y \sin(\omega t + \theta)$*

11.1.2 Complex numbers

In the discussions that follow in this chapter, complex numbers are used. The term *complex number* is used for the sum of a real number and an imaginary number, e.g. $a + jb$ with a being the real part of the complex number and jb the imaginary part; $j = -\sqrt{1}$ and so $j^2 = -1$. The manipulation rules for handling complex numbers are:

1   To *add* complex numbers we separately add the real parts and add the imaginary parts:

$$(a + jb) + (c + jd) = (a + c) + j(b + d)$$

2   To *subtract* complex numbers we separately subtract the real parts and subtract the imaginary parts:

$$(a + jb) - (c + jd) = (a - c) + j(b - d)$$

3   In the *multiplication* of the two complex numbers $z_1 = a + jb$ and $z_2 = c + jd$, the product z is given by:

$$z = (a + jb)(c + jd) = ac + j(ad + bc) + j^2bd$$

$$= ac + j(ad + bc) - bd$$

4   If $z = a + jb$ then the term *complex conjugate* is used for the complex number given by $z^* = a - jb$. The imaginary part of the complex number changes sign to give the conjugate, conjugates being denoted as z^*. Consider now the product of a complex number and its conjugate:

$$zz^* = (a + jb)(a - jb) = a^2 - j^2 b = a^2 + b^2$$

The product of a complex number and its conjugate is a real number.

5   Consider the *division* of $z_1 = a + jb$ by $z_2 = c + jd$, i.e.

$$z = \frac{z_1}{z_2} = \frac{a + jb}{c + jd}$$

To divide one complex number by another we have to convert the denominator into a real number. This can be done by multiplying it by its conjugate. Thus:

$$z = \frac{a + jb}{c + jd} \times \frac{c - jd}{c - jd} = \frac{(a + jb)(c - jd)}{c^2 + d^2}$$

The effect of multiplying a real number by (-1) is to move the point from one side of the origin on a graph to the other (Figure 11.3). We can think of the positive number line radiating out from the origin being rotated through 180° to its new position after being multiplied by (-1). But $(-1) = j^2$. Thus, multiplication by j^2 is equivalent to a 180° rotation. Multiplication by j^4 is a multiplication by $(+1)$ and so is equivalent to a rotation through 360°. On this basis it seems reasonable to take a multiplication by j to be equivalent to a rotation through 90° and a multiplication by j^3 a rotation through 270°. This concept of multiplication by j as involving a rotation is the basis of the use of complex numbers to represent phasors in alternating current circuits.

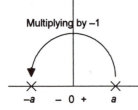

Figure 11.3   *Multiplying by –1*

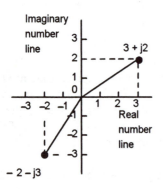

Figure 11.4   *Argand diagram*

Since rotation by 90° from the x-axis on a graph gives the y-axis, the y-axis is used for imaginary numbers and the x-axis for real numbers (Figure 11.4). The resulting graph is called the *Argand diagram* and is used to represent complex numbers.

Figure 11.4 shows how we represent the complex numbers $3 + j2$ and $-2 - j3$ on such an Argand diagram. The line joining the number to the origin is taken as the graphical representation of the complex number.

If the complex number $z = a + jb$ is represented on an Argand diagram by the line OP, as in Figure 11.5, then the length of the line OP, denoted by $|z|$, is called the *modulus* of the complex number and its inclination θ to the real number axis is termed the *argument* of the complex number and denoted by θ or arg z.

Using Pythagoras' theorem:

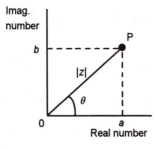

Figure 11.5   *Modulus and argument*

$$|z| = \sqrt{a^2 + b^2}$$

and, since $\tan \theta = b/a$:

$$\arg z = \tan^{-1}\left(\frac{b}{a}\right)$$

11.2 Sinusoidal inputs

In discussing, in Chapter 10, the responses of systems to impulse and step inputs we were interested in both the transient and steady-state responses. However, when considering sinusoidal inputs, we are normally only interested in the steady-state response.

> The *frequency response* of a system (Figure 11.6) is the steady-state response of the system to a sinusoidal input signal. The steady-state output is a sinusoidal signal of the same frequency as the input signal, differing only in amplitude and phase angle.

Application
The frequency response of a system can be found by applying a sinusoidal input signal to the input of the system and measuring the output so that the gain and the phase can be determined. Such measurements can be carried out for a number of frequencies.

As we will see in this chapter, an alternative way of establishing the frequency response is to compute it from the transfer function of the system.

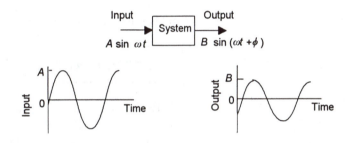

Figure 11.6 *Frequency response of a system*

11.2.1 Frequency response function

For a system with a transfer function $G(s)$, the output transform $Y(s)$ is related to the input transform $X(s)$ by $Y(s) = G(s) X(s)$. If we consider a unit amplitude sinusoidal input then $X(s) = \omega/(s^2 + \omega^2)$, see item 12 in Table 10.1. Thus we have:

$$Y(s) = G(s)\left[\frac{\omega}{s^2 + \omega^2}\right]$$

We can write this as:

$$Y(s) = G(s)\left[\frac{\omega}{(s - j\omega)(s + j\omega)}\right]$$

Using partial fractions, we can write this equation in the form:

$$Y(s) = \frac{A}{s + p_1} + \frac{B}{s + p_2} + \dots + \frac{P}{s + j\omega} + \frac{Q}{s - j\omega}$$

where the poles p_1, p_2, ... arise from the $G(s)$ term. Thus the poles of $Y(s)$ will be the poles of $G(s)$ plus the two imaginary poles at $s = j\omega$ and $s = -j\omega$. Provided the system is stable, all the poles of $G(s)$ will be in the left-hand half of the s-plane and so all the terms arising from them in the output response of the system will decay to zero with increasing time. Thus we do not have to consider them when we want to find the steady-state response of the system.

> When we want to find the steady-state response of a system to a sinusoidal input, we only need to consider its response for when $s = j\omega$.

So for a stable system we have a steady-state response given by the transform:

$$Y(s) = \frac{P}{s+j\omega} + \frac{Q}{s-j\omega}$$

and hence a steady-state response y_{ss} of the form:

> $$y_{ss} = |G(j\omega)| \sin(\omega t + \phi)$$

where $|G(j\omega)|$ is the magnitude of the output and ϕ its phase when we have the condition $s = j\omega$.

As a consequence, the function $G(j\omega)$, which is the transfer function $G(s)$ evaluated for when $s = j\omega$, is known as the *frequency response function*.

> The frequency response function is obtained from the transfer function by replacing s by $j\omega$.

There is another way of looking at the above. Consider a simple system with a sinusoidal input and a steady-state sinusoidal output with the input x related to the output y by the differential equation:

$$\tau\frac{dy}{dt} + y = kx$$

Let us take the output to be $y = \sin \omega t$ and so, since $dy/dt = \omega \cos \omega t = \omega \sin(\omega t + 90°)$, the left-hand side of the equation can be written as:

$$\tau\omega \sin(\omega t + 90°) + \sin \omega t$$

We can represent sinusoidal signals by phasors and describe them by complex numbers. Thus, if $\sin \omega t$ is the output phasor Y, then $\sin(\omega t + 90°)$ is Y with a phase angle of 90° (Figure 11.7) and so can be

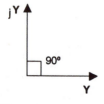

Figure 11.7 *The phasors*

represented as j**Y**. Thus, if **X** is the input phasor, the above equation can be written in terms of phasors as:

$$j\tau\omega\mathbf{Y} + \mathbf{Y} = k\mathbf{X}$$

Hence, we can write:

$$\frac{\text{output phasor } \mathbf{Y}}{\text{input phasor } \mathbf{X}} = \frac{k}{1 + j\omega\tau}$$

> We can define the *frequency response function* as being the output phasor divided by the input phasor.

We can compare the above equation with the differential equation written in its Laplace transform format as:

$$\tau sY(s) + Y(s) = kX(s)$$

and the resulting transfer function:

$$G(s) = \frac{Y(s)}{X(s)} = \frac{k}{1 + \tau s}$$

The frequency response function equation is of the same form as the transfer function if we replace s by $j\omega$. Hence the frequency response function is denoted by $G(j\omega)$. In general we can state:

> The frequency response function is obtained from the transfer function by replacing s by $j\omega$.

Example

Determine the frequency response function for a system having a transfer function of $G(s) = 5/(2 + s)$.

Replacing s by $j\omega$ gives the frequency response function of $G(j\omega) = 5/(2 + j\omega)$.

Example

Determine the frequency response function for a system having a transfer function of:

$$G(s) = \frac{2s + 1}{s^2 + 4s + 3}$$

Replacing s by $j\omega$ gives the frequency response function of:

$$G(j\omega) = \frac{j2\omega + 1}{(j\omega)^2 + j4\omega + 3} = \frac{j2\omega + 1}{-\omega^2 + j4\omega + 3}$$

11.2.2 Frequency response for first-order systems

Consider a first-order system and the determination, from the frequency response function, of the magnitude and phase of the steady-state output when it is subject to a sinusoidal input. In general, a first-order system has a transfer function of the form:

$$G(s) = \frac{1}{1 + \tau s}$$

where τ is the time constant of the system. The frequency response function $G(j\omega)$ can be obtained by replacing s by $j\omega$. Hence:

$$G(j\omega) = \frac{1}{1 + j\omega\tau}$$

We can put this into the form $a + jb$ by multiplying the top and bottom of the expression by $(1 - j\omega\tau)$ to give:

$$G(j\omega) = \frac{1}{1 + j\omega\tau} \times \frac{1 - j\omega\tau}{1 - j\omega\tau} = \frac{1 - j\omega\tau}{1 + j^2\omega^2\tau^2}$$

But $j^2 = -1$, thus we can write this equation as:

$$G(j\omega) = \frac{1}{1 + \omega^2\tau^2} - j\frac{\omega\tau}{1 + \omega^2\tau^2}$$

The frequency response function has thus a real element of $1/(1 + \omega^2\tau^2)$ and an imaginary element of $-\omega\tau/(1 + \omega^2\tau^2)$. This is the steady-state response of the system to a sinusoidal input and so we have:

$$y_{ss} = |G(j\omega)| \sin (\omega t + \phi)$$

The magnitude of a complex number $a + jb$ is given by $\sqrt{(a^2 + b^2)}$ and so:

$$|G(j\omega)| = \sqrt{\left(\frac{1}{1 + \omega^2\tau^2}\right)^2 + \left(\frac{\omega\tau}{1 + \omega^2\tau^2}\right)^2}$$

Hence:

$$|G(j\omega)| = \frac{1}{\sqrt{1 + \omega^2\tau^2}}$$

$|G(j\omega)|$ is often referred to as the *gain* of the system. Since $G(j\omega)$ is the output phasor divided by the input phasor, then the output phasor has a magnitude bigger than that of the input phasor by the factor $|G(j\omega)|$.

The phase of a complex number is given by tan ϕ = b/a, so the phase difference ϕ between the output and the input is given by:

$$\tan \phi = -\omega\tau$$

The negative sign indicates that the output signal lags behind the input signal by this angle. Thus:

The gain and phase of a system when subject to a sinusoidal input is obtained by putting the frequency response function in the form $a + jb$ and then the gain is $\sqrt{(a^2 + b^2)}$ and the phase is $\tan^{-1}(b/a)$.

Example

Determine the magnitude and phase of the output from a system when subject to a sinusoidal input of 2 sin 3t if it has a transfer function of $G(s) = 2/(s + 1)$.

The frequency response function is obtained by replacing s by jω:

$$G(j\omega) = \frac{2}{j\omega + 1}$$

Multiplying top and bottom of the equation by $(-j\omega + 1)$ gives:

$$G(j\omega) = \frac{-j2\omega + 2}{\omega^2 + 1} = \frac{2}{\omega^2 + 1} - j\frac{2\omega}{\omega^2 + 1}$$

The magnitude is thus:

$$|G(j\omega)| = \sqrt{\frac{2^2}{(\omega^2 + 1)^2} + \frac{2^2\omega^2}{(\omega^2 + 1)^2}}$$

$$= \frac{2}{\sqrt{\omega^2 + 1}}$$

and the phase angle is given by:

$$\tan \phi = -\omega$$

For the specified input we have ω = 3 rad/s. The magnitude is thus:

$$|G(j\omega)| = \frac{2}{\sqrt{3^2 + 1}} = 0.63$$

and the phase is given by tan ϕ = −3 as ϕ = −72°. This is the phase angle between the input and the output. Thus, the output is the sinusoidal signal of the same frequency as the input signal and described by $2 \times 0.63 \sin (3t - 72°) = 1.26 \sin (3t - 72°)$.

11.2.3 Frequency response for second-order systems

Consider a second-order system and the determination, from the frequency response function, of the magnitude and phase of the steady-state output when it is subject to a sinusoidal input. In general, a second-order system has a transfer function of the form:

$$G(s) = \frac{\omega_n^2}{s^2 + 2\zeta\omega_n s + \omega_n^2}$$

where ω_n is the natural angular frequency and ζ the damping ratio. The frequency response function is obtained by replacing s by $j\omega$. Thus:

$$G(j\omega) = \frac{\omega_n^2}{-\omega^2 + j2\zeta\omega\omega_n + \omega_n^2} = \frac{\omega_n^2}{(\omega_n^2 - \omega^2) + j2\zeta\omega_n}$$

$$= \frac{1}{\left[1 - \left(\frac{\omega}{\omega_n}\right)^2\right] + j2\zeta\left(\frac{\omega}{\omega_n}\right)}$$

Multiplying the top and bottom of the expression by:

$$\left[1 - \left(\frac{\omega}{\omega_n}\right)^2\right] - j2\zeta\left(\frac{\omega}{\omega_n}\right)$$

gives:

$$G(j\omega) = \frac{\left[1 - \left(\frac{\omega}{\omega_n}\right)^2\right] - j2\zeta\left(\frac{\omega}{\omega_n}\right)}{\left[1 - \left(\frac{\omega}{\omega_n}\right)^2\right]^2 + \left[2\zeta\left(\frac{\omega}{\omega_n}\right)\right]^2}$$

This is of the form $a + jb$ and so we have the magnitude of the output, i.e. the gain, as:

$$|G(j\omega)| = \frac{1}{\sqrt{\left[1 - \left(\frac{\omega}{\omega_n}\right)^2\right]^2 + \left[2\zeta\left(\frac{\omega}{\omega_n}\right)\right]^2}}$$

The phase ϕ difference between the input and output is given by:

$$\tan \phi = -\frac{2\zeta\left(\frac{\omega}{\omega_n}\right)}{1 - \left(\frac{\omega}{\omega_n}\right)^2}$$

The minus sign indicates that the output phasor lags the input phasor.

Example

Determine the magnitude and phase of the output from a system when subject to a sinusoidal input of 3 sin 2*t* if it has a transfer function of $G(s) = 2/(s^2 + 4s + 1)$.

The frequency response function is obtained by replacing *s* by jω and so we have:

$$G(j\omega) = \frac{2}{-\omega^2 + j4\omega + 1}$$

With $\omega = 2$ this becomes:

$$G(j2) = \frac{2}{-4 + j8 + 1} = \frac{2}{-3 + j8}$$

Multiplying the top and the bottom of the equation by $-3 - j8$ gives:

$$G(j2) = \frac{2}{-3 + j8} \frac{-3 - j8}{-3 - j8} = \frac{-6 - j8}{9 + 64} = -0.082 - j0.110$$

This has a magnitude of $\sqrt{(0.082^2 + 0.110^2)} = 0.137$ and a phase of $\tan^{-1}(0.110/0.082) = 53.3°$. Thus the output for an input of amplitude 3 is $0.411 \sin(2t + 53.3°)$.

11.3 Bode plots

The frequency response of a system is described by values of the gain and the phase angle which occur when the sinusoidal input signal is varied over a range of frequencies. The term *Bode plot* is used for the pair of graphs of the logarithm to base 10 of the gain plotted against the logarithm to base 10 of the frequency and the phase angle plotted against the logarithm to base 10 of the frequency. The reason for the graphs being in this form is that it enables plots for complex frequency response functions to be obtained by merely adding together the plots obtained for each constituent element.

Suppose we have a sinusoidal input to a number of systems in series (Figure 11.8). The first produces a gain of $|G_1(j\omega)|$ and a phase angle shift of ϕ_1, the second produces a gain of $|G_2(j\omega)|$ and a phase angle shift of ϕ_2, and the third produces a gain of $|G_3(j\omega)|$ and a phase angle shift of ϕ_3. The overall gain of the system will be the products of the gains of each of the systems and thus be:

$$|G(j\omega)| = |G_1(j\omega)||G_2(j\omega)||G_3(j\omega)|$$

If we take the logarithms of the gain equation we obtain:

$$\lg |G(j\omega)| = \lg |G_1(j\omega)| + \lg |G_2(j\omega)| + \lg |G_3(j\omega)|$$

Thus,

System 1 System 2 System 3

Figure 11.8 *Systems in series*

> When we use the logarithms of the gains, we obtain the overall gain by just adding the logarithms of the gains of the individual elements.

This enables us to consider any frequency response function as being made up of a number of simple elements and so obtain the response by adding the logarithms of the gains of the simple elements. For example, we can think of the frequency response function $5/(2 + j\omega)$ as being two elements, one with frequency response function 5 and the other with frequency response function $1/(2 + j\omega)$.

The total phase shift will be:

$$\phi = \phi_1 + \phi_2 + \phi_3$$

> The total phase is the sum of the phases of the individual elements.

Application
A MATLAB program that can be used to plot the Bode plot for a transfer function. e.g. $(6s + 1)/(2s^2 + 3s + 5)$ is:

```
num=[0 6 1];
den=[2 3 5];
bode(num,den)
```

Because we can obtain the Bode plot for a system by considering the plots for the constituent elements of its frequency response function and adding, it is only necessary to know the form of the Bode plots for a small number of transfer function terms. The following sections show the Bode plots for common terms.

It is usual to express the gain in decibels (dB), with:

> gain in dB = $20 \lg |G(j\omega)|$

Example

What is the gain in dB of the following gains (a) 1, (b) 0.707, (c) 3, (d) 10?

(a) A gain of 1 is a gain in dB of $20 \lg 1 = 0$ dB.
(b) A gain of 0.707 is a gain in dB of $20 \lg 0.707 = -3$ dB.
(c) A gain of 3 is a gain in dB of $20 \lg 3 = 9.5$ dB.
(d) A gain of 10 is a gain in dB of $20 \lg 10 = 20$ dB.

11.3.1 Transfer function a constant K

This has $G(s) = K$ and so $G(j\omega) = K$. The gain is $|G(j\omega)| = K$ and in decibels this is $20 \lg K$ dB. This does not change when the frequency changes. The phase $= 0°$ if K is positive and $-180°$ if K is negative. Again, it does not change when the frequency changes. The form of the Bode plot is thus as shown in Figure 11.9. Thus, for $K = 10$ the gain is a constant line of $20 \lg 10 = 20$ dB and the phase a constant line of $0°$.

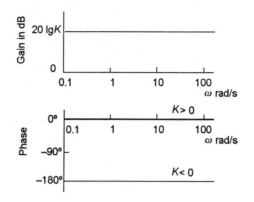

Figure 11.9 *Bode plot for constant transfer function*

11.3.2 Transfer function $1/s^n$

For $G(s) = 1/s$, i.e. on the s-plane (see Section 10.6.1) the pole is at the origin as we have a pole with $s = 0$, we have the frequency response function of $G(j\omega) = 1/(j\omega) = -(j/\omega)$. For such a system the gain is $|G(j\omega)| = 20 \lg(1/\omega) = -20 \lg \omega$ dB and the phase $-90°$. The plots are thus straight lines (Figure 11.10). When $\omega = 1$ rad/s then the gain is 0. When $\omega = 10$ rad/s it is -20 dB. When $\omega = 100$ rad/s it is -40 dB. Thus the slope of the gain line for a transfer function of $1/s$ is -20 dB for each tenfold increase in frequency (this is termed a decade).

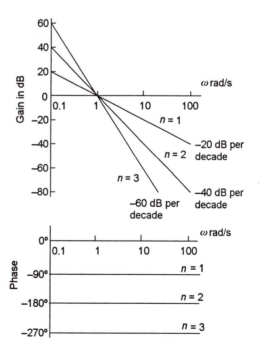

Figure 11.10 *Bode plot for transfer functions having $1/s^n$ terms*

For $G(s) = 1/s^n$ we have $G(j\omega) = 1/(j\omega)^n = -(j/\omega)^n$. For such a system the gain is $|G(j\omega)| = 20 \lg(1/\omega)^n = -20n \lg \omega$ dB and the phase $-n90°$. When $\omega = 1$ rad/s then the magnitude is 0. When $\omega = 10$ rad/s it is $-20n$ dB. When $\omega = 100$ rad/s it is $-40n$ dB. Thus the slope of the gain line for a transfer function of $1/s^n$ is $-20n$ for each tenfold increase in frequency. Figure 11.10 shows the Bode plots.

11.3.3 Transfer function s

For a transfer function $G(s) = s$ (note that values of s which make the numerator of a transfer function zero are termed zeros, the values which make the denominator zero being termed poles) we have a frequency response function of $G(j\omega) = (j\omega)$. For such a system the gain is $|G(j\omega)| = 20 \lg \omega = 20 \lg \omega$ dB and the phase 90°. The Bode plots are thus straight lines. When $\omega = 1$ rad/s then the gain is 0. When $\omega = 10$ rad/s it is 20 dB. When $\omega = 100$ rad/s it is 40 dB. Thus the slope of the line is 20 dB for each tenfold increase in frequency.

For a transfer function $G(s) = s^m$ and so a frequency response function $G(j\omega) = (j\omega)^m$, the gain is $|G(j\omega)| = 20 \lg \omega^m = 20m \lg \omega$ dB and the phase $m90°$. When $\omega = 1$ rad/s then the magnitude is 0; when $\omega = 10$ rad/s it is $20m$ dB. Thus the slope of the line is $20m$ dB for each tenfold increase in frequency. Figure 11.11 shows the Bode plot.

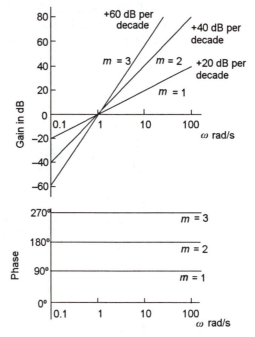

Figure 11.11 *Bode plot for transfer functions s^m*

11.3.4 Transfer function $1/(1 + \tau s)$

For $G(s) = 1/(1 + \tau s)$, i.e. a pole at $s = -1/\tau$ where τ is the time constant, the frequency response is:

$$G(j\omega) = \frac{1}{1+j\omega\tau} = \frac{1-j\omega\tau}{1+\omega^2\tau^2}$$

This means a gain in dB of:

$$\text{gain} = 20\lg\left(\frac{1}{\sqrt{1+\omega^2\tau^2}}\right) = -10\lg(1+\omega^2\tau^2)$$

and a phase of $\tan^{-1} -\omega\tau$.

When $\omega \ll 1/\tau$ then $\omega^2\tau^2$ is negligible compared with 1 and so the gain is 0 dB. Thus, at low frequencies the Bode plot is a straight line with a constant value of 0 dB. For high frequencies when $\omega \gg 1/\tau$ then $\omega^2\tau^2$ is much greater than 1 and so we can neglect the 1 and the magnitude is $-20\lg\omega\tau$. This is a straight line of slope -20 dB per decade which intersects with the zero decibel line when $\omega\tau = 1$. Figure 11.12 shows these lines for the low and high frequencies, their intersection being when $\omega = 1/\tau$; this intersection is termed the *break point* or *corner frequency*. The two lines are called the *asymptotic approximation* to the true plot. The true plot, when we do not make these approximations, differs slightly from the approximate plot and has a maximum error of 3 dB at the break point. Table 11.1 gives the errors in using the asymptotes.

At low frequencies, when ω is less than about $0.1/\tau$, the phase angle is virtually $0°$. At high frequencies, when ω is more than about $10/\tau$, it is virtually $-90°$. Between these frequencies the phase angle can be considered to give a reasonable straight line. The maximum error in assuming a straight line is $5\frac{1}{2}°$. When $\omega = 1/\tau$ then the phase angle is $45°$. Table 11.1 gives the errors in using the asymptotes.

Table 11.1 *Asymptote errors for transfer function $1/(1 + \tau s)$*

ω	Magnitude error dB	Phase error
$0.10/\tau$	-0.04	$-5.7°$
$0.20/\tau$	-0.2	$+2.3°$
$0.50/\tau$	-1.0	$+4.9°$
$1.00/\tau$	-3.0	$0°$
$2.00/\tau$	-1.0	$-4.9°$
$5.00/\tau$	-0.2	$-2.3°$
$10.0/\tau$	-0.04	$+5.7°$

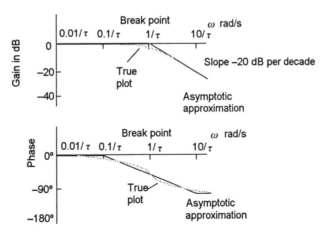

Figure 11.12 *Bode plot for transfer function $1/(1 + \tau s)$*

Example

Sketch the asymptotes of the Bode plots for a system having a transfer function of $100/(s + 100)$.

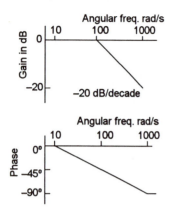

Figure 11.13 *Example*

The transfer function can be put in the form $1/(1 + s/100)$ and so the Bode plot is of the form shown in Figure 11.12. Since the time constant $(1/100)$ then the break point is at $\omega = 100$ rad/s. At higher frequencies the slope of the gain asymptote will be -20 dB/decade and thus the gain plot is as shown in Figure 11.13. The phase is $\tan^{-1} -(\omega/100)$ and is thus $0°$ at low frequencies, $-45°$ at the break point and $90°$ at high frequencies; Figure 11.13 shows the plot.

11.3.5 Transfer function $(1 + \tau s)$

For $G(s) = (1 + \tau s)$, i.e. a zero at $s = 1/\tau$ where τ is the time constant, the frequency response is:

$$G(j\omega) = 1 + j\omega\tau$$

This means a gain in dB of:

$$\text{gain} = 20\lg \sqrt{1 + \omega^2\tau^2} = 10\lg(1 + \omega^2\tau^2)$$

and a phase of $\tan^{-1} \omega\tau$.

When $\omega \ll 1/\tau$ then $\omega^2\tau^2$ is negligible compared with 1 and so the gain is 0 dB. Thus at low frequencies the Bode plot is a straight line with a constant value of 0 dB. For high frequencies when $\omega \gg 1/\tau$ then $\omega^2\tau^2$ is much greater than 1 and so we can neglect the 1 and the magnitude is $20\lg \omega\tau$. This is a straight line of slope $+20$ dB per decade which intersects with the zero decibel line when $\omega\tau = 1$. Figure 11.14 shows these lines, the *asymptotic approximation* to the true plot, for the low and high frequencies, their intersection being the *break point* or *corner frequency* of $\omega = 1/\tau$. The true plot has a maximum error which is 3 dB at the break point. The errors are the same as in Table 11.1.

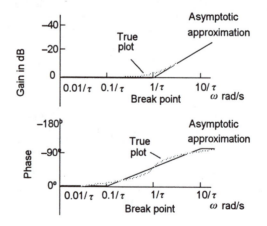

Figure 11.14 *Bode plot for transfer function $1/(1 + \tau s)$*

At low frequencies, when ω is less than about $0.1/\tau$, the phase angle is virtually $0°$. At high frequencies, when ω is more than about $10/\tau$, it is virtually $+90°$. Between these frequencies the phase angle can be considered to give a reasonable straight line. The maximum error in assuming a straight line is 5½°. When $\omega\tau = 1$ then the phase angle is 45°. Table 11.1 gives the errors.

11.3.6 Transfer function $\omega_n^2/(s^2 + 2\zeta\omega_n s + \omega_n^2)$

For a system having a transfer function:

$$G(s) = \frac{\omega_n^2}{s^2 + 2\zeta\omega_n s + \omega_n^2}$$

i.e. a pair of complex poles, the frequency response function is:

$$G(j\omega) = \frac{\omega_n^2}{-\omega^2 + j2\zeta\omega_n\omega + \omega_n^2}$$

$$= \frac{1}{[1 - (\omega/\omega_n)^2] + j[2\zeta(\omega/\omega_n)]}$$

$$= \frac{[1 - (\omega/\omega_n)^2] - j[2\zeta(\omega/\omega_n)]}{[1 - (\omega/\omega_n)^2]^2 + [2\zeta(\omega/\omega_n)]^2}$$

Thus the gain in decibels is:

$$\text{gain} = 20\lg\sqrt{\frac{1}{[1 - (\omega/\omega_n)^2]^2 + [2\zeta(\omega/\omega_n)]^2}}$$

$$= -10\lg\{[1 - (\omega/\omega_n)^2]^2 + [2\zeta(\omega/\omega_n)]^2\}$$

and the phase is:

$$\text{phase} = -\tan^{-1}\frac{2\zeta(\omega/\omega_n)}{1 - (\omega/\omega_n)^2}$$

For $\omega/\omega_n \ll 1$ the magnitude approximates to 0 dB and thus at low frequencies the asymptotic approximation is a straight line of 0 dB. For $\omega/\omega_n \gg 1$ the magnitude approximates to $-40\lg(\omega/\omega_n)$. Thus, at high frequencies the asymptotic approximation is a straight line of slope -40 dB per decade. The intersection of these two lines is a break point of $\omega/\omega_n = 1$. The true value, however, depends on the damping ratio. Figure 11.15 shows the asymptotes and some true plots at different damping factors.

Table 11.2 gives the errors, for a number of damping ratios, between the asymptote lines and the true magnitude plot.

The phase is approximately constant at 0° for $\omega/\omega_n \ll 1$ and for $\omega/\omega_n \gg 1$ approximately $-180°$. Usually an asymptote line is drawn through the points $\omega/\omega_n = 0.2$ as 0° and $\omega/\omega_n = 5$ as $-180°$. The discrepancy between this line and the true phase plots is shown in Table 11.3.

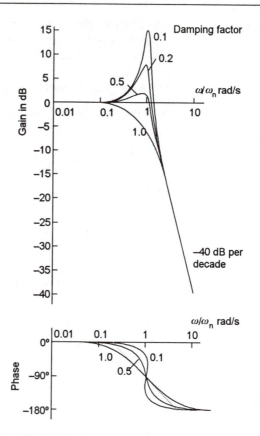

Figure 11.15 *Bode plot for a transfer function $\omega_n^2/(s^2 + 2\zeta\omega_n s + \omega_n^2)$*

Table 11.2 *Asymptote errors for gain in dB*

ζ				ω/ω_n			
	0.1	0.2	0.5	1.0	2.0	5.0	10.0
1.0	−0.09	−0.34	−1.94	−6.0	−1.92	−0.34	−0.09
0.7	0	−0.01	−0.26	−3.0	−0.26	−0.01	0
0.5	+0.04	+0.17	+0.90	0	+0.90	+0.17	+0.04
0.3	+0.07	+0.29	+1.85	+4.4	+1.85	+0.29	+0.07
0.2	+0.08	+0.33	+2.2	+8.0	+2.2	+0.33	+0.08

Table 11.3 *Asymptote errors in degrees for phase*

ζ				ω/ω_n			
	0.1	0.2	0.5	1.0	2.0	5.0	10.0
1.0	−11.4	−4.6	−9.8	0	+9.8	+4.6	+11.4
0.7	−8.1	−10.7	−19.6	0	+19.6	+10.7	+ 8.1
0.5	−5.8	−15.3	−29.2	0	+29.2	+15.3	+5.8
0.3	−3.5	−20.0	−41.1	0	+41.1	+20.0	+3.5
0.2	−2.3	−22.3	−48.0	0	+48.0	+22.3	+2.3

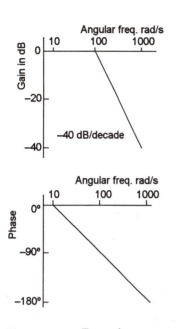

Figure 11.16 *Example*

Example

Sketch the asymptotes of the Bode plots for a system having a transfer function of $100/(s^2 + 4s + 100)$ and indicate the size of the error from the true plot at the break point.

The system has $\omega_n = 10$ rad/s and, since $2\zeta\omega_n = 4$, a damping factor ζ of 0.2. The break point is when $\omega = \omega_n = 10$ rad/s. For lower frequencies the gain asymptote will be 0 and at the break point will become –40 dB/decade. Figure 11.16 shows the plot. The phase will be –90° at the break point and effectively 0° one decade down from the break point and –180° one decade up from the break point. Figure 11.16 shows the plot.

With a damping factor of 0.2, the error at the break point for the asymptote gain plot is given by Table 11.2 as +8.0 dB and for the asymptote phase plot by Table 11.3 as zero.

11.3.7 Transfer function $(s^2 + 2\zeta\omega_n s + \omega_n^2)/\omega_n^2$

For a system having a transfer function:

$$G(s) = \frac{s^2 + 2\zeta\omega_n s + \omega_n^2}{\omega_n^2}$$

i.e. a pair of complex zeros, the frequency response is:

$$G(j\omega) = \frac{-\omega^2 + j2\zeta\omega_n\omega + \omega_n^2}{\omega_n^2}$$

$$= [1 - (\omega/\omega_n)^2] + j[2\zeta(\omega/\omega_n)]$$

The gain in decibels is thus:

$$\text{gain} = 20\lg\sqrt{[1 - (\omega/\omega_n)^2]^2 + [2\zeta(\omega/\omega_n)]^2}$$

$$= 10\lg\{[1 - (\omega/\omega_n)^2]^2 + [2\zeta(\omega/\omega_n)]^2\}$$

and the phase is:

$$\text{phase} = \tan^{-1}\left[\frac{2\zeta(\omega/\omega_n)}{1 - (\omega/\omega_n)^2}\right]$$

The gain differs only from that in Section 11.3.5 in being positive rather than negative. Thus the magnitude plot is just the mirror image of Figure 11.15 about the 0 dB line. The phase differs from that in Section 11.3.5 in being positive rather than negative. Thus the phase plot is just the mirror image of Figure 11.15 about the 0° line. The differences of the true plots from the asymptote lines is the same as in Tables 11.2 and 11.3.

Example

Determine the asymptote Bode plot for the system having the transfer function:

$$G(s) = \frac{50(s+2)}{s(s+10)}$$

This can be considered to the multiplication of four elements:

$$G(s) = 10 \times (1 + \tfrac{1}{2}s) \times \frac{1}{s} \times \frac{1}{1 + s/10}$$

We can draw the Bode plots for each of these elements and then sum them to obtain the overall plot. Figure 11.17 shows the result.

1 For $G_1(s) = 10$ we have a straight line of magnitude 20 lg 10 = 0 dB and a constant phase of 0°.

2 For $G_2(s) = 1 + \tfrac{1}{2}s$ we have a magnitude of 0 dB when $\omega\tau \ll 1$ and a line of slope +20 lg $\omega\tau$ = +20 dB per decade when $\omega\tau \gg$ 1. The break point is $\omega = 1/\tau = 2$ rad/s. The phase is effectively 0° up to $0.1/\tau = 0.2$ rad/s and +90° for frequencies greater than $10/\tau = 20$ rad/s.

3 For $G_3(s) = 1/s$ we have a straight line of slope –20 dB per decade passing through the 0 dB point at $\omega = 1$ rad/s. The phase is a constant –90°.

4 For $G_4 = 1/(1 + s/10)$ we have a magnitude of 0 dB when $\omega\tau \ll$ 1 and a line of slope –20 lg $\omega\tau$ = –20 dB/decade when $\omega\tau \gg$ 1. The break point is $\omega = 1/\tau = 10$ rad/s. The phase is effectively 0° up to $0.1/\tau = 1$ rad/s and –90° for frequencies greater than $10/\tau = 100$ rad/s.

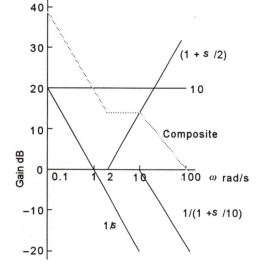

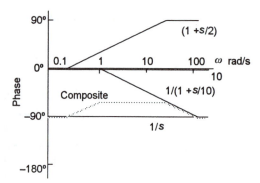

Figure 11.17 *Composite Bode plot*

11.4 System identification

In Chapter 8, methods were indicated by which models, i.e. differential equations describing the input–output relationship or transfer function, can be devised for systems by considering them to be made up of simple elements. An alternative way of developing a model for a real system is determine its response to some input and then find the model that fits the response; this process of determining a mathematical model is known as *system identification*.

A particularly useful method of system identification is to use a sinusoidal input and determine the output over a range of frequencies. Bode plots are then plotted with this experimental data. We then find the Bode plot elements that fit the experimental plot by drawing asymptotes on the gain Bode plot and considering their gradients.

1   If the gradient at low frequencies prior to the first corner frequency is zero then there is no s or $1/s$ element in the transfer function. The K element in the numerator of the transfer function can be obtained from value of the low frequency gain since the gain in dB = $20 \lg K$.

2   If the initial gradient at low frequencies is –20 dB/decade then the transfer function has a $1/s$ element.

3   If the gradient becomes more negative at a corner frequency by 20 dB/decade, there is a $(1 + s/\omega_c)$ term in the denominator of the transfer function, with ω_c being the corner frequency at which the change occurs. Such terms can occur for more than one corner frequency.

4   If the gradient becomes more positive at a corner frequency by 20 dB/decade, there is a $(1 + s/\omega_c)$ term in the numerator of the transfer function, with ω_c being the frequency at which the change occurs. Such terms can occur for more than one corner frequency.

5   If the gradient at a corner frequency becomes more negative by 40 dB/decade, there is a $(s^2/\omega_c^2 + 2\zeta s/\omega_c + 1)$ term in the denominator of the transfer function. The damping ratio ζ can be found   from considering the behaviour of the system to a unit step input.

6   If the gradient at a corner frequency becomes more positive by 40 dB/decade, there is a $(s^2/\omega_c^2 + 2\zeta s/\omega_c + 1)$ term in the numerator of the transfer function. The damping ratio ζ can be found   from considering the detail of the Bode plot at a corner frequency.

7   If the low-frequency gradient is not zero, the K term in the numerator of the transfer function can be determined by considering the value of the low-frequency asymptote. At low frequencies, many terms in transfer functions can be neglected and the gain in dB approximates to $20 \lg (K/\omega^2)$. Thus, at $\omega = 1$ the gain in dB approximates to $20 \lg K$.

The phase angle curve is used to check the results obtained from the magnitude analysis.

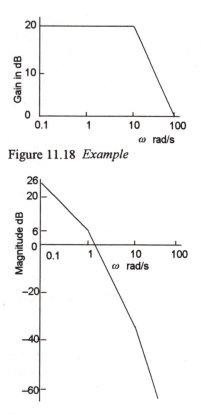

Figure 11.18 *Example*

Figure 11.19 *Example*

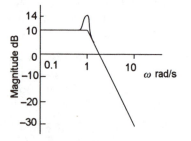

Figure 11.20 *Example*

11.5 Stability

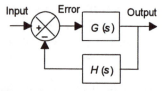

Figure 11.21 *Closed-loop system*

Example

Determine the transfer function of the system giving the Bode magnitude plot shown in Figure 11.18.

The initial gradient is 0 and so there is no $1/s$ or s term in the transfer function. The initial gain is 20 and thus $20 = 20 \lg K$ and so we have $K = 10$. The gradient changes by -20 dB/decade at a frequency of 10 rad/s. Hence there is a $(1 + s/10)$ term in the denominator. The transfer function is thus $10/(1 + 0.1s)$.

Example

Determine the transfer function of the system giving the Bode magnitude plot shown in Figure 11.19.

There is an initial slope of -20 dB/decade and so a $1/s$ term. At the corner frequency 1.0 rad/s there is a -20 dB/decade change in gradient and so a $1/(1 + s/1)$ term. At the corner frequency 10 rad/s there is a further -20 dB/decade change in gradient and so a $1/(1 + s/10)$ term. At $\omega = 1$ the magnitude is 6 dB and so $6 = 20 \lg K$ and $K = 10^{6/20} = 2.0$. The transfer function is thus $2.0/s(1 + s)(1 + 0.1s)$.

Example

Determine the transfer function of the system giving the Bode magnitude plot shown in Figure 11.20. This shows both the asymptotes and the departure of the true plot from them in the vicinity of the break point.

The gain Bode plot has an initial zero gradient. Since the initial magnitude is 10 dB then $10 = 20 \lg K$ and so $K = 10^{0.5} = 3.2$. The change of -40 dB/decade at 1 rad/s means there is a $1/(s^2 + 2\zeta s + 1)$ term. The transfer function is thus $3.2/(s^2 + 2\zeta s + 1)$.

The damping factor ζ can be obtained by considering the departure of the true Bode plot from the asymptotes at the break point. Since it rises by about 4 dB, Table 11.2 indicates that this corresponds to a damping factor of about 0.3. The transfer function is thus $3.2/(s^2 + 0.6s + 1)$.

Consider the stability of a closed-loop control system (Figure 11.21) when we have a brief input of a pulse. We will regard the pulse as essentially half of a sinusoidal signal at a particular frequency (Figure 11.22(a)). This passes through $G(s)$ to give an output which is then fed back through $H(s)$. Suppose it arrives back with amplitude unchanged from that of the input but with a phase such that when it is subtracted from the now zero input signal we have a resulting error signal which just continues the initial half-rectified pulse (Figure 11.22(b)). This then continues round the feedback loop to once again arrive just in time to continue the signal. There is a self-sustaining oscillation.

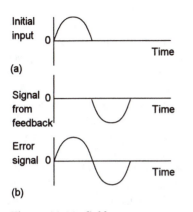

Figure 11.22 *Self sustaining signal: (a) input to system, (b) fed back signal, (c) resulting error signal*

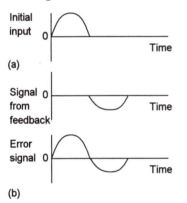

Figure 11.23 *Stable system: (a) input to system, (b) fed back signal, (c) resulting error signal*

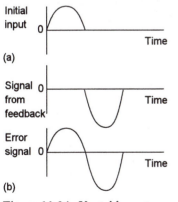

Figure 11.24 *Unstable system: (a) input to system, (b) fed back signal, (c) resulting error signal*

If the fed back signal has an amplitude which was smaller than the initial pulse amplitude then the signal would die away with time (Figure 11.23). The system is *stable*. If the fed back signal had an amplitude which was just the same as that of the initial pulse amplitude then the oscillation would continue with constant amplitude and the system is said to be *marginally stable* (Figure 11.22). If the fed back signal had a larger amplitude than that of the initial pulse amplitude (Figure 11.24) then the oscillation would continue with an increasing amplitude and the system would be unstable. Thus the condition for *instability* is that the gain resulting from a signal fed though the series arrangement of $G(s)$ and $H(s)$ should be greater than 1 and the signal fed back into $G(s)H(s)$ must have suffered a phase change of $-180°$.

The transfer function for $G(s)$ in series with $H(s)$ is called the *open-loop transfer function*. The stability criterion can thus be stated as:

> The critical point which separates stable from unstable systems is when the open-loop phase shift is $-180°$ and the open-loop gain is 1.

A good stable control system usually has an open-loop gain significantly less than 1, typically about 0.4 to 0.5, when the phase shift is $-180°$ and an open-loop phase shift of between $-115°$ to $-125°$ when the gain is 1. Such values give a slightly underdamped system which gives, with a step input, about a 20 to 30% overshoot.

11.5.1 Stability measures

Measures of the stability of systems that are used in the frequency domain are:

1 *Phase crossover frequency*
The phase crossover frequency is the frequency at which the phase angle first reaches $-180°$.

2 *Gain margin*
This is the factor by which the gain must be multiplied at the phase crossover to have the value 1. A good stable control system usually has an open-loop gain significantly less than 1, typically about 0.4 to 0.5, when the phase shift is $-180°$ and so a gain margin of 1/0.5 to 1/0.4, i.e. 2 to 2.5.

3 *Gain crossover*
This is the frequency at which the open-loop gain first reaches 1.

4 *Phase margin*
This is the number of degrees by which the phase angle is smaller than $-180°$ at the gain crossover. A good stable control system usually has typically an open-loop phase shift of between $-115°$ to $-125°$ when the gain is 1; thus, the phase margin is between $45°$ and $65°$.

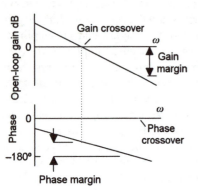

Figure 11.25 *Stability and Bode plots*

The above parameters enable the questions of how much change in gain and phase of the $G(s)H(s)$ product can be tolerated before a system becomes unstable. They are thus useful in the design of stable systems.

The Bode plot for the open-loop transfer function, i.e. $G(s)H(s)$, gives a convenient way to determine the above parameters and hence the stability of a system. An open-loop gain of 1 is, on the log scale of dB, a gain of 20 lg 1 = 0 dB. Figure 11.25 shows the parameters on a Bode plot. Note that on the Bode gain plot we are working with the log of the gain and so the gain margin is the additional dB that is necessary to make the gain of the signal unity at the phase crossover frequency.

Figure 11.26(a) is an example of Bode plot for a stable system with Figure 11.26(b) being for an unstable system. In (a), the open-loop gain is less than 0 dB, i.e. a gain of 1, when the phase is −180°. In (b), the open-loop gain is greater than 0 dB, i.e. greater than 1, when the phase is −180°.

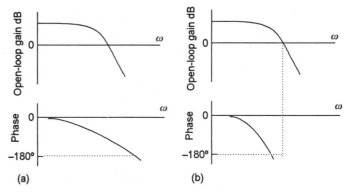

Figure 11.26 *(a) Stable system, (b) unstable system*

Example

Determine the gain margin and the phase margin for a system that gave the following open-loop experimental frequency response data: at frequency 0.005 Hz a gain of 1.00 and phase −120°, at 0.010 Hz a gain of 0.45 and phase −180°.

The gain margin is the factor by which the gain must be multiplied at the phase crossover to have the value 1. The phase crossover occurs at 0.010 Hz and so the gain margin is 1.00/0.45 = 2.22. The phase margin is the number of degrees by which the phase angle is smaller than −180° at the gain crossover. The gain crossover is the frequency at which the open-loop gain first reaches the value 1 and so is 0.005 Hz. Thus, the phase margin is 180° − 120° = 60°.

Example

For the Bode plot shown in Figure 11.27, determine (a) whether the system is stable, (b) the gain margin, (c) the phase margin.

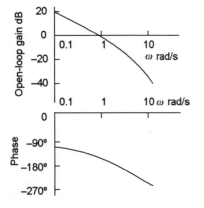

Figure 11.27 *Example*

(a) The system is stable because it has an open-loop gain less than 1 when the phase is −180°.
(b) The gain margin is about 12 dB.
(c) The phase margin is about 30°.

11.6 Compensation

The term *compensation* is used for the modification of the performance characteristics of a system so that the required characteristics are obtained. A *compensator* is thus an additional component which is added into a control system to modify the closed-loop performance and compensate for a deficient performance. A compensator placed in the forward path is called a *cascade* or *series* compensator (Figure 11.28).

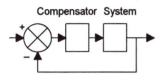

Figure 11.28 *Cascade compensation*

11.6.1 Changing the gain

Consider the effects of adjusting the performance of a control system by changing the gain in the forward path (Figure 11.29). The effect of increasing the gain is to shift upwards the gain–frequency Bode plot by an equal amount over all the frequencies; this is because we are adding a constant gain element to the Bode plot. Figure 11.30 illustrates this. There is no effect on the phase–frequency Bode plot. Increasing the gain thus shifts the 0 dB crossing point of the gain plot to the right and so to a higher frequency. This decreases the stability of the system since it decreases the gain margin and the phase margin. Hence, if an increase in stability is required then the gain needs to be reduced.

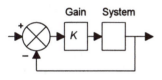

Figure 11.29 *Adjusting the performance of a system by introducing gain*

Example

For the control system giving the open-loop Bode plot of Figure 11.31 we have a control system with a phase margin of 35°. By how much should the gain of the system be changed if a gain margin of 45° is required?

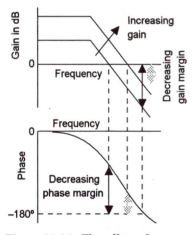

Figure 11.30 *The effect of increasing the gain*

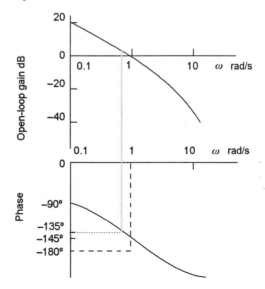

Figure 11.31 *Example*

The phase margin of 35° means that the phase is 180° − 35° = 145° when the gain is 0 dB. For a phase margin of 45° the phase must be 180° − 45° = 135° when the gain is 0 dB. At present, when the phase is 135° the gain is about +2 dB. If follows that if the gain is reduced by 2 dB that the gain–frequency line will be shifted downwards by 2 dB and give the required phase margin. Since the change in gain ΔK is given by 2 = 20 lg ΔK then ΔK = 1.3. This is the factor by which the gain has to be reduced.

11.6.2 Phase-lead compensation

The transfer function of a phase-lead compensator is of the form:

$$G(s) = \frac{1 + a\tau s}{1 + \tau s}$$

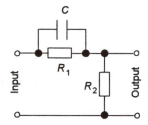

Figure 11.32 *Phase-lead compensator*

with $a > 1$. Such a compensator can be provided by the circuit shown in Figure 11.32 ($a = (R_1 + R_2)/R_2$ and $\tau = R_1R_2C/(R_1 + R_2)$). Figure 11.33 shows the Bode plot for a phase-lead compensator (it can be obtained by adding the Bode plots for the numerator term and the denominator term). The term *phase-lead* is used because the compensator has a positive phase and so is used to add phase to an uncompensated system. The maximum value of the phase occurs at a frequency ω_m which is midway between the frequencies of $1/\tau$ and $1/a\tau$ on the logarithmic scale and so:

$$\lg \omega_m = \frac{1}{2}\left(\lg \frac{1}{\tau} + \lg \frac{1}{a\tau}\right)$$

and hence:

$$\omega_m = \frac{1}{\tau\sqrt{a}}$$

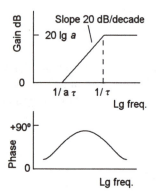

Figure 11.33 *Bode plot for a phase-lead compensator*

The value of this maximum phase angle ϕ_m is given by:

$$\sin \phi_m = \frac{a - 1}{a + 1}$$

With a cascade phase-lead compensator we add its Bode plot to that of the system being compensated in order to obtain the required specification. As a consequence we can increase the phase margin. The cascade phase-lead compensator is thus used to provide a satisfactory phase margin for a system. The procedure to determine the required values of a and τ for the compensator is to:

1   Determine the phase margin of the open-loop uncompensated system and so ascertain the addition amount of phase required to give the desired phase margin.

2   The above equation is then used to determine the value of a to give this additional phase.

3 The high frequency gain of the compensator is 20 lg a dB and the low frequency gain is 0 dB; hence, the gain at the maximum phase is ½ 20 lg a = 10 lg a dB. From the value of a obtained in item 2 we can determine this gain.

4 The compensator is used to give the new gain crossover at the frequency at which the phase is a maximum and thus we need to place this frequency where the gain of the uncompensated system is −10 lg a dB. In doing this we obtain the required value of ϕ_m and so can determine the required value of τ.

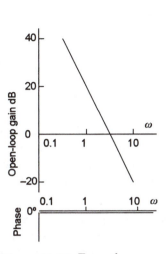

Figure 11.34 *Example*

Example

Determine the transfer function of the cascade lead-compensator that can be used with a system having an open-loop transfer function of $10/s^2$ in order to give a phase margin of 45°.

Figure 11.34 shows the Bode plot for the uncompensated system, the uncompensated phase being 0° at all frequencies. Thus, the phase margin of the uncompensated system is 180°. To obtain a phase margin of 45° we require:

$$\sin 45° = \frac{a-1}{a+1}$$

$$0.71(a + 1) = a - 1$$

$$0.29a = 1.71$$

and thus a is 5.9. The gain of the compensator at the maximum compensator phase is 10 lg a = 7.7 dB. The uncompensated system has a gain of −7.7 dB at a frequency of about 5 rad/s and so $\omega_m = 5 = 1/\tau\sqrt{a} = 1/2.4\tau$ and thus $\tau = 0.083$. The required compensator has thus a transfer function of:

$$G(s) = \frac{1+a\tau s}{1+\tau s} = \frac{1+0.49s}{1+0.083s}$$

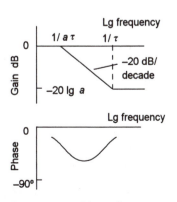

Figure 11.35 *Phase-lag compensator*

11.6.3 Phase-lag compensation

The phase-lag compensator has a negative phase angle and so is used to subtract phase from an uncompensated system. A phase-lag compensator has a transfer function of the form:

$$G(s) = \frac{1+\tau s}{1+a\tau s}$$

Figure 11.36 *Phase-lag compensator*

where a is greater than 1. Figure 11.35 shows the type of circuit that can be used ($a = (R_1 + R_2)/R_2$ and $\tau = R_2C$). Figure 11.36 shows the Bode plot. Because the phase-lag compensator adds a negative phase angle to a system, the phase lag is not a useful effect of the compensation and does not provide a direct means of improving the phase margin. The phase-lag compensator does, however, reduce the gain and so can be used to lower the crossover frequency. A consequence of this is that, as usually the phase margin of the system is higher at the lower frequency, the stability can be improved. The procedure that can be adopted to design with a cascade phase-lag compensator is:

1   Determine the frequency where the required phase margin would be obtained if the gain plot crossed the 0 dB line at this frequency. Allow for 5° phase lag from the phase-lag compensator when determining the new crossover frequency.

2   One decade below the new crossover frequency is taken to be $1/\tau$. It is necessary to ensure that the phase minimum occurs at a frequency which is well below the crossover frequency of the uncompensated system so that the phase lag introduced by the compensator does not significantly destabilise the system.

3   Determine the gain required at the new crossover frequency to ensure that the compensated system gain plot crosses at this frequency. Since the gain produced by the phase-lag compensator at this frequency is $-20 \lg a$, we can calculate a.

Example

Determine the transfer function of the cascade phase-lag compensator that can be used with a system having an open-loop Bode plot of Figure 11.37 in order to give a phase margin of 40°.

The uncompensated system has a phase margin of about 20°. The frequency where the phase margin is the required 40° + 5° is 2 rad/s and this is to become the new crossover frequency. One decade below this is 0.2 rad/s. Thus $1/\tau = 0.2$ and so $\tau = 5$. The gain of the uncompensated system at the new crossover frequency is 20 dB and so $20 \lg a = 20$ and hence $a = 10$. Thus, the required transfer function of the compensator is $(1 + 5s)/(1 + 50s)$.

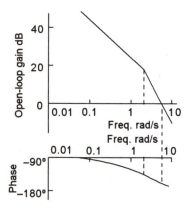

Figure 11.37 *Example*

Problems   *Questions 1 to 7 have four answer options: A, B, C or D. Choose the correct answer from the answer options.*

1   The phase difference between the two signals $y_1 = 2 \sin (3t + \pi/2)$ and $y_2 = -5 \sin (3t + \pi/2)$ is:

A $\pi/2$
B $-\pi/2$
C π
D $-\pi$

2   Decide whether each of these statements is True (T) or False (F).

A system with a transfer function of $1/(1 + 3s)$ has, for a sinusoidal input of $\sin \omega t$:
(i)   A frequency response function of $1/(1 + j3\omega)$.

(ii) A gain of $1/\sqrt{(1 + 9\omega^2)}$.

A   (i) T (ii) T
B   (i) T (ii) F
C   (i) F (ii) T
D   (i) F (ii) F

3 Decide whether each of these statements is True (T) or False (F).

A system with a transfer function of $1/(1 + 5s)$ has, for a sinusoidal input of sin ωt:
(i) A phase difference between the output and input of $\tan^{-1} 5\omega$.
(ii) A gain of $1/\sqrt{(1 + 5\omega^2)}$.

A   (i) T (ii) T
B   (i) T (ii) F
C   (i) F (ii) T
D   (i) F (ii) F

4 Decide whether each of these statements is True (T) or False (F).

At $\omega = 2$ rad/s, a system gives an output with a gain of 5 and a phase shift of 30° for a sinusoidal input. For a sinusoidal input of $2 \sin (3t - 30°)$, the output has:

(i) A phase of 0°.
(ii) A magnitude of 5.

A   (i) T (ii) T
B   (i) T (ii) F
C   (i) F (ii) T
D   (i) F (ii) F

5 The gain of a system is $|G(j\omega_1)| = 0$ dB. For an input of a sinusoidal signal of frequency ω_1, the output will have a magnitude which is:

A   Zero
B   Reduced in size compared with input
C   Same size as input
D   Larger in size than input

6 Decide whether each of these statements is True (T) or False (F).

For a system to be stable, it must have:

(i) An open-loop gain greater than 1.
(ii) A phase shift between 0°C and 180°.

A   (i) T (ii) T
B   (i) T (ii) F
C   (i) F (ii) T
D   (i) F (ii) F

7 The phase crossover frequency of a system is the frequency at which the phase angle first reaches:

A −180°
B −90°
C 0°
D +180°

8 What are the frequency response functions for systems with transfer functions (a) $1/(s + 5)$, (b) $7/(s + 2)$, (c) $1/[(s + 10)(s + 2)]$?

9 Determine the magnitude and the phase of the response of a system with transfer function $3/(s + 2)$ to sinusoidal inputs of angular frequency (a) 1 rad/s, (b) 2 rad/s.

10 Sketch the asymptotes for the Bode plots of systems with the transfer functions (a) 100, (b) $1000/(s + 1000)$, (c) $4/(s^2 + s + 4)$.

11 Sketch the asymptotes for the Bode plots of systems with the transfer function (a) $10/s^2$, (b) $(s − 10)/(s + 10)$, (c) $s/(s^2 + 20s + 100)$.

12 Obtain the transfer functions of the systems giving the Bode gain plots in Figure 11.38.

13 The following are experimentally determined frequency response data for a system. By plotting the Bode gain diagram, determine the transfer function of the system.

Freq. Hz	0.16	0.47	1.3	2.5	4.8	10.0	16.0	20.0	24.0
Gain dB	24.0	24.0	23.6	23.0	17.0	14.0	9.5	8.0	5.0

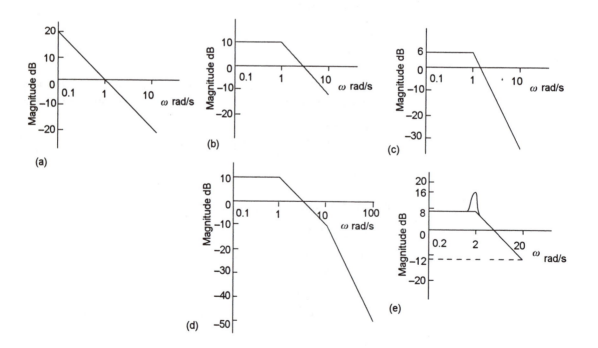

Figure 11.38 *Problem 12*

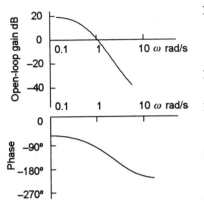

Figure 11.39 *Problem 15*

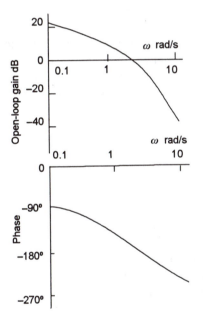

Figure 11.40 *Problem 18*

14 Determine the gain margin and the phase margin for a system that gave the following open-loop experimental frequency response data: at frequency 0.01 Hz a gain of 1.00 and phase −130°, at 0.02 Hz a gain of 0.55 and phase −180°.

15 For the Bode plot shown in Figure 11.39, determine (a) whether the system is stable, (b) the gain margin, (c) the phase margin.

16 A system has an open-loop transfer function of $1/[s(1 + s)(1 + 0.2s)]$. By plotting the Bode diagrams, determine (a) whether the system is stable, (b) the gain margin, (c) the phase margin.

17 A system has an open-loop transfer function of $1/[s(1 + 0/0.02s)(1 + 0.2s)]$. By plotting the Bode diagrams, determine (a) whether the system is stable, (b) the gain margin, (c) the phase margin.

18 For the control system giving the open-loop Bode plot of Figure 11.40 we have a control system with a phase margin of 10°. By how much should the gain of the system be changed if a gain margin of 40° is required?

19 The following are experimentally determined open-loop frequency response data for a system. Determine the phase margin when uncompensated and the change in the gain of the system that is necessary from a phase-lead compensator, when added in cascade, to increase the phase margin by 5°.

Frequency rad/s	0.6	0.8	1.0	2.0	4.0	6.0
Gain dB	13.2	10.3	8.0	0.0	−10.7	−18.0
Phase deg.	−110	−116	−122	−146	−175	−192

20 Determine the maximum phase lead introduced by a phase-lead compensator with a transfer function of $(1 + 0.2s)/(1 + 0.02s)$.

21 A system has an open-loop transfer function of $12/[s(s + 1)]$. What will be the transfer function of a phase-lead compensator which, when added in cascade, will increase the uncompensated phase margin of 15° to 40°?

22 A system has an open-loop transfer function of $4/[s(2s + 1)]$. What will be the transfer function of a phase-lag compensator which, when added in cascade, will give a phase margin of 40°?

12 Nyquist diagrams

12.1 Introduction

This chapter follows on from Chapter 11 and presents another method of describing the frequency response of systems and their stability. The method uses *Nyquist diagrams*; in these diagrams the gain and the phase of the open-loop transfer function, i.e. the product of the forward path and the feedback path transfer functions, are plotted as polar graphs for various values of frequency. With Cartesian graphs the points are plotted according to their x and y coordinates from the origin; with polar graph the points are plotted from the origin according to their radial distance from it and their angle to the reference axis (Figure 12.1).

12.2 The polar plot

The *polar plot* of the frequency response of a system is the line traced out as the frequency is changed from 0 to infinity by the tips of the phasors whose lengths represent the magnitude, i.e. amplitude gain, of the system and which are drawn at angles corresponding to their phase (Figure 12.2).

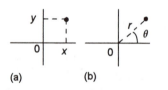

(a)　　　　　(b)

Figure 12.1 *(a) Cartesian graph with points specified by x and y values, (b) polar graph with points specified by r and θ values*

Example

Draw the polar diagram for the following frequency response data.

Freq. rad/s	1.4	2.0	2.6	3.2	3.8
Magnitude	1.6	1.0	0.6	0.4	0.2
Phase deg.	−150	−160	−170	−180	−190

Note that the above data gives the phases with negative signs. This means they are lagging behind the 0° line by the amounts given. Figure 12.3 shows the polar plot obtained.

12.2.1 Nyquist diagrams

The term *Nyquist diagram* is used for:

> The Nyquist diagram is the line joining the series of points plotted on a polar graph when each point represents the magnitude and phase of the open-loop frequency response corresponding to a particular frequency.

To plot the Nyquist diagram from the open-loop transfer function of a system we need to determine the magnitude and the phase as functions of frequency.

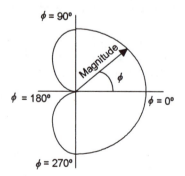

Figure 12.2　*Polar plot with the plot as the line traced out by the tips of the phasors as the frequency is changed from zero to infinity*

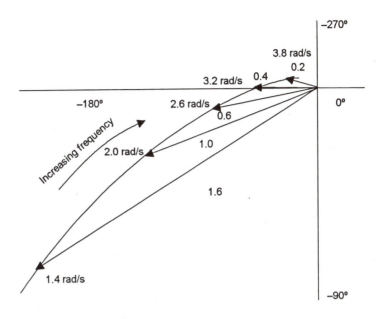

Figure 12.3   *Example*

Application

A MATLAB program that can be used to plot a Nyquist diagram for, say, the open loop transfer function 4/(s² + 5s + 2) is:

num = 4;
den = [1 5 2];
subplot (211), nyquist(num,den)

Note: subplot(211) divides the screen into two equal halves and puts the current plot into the top half of the screen.

Example

Determine the Nyquist diagram for a first-order system with an open-loop transfer function of $1/(1+\tau s)$.

The frequency response is:

$$\frac{1}{1+j\omega\tau} = \frac{1}{1+j\omega\tau}\frac{1-j\omega\tau}{1-j\omega\tau} = \frac{1}{1+\omega^2\tau^2} - j\frac{\omega\tau}{1+\omega^2\tau^2}$$

The magnitude is thus:

$$\text{magnitude} = \frac{1}{\sqrt{1+\omega^2\tau^2}}$$

and the phase is:

$$\text{phase} = -\tan^{-1}\omega\tau$$

At zero frequency the magnitude is 1 and the phase 0°. At infinite frequency the magnitude is zero and the phase is −90°. When $\omega\tau = 1$ the magnitude is $1/\sqrt{2}$ and the phase is −45°. Substitution of other values leads to the result shown in Figure 12.4 of a semicircular plot.

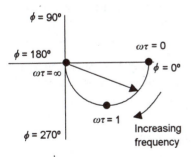

Figure 12.4   *Nyquist diagram for a first-order system*

Example

Determine the Nyquist plot for the system having the open-loop transfer function of $1/s(s + 1)$.

The frequency response is:

$$G(j\omega) = \frac{1}{j\omega(j\omega + 1)} = \frac{1}{j\omega - \omega^2} = \frac{1}{j\omega - \omega^2} \frac{-j\omega - \omega^2}{-j\omega - \omega^2}$$

$$= \frac{-\omega^2}{\omega^2 + \omega^4} - j\frac{\omega}{\omega^2 + \omega^4}$$

the magnitude and phase are thus:

$$\text{magnitude} = \frac{1}{\omega\sqrt{\omega^2 + 1}}$$

$$\text{phase} = \tan^{-1} -1/(-\omega) = 180° + \tan^{-1} 1/\omega$$

When $\omega = \infty$ then the magnitude is 0 and the phase is 0°. As ω tends to 0 then the magnitude tends to infinity and the phase to 270° or −90°. Figure 12.5 shows the polar plot.

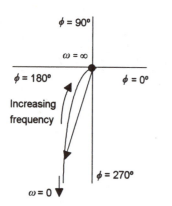
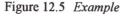

Figure 12.5 *Example*

12.3 Stability

As indicated in Section 11.5: the critical point which separates stable from unstable systems is when the open-loop phase shift is −180° and the open-loop magnitude is 1. If a Nyquist diagram of the open-loop frequency response is plotted then for the system to be stable there must not be any phasor with length greater than 1 and phase −180°. Thus the line traced by the tips of the phasors, the so-termed loci, must not enclose the −1 point. If an open-loop system is stable, then the corresponding unity feedback closed-loop system is stable.

> *Nyquist stability criterion*: Closed-loop systems whose open-loop frequency response $G(j\omega)H(j\omega)$, as ω goes from 0 to ∞, does not encircle the −1 point will be *stable*, those which encircle the −1 point are *unstable* and those which pass through the −1 point are *marginally stable*. Encircling the point may be taken as passing to the left of the point.

Figure 12.6 illustrates the above with examples of stable, marginally stable and unstable systems. The Nyquist plots, not to scale, correspond to the open-loop frequency response of:

$$G(j\omega)H(j\omega) = \frac{K}{(1 + j\omega 0.2)(1 + j\omega)(1 + j\omega 10)}$$

with $K = 10$ for the stable plot, $K = 137$ for the marginally stable plot and $K = 500$ for the unstable plot.

The vertical axis of the Nyquist plot corresponds to the phase equal to 90° and so is the imaginary part of the open-loop frequency response. The horizontal axis corresponds to the phase equal to 0° and so is the real part of the open-loop frequency response.

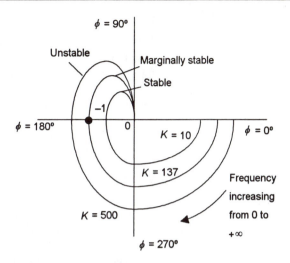

Figure 12.6   *Stability and the Nyquist plot*

Figures 12.7 to 12.10 show examples of Nyquist plots for common forms of open-loop transfer functions and their conditions for stability.

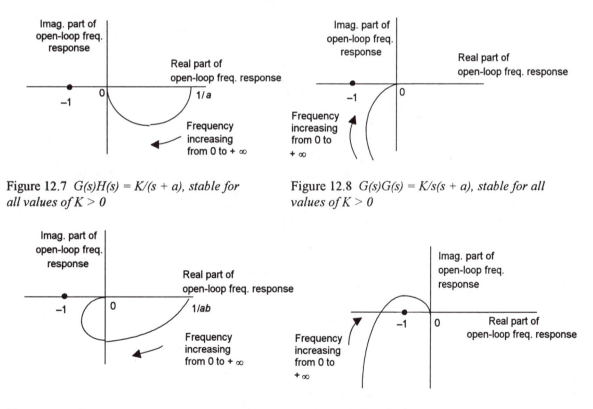

Figure 12.7 $G(s)H(s) = K/(s + a)$, *stable for all values of K > 0*

Figure 12.8 $G(s)G(s) = K/s(s + a)$, *stable for all values of K > 0*

Figure 12.9 $G(s)H(s) = K/(s + a)(s + b)$, *stable for all of K > 0*

Figure 12.10 $G(s)H(s) = K/s(s + a)(s + b)$; *this is values unstable with large K but can become stable if K is reduced, the point at which the plot crosses the axis being $-K/(a + b)$ and so stability is when $-K/(a + b) > -1$*

Example

Plot the Nyquist diagram for a system with the open-loop transfer function $K/(s + 1)(s + 2)(s + 3)$ and consider the value of K needed for stability.

The open-loop frequency response is:

$$\frac{K}{(j\omega + 1)(j\omega + 2)(j\omega + 3)}$$

The magnitude and phase are:

$$\text{magnitude} = \frac{K}{\sqrt{(\omega^2 + 1)(\omega^2 + 4)(\omega^2 + 9)}}$$

$$\text{phase} = \tan^{-1}\left(\frac{\omega}{1}\right) + \tan^{-1}\left(\frac{\omega}{2}\right) + \tan^{-1}\left(\frac{\omega}{3}\right)$$

When $\omega = 0$ then the magnitude is $K/6$ and the phase is $0°$. When $\omega = \infty$ then the magnitude is 0 and the phase is $270°$. We can use these, and other points to plot the polar graph.

Alternatively we can consider the frequency response in terms of real and imaginary parts. We can write the open-loop frequency function as:

$$\frac{6K(1 - \omega^2)}{(\omega^2 + 1)(\omega^2 + 4)(\omega^2 + 9)} + j\frac{\omega K(\omega^2 - 11)}{(\omega^2 + 1)(\omega^2 + 4)(\omega^2 + 9)}$$

When $\omega = 0$ then the imaginary part is zero and the real part is $K/6$. When $\omega = \infty$ then the imaginary part is zero and the real part is 0. The imaginary part will be zero when $\omega = \sqrt{11}$. This is a real part, and hence magnitude, of $-K/60$ and is the point at which the plot crosses the real axis. Thus for a stable system we must have $-K/60$ less than -1, i.e. K must be less than 60. Figure 12.11 shows the complete Nyquist plot (not to scale).

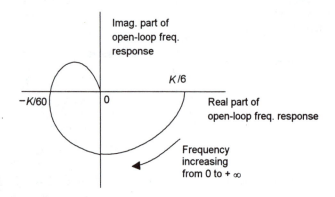

Figure 12.11 *Example*

12.4 Relative stability

The use of gain margin and phase margin was introduced in Section 11.5.1 to discuss the relative stability of a system in the frequency domain when described by a Bode plot. With Nyquist plots:

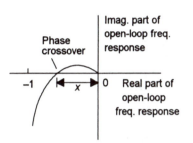

Figure 12.12. *Phase crossover and gain margin*

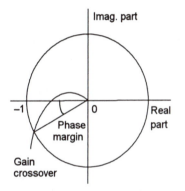

Figure 12.13 *Phase margin: the angle through which the gain crossover line must be rotated to reach the real axis and pass through the (–1, j0) point*

1 *Gain margin*

The phase crossover frequency is the frequency at which the phase angle first reaches –180° and thus is the point where the Nyquist plot crosses the real axis (Figure 12.12). On a Nyquist plot the (–1, j0) point is the point separating stability from instability. The gain margin is the amount by which the actual gain must be multiplied before the onset of instability. Thus if the plot cuts the negative real axis at –x (Figure 12.12), it has to be multiplied by 1/x to give the value –1 and so the gain margin, which is expressed in dB, is 20 lg(1/x).

When the open-loop plot goes through the (–1, j0) point the gain margin is 0 dB, the system being on the margin of instability. When the open-loop plot goes to the left of (–1, j0) point the gain margin is negative in dB, the system being unstable. When the open-loop plot goes to the right of (–1, j0) point the gain margin is positive in dB, the system being stable. When the open-loop plot does not intersect the negative real axis the gain margin is infinite in dB.

2 *Phase margin*

The phase margin is defined as the angle in degrees by which the phase angle is smaller than –180° at the gain crossover, the gain crossover being the frequency at which the open-loop gain first reaches 1. Thus, with a Nyquist plot, if we draw a circle of radius 1 centred on the origin, then the point at which it intersects the Nyquist line gives the gain crossover. The phase margin is the angle through which this gain crossover line must be rotated about the origin to reach the real axis and pass through the (–1, j0) point (Figure 12.13).

Example

Determine the gain margin and the phase margin for a system with the open-loop transfer function $K/(s + 1)(s + 2)(s + 3)$ with $K = 20$. This system was discussed earlier in this chapter (see Figure 12.11 for the Nyquist plot).

The open-loop frequency response is:

$$\frac{K}{(j\omega + 1)(j\omega + 2)(j\omega + 3)}$$

and this can be rearranged to give:

$$\frac{6K(1 - \omega^2)}{(\omega^2 + 1)(\omega^2 + 4)(\omega^2 + 9)} + j\frac{\omega K(\omega^2 - 11)}{(\omega^2 + 1)(\omega^2 + 4)(\omega^2 + 9)}$$

The imaginary part will be zero when $\omega = \sqrt{11}$ and thus the real part is $-K/60$ and is the point at which the plot crosses the real axis. Hence, if we have $K = 20$ then the plot intersects the negative real axis at $-20/60 = -1/3$. The gain can thus be increased by a factor of 3 in order to reach the -1 point. The gain margin is thus $20 \lg 3 = 9.5$ dB.

The magnitude is:

$$\text{magnitude} = \frac{K}{\sqrt{(\omega^2 + 1)(\omega^2 + 4)(\omega^2 + 9)}}$$

Thus, for $K = 20$, the magnitude is 1 when $\omega = 1.84$ rad/s. The phase is given by:

$$\text{phase} = \tan^{-1}\left(\frac{\omega}{1}\right) + \tan^{-1}\left(\frac{\omega}{2}\right) + \tan^{-1}\left(\frac{\omega}{3}\right)$$

and so, at this frequency, the phase is $-135.5°$. Thus the phase margin is $44.5°$.

Problems

Questions 1 to 3 have four answer options: A, B, C or D. Choose the correct answer from the answer options.

1   The system giving the Nyquist diagram shown in Figure 12.14 is:

   A   Stable for all frequencies
   B   Stable only at low frequencies
   C   Unstable at all frequencies
   D   Marginally stable

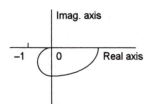

Figure 12.14   *Problem 1*

2   The system giving the Nyquist diagram shown in Figure 12.15 has a value of $K/10$ where it cuts the negative real axis and so is:

   A   Stable when K is greater than 10
   B   Stable when K is equal to 10
   C   Stable when K is less than 10
   D   Stable for all values of K

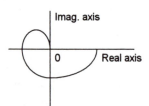

Figure 12.15   *Problem 2*

3   The system giving the Nyquist diagram shown in Figure 12.16 is:

   A   Stable for all frequencies
   B   Stable only at low frequencies
   C   Unstable at all frequencies
   D   Marginally stable

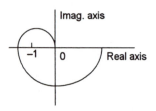

Figure 12.16   *Problem 3*

4 Sketch the Nyquist diagram for a system having an open-loop transfer function of $1/[s(s + 1)]$.

5 With a Nyquist diagram for the open-loop frequency response for a system, what is the condition for the system to be stable?

6 Determine the gain margin and the phase margin for a system which gave the following open-loop frequency response:

Freq. rad/s	1.4	2.0	2.6	3.2	3.8
Magnitude	1.6	1.0	0.6	0.4	0.2
Phase deg.	−150	−160	−170	−180	−190

7 Determine the gain margin and the phase margin for a system which gave the following open-loop frequency response:

Freq. rad/s	4	5	6	8	10
Gain	3.2	2.3	1.7	1.0	0.6
Phase in deg.	−140	−150	−157	−170	−180

8 Determine the gain margin and the phase margin for a system having an open-loop transfer function of:

$$\frac{K}{s(s+1)(s+2)}$$

when $K = 4$.

9 Determine the gain margin and the phase margin for a system having an open-loop transfer function of :

$$\frac{1}{s(0.2s+1)(0.05s+1)}$$

13 Controllers

13.1 Introduction

The design of a control system involves the need to be able to predict the performance of the system and, as a consequence, select the most appropriate control law to be used for the controller in order that the system is able to meet the required performance specification. The performance of the system can be defined in terms of its static characteristics, dynamic response as a function of time and frequency response. For example, the steady-state accuracy can be specified for when all transient components of the response have decayed away, how the performance is to vary with time for, say, a step input and the response to a sinusoidal input after all transient components have decayed away.

This chapter starts with a discussion of controllers in terms of transfer functions and then considers how the form of control law determines the way control systems respond.

13.2 Controllers

See Chapter 5 for a preliminary discussion of controllers. With *proportional control* (Figure 13.1), the controller produces a control action that is proportional to the error. There is a constant gain K_p acting on the error signal e and so:

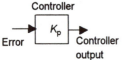

Controller

$$\text{controller output} = K_p e$$

Figure 13.1 *Proportional controller*

Increasing the gain speeds up the response to changes in inputs (Figure 13.2) and decreases, but does not eliminate, the output offset (see Section 5.3.2).

With *derivative control*, the controller produces a control action that is proportional to the rate at which the error is changing. Derivative control is not used alone but always in conjunction with proportional control and, often, also integral control. Figure 13.3 shows the basic form of a proportional plus derivative controller. The proportional element has an input of the error e and an output of $K_p e$. The derivative element has an input of e and an output which is proportional to the derivative of the error with time, i.e.

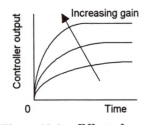

Figure 13.2 *Effect of increasing the gain for a proportional controller when there is a step input*

$$K_d \frac{de}{dt}$$

where K_d is the derivative gain. Thus the controller output is:

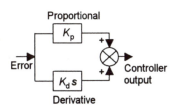

Figure 13.3 *PD controller*

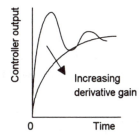

Figure 13.4 *Increasing the derivative gain increases the damping ratio, the graph illustrates this for a step input*

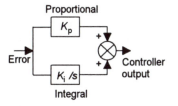

Figure 13.5 *PI controller*

$$\text{controller output} = K_\text{p}e + K_\text{d}\frac{de}{dt}$$

In terms of the Laplace transform we have:

$$\text{controller output } (s) = (K_\text{p} + K_\text{d}s)E(s)$$

This can be written as:

$$\boxed{\text{controller output } (s) = K_\text{P}(1 + T_\text{d}s)E(s)}$$

where $T_\text{d} = K_\text{d}/K_\text{P}$ and is called the *derivative time constant*.

Derivative control has the controller effectively anticipating the way an error signal is growing and responding as the error signal begins to change. A problem with this is that noise can lead to quite large responses. Adding derivative control to proportional control still leaves the output steady-state offset. Changing the amount of derivative control in a closed-loop system will change the damping ratio since increasing K_D increases the damping ratio. Figure 13.4 illustrates this.

With PI control, the proportional element is augmented with an additional element (Figure 13.5) which gives an output proportional to the *integral* of the error with time. The proportional element has an input of the error e and an output of $K_\text{p}e$. The integral element has an input of e and an output which is proportional to the integral of the error with time, i.e.

$$K_\text{i} \int e \, dt$$

where K_i is the integrating gain. Thus the controller output is:

$$\text{controller output} = K_\text{p}e + K_\text{i} \int e \, dt$$

In terms of the Laplace transform:

$$\text{controller output } (s) = \left(K_\text{p} + \frac{K_\text{i}}{s}\right)E(s)$$

This can be written as:

$$\boxed{\text{controller output } (s) = \frac{K_\text{p}}{s}\left(s + \frac{1}{T_\text{i}}\right)E(s)}$$

where $T_\text{i} = K_\text{p}/K_\text{i}$ and is called the *integral time constant*.

The presence of integral control eliminates steady-state offsets and this is generally an important feature required in a control system.

Figure 13.6 shows the basic form of a PID, i.e. *three-term controller*. The controller output is:

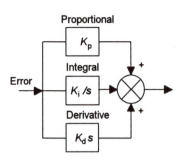

Figure 13.6 *PID controller*

$$\text{output} = K_{\text{p}}e + K_{\text{i}} \int e \, dt + K_{\text{d}} \frac{de}{dt}$$

Taking Laplace transforms gives:

$$\text{output }(s) = K_{\text{p}}\left(1 + \frac{K_{\text{i}}}{K_{\text{p}}s} + \frac{K_{\text{d}}}{K_{\text{p}}}s\right)E(s)$$

and so:

$$\text{output }(s) = K_{\text{p}}\left(1 + \frac{1}{T_{\text{i}}s} + T_{\text{d}}s\right)E(s)$$

Figure 13.7 illustrates the type of issues that determine whether P, PI or PID control is selected for a particular system, the choice generally being restricted to these three types of control law.

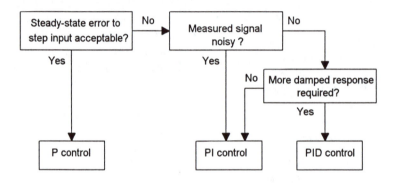

Figure 13.7 *Selection chart*

Example

Determine the open-loop transfer function of the system shown in Figure 13.8 if the controller to be used is PD and has a transfer function of $K_{\text{p}} + K_{\text{d}}s$ and $G(s) = \omega_n^2/(s^2 + 2\zeta\omega_n s)$.

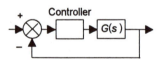

Figure 13.8 *Example*

The forward path gain, i.e. the open loop gain, is:

$$(K_{\text{p}} + K_{\text{d}}s)\left(\frac{\omega_n^2}{s^2 + 2\zeta\omega_n}\right)$$

Example

A proportional control system for angular motion of a shaft consists of a d.c. motor with transfer function $1/[s(s + 4)]$ and a negative feedback loop with a measurement system with a transfer function of 0.1. Determine the value of the proportional gain which will give a critically damped system.

The overall transfer function of the system is:

$$\frac{G(s)}{1+G(s)H(s)} = \frac{K/[s(s+4)]}{1+0.1K/[s(s+4)]}$$

This can be simplified to:

$$\frac{K}{s(s+4)+0.1K} = \frac{K}{s^2+4s+0.1K}$$

For a critically damped system we require the roots of the denominator to be real and equal. The roots are given by the usual equation for the roots of a quadratic equation as:

$$s = \frac{-4 \pm \sqrt{16-0.4K}}{2}$$

For the roots to be equal and real, we must have $0.4K = 16$ and so $K = 40$.

13.2.1 Proportional steady-state offset

The *steady-state offset* is the value of the error signal needed with a proportional control system to maintain the output at some value other than that it had when the error was zero. In order to see how the value of the proportional controller gain affects the steady-state offset, consider a closed-loop control system with a proportional controller of gain K_P, a process of transfer function $G(s)$ and a unity feedback loop (Figure 13.9), the output $Y(s)$ is related to the input $X(s)$ by:

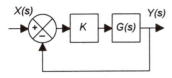

Figure 13.9   *Proportional control system*

$$Y(s) = \frac{G(s)K_P}{1+G(s)K_P}X(s)$$

A simple way of determining the value of the output as the time t tends to an infinite value, i.e. the steady-state value y_{ss} is by using the *final value theorem* (see Appendix C). This states that if a function of time has a Laplace transform $F(s)$ then in the limit as the time tends to infinity, the value of the function of time is given by the limit as s tends to zero of $sF(s)$. Thus, for the above we have:

$$y_{ss} = \lim_{s \to 0} s\frac{G(s)K_P}{1+G(s)K_P}X(s)$$

Suppose we have a unit step input, i.e. $X(s) = 1/s$, and a system with $G(s) = 1/(s+1)$. Then we have:

$$y_{ss} = \lim_{s \to 0} s\frac{K_P/(s+1)}{1+K_P/(s+1)}\frac{1}{s}$$

$$= \lim_{s \to 0} \frac{K_P}{s+1+K_P}$$

Thus, in the limit a s tends to zero, we have:

$$y_{ss} = K_P/(1 + K_P)$$

When $K_P = 0$ then the steady-state output is 1, when K_P is 1 it is 0.5, when $K_P = 4$ it is 0.8, when K_P is 10 it is 0.91 and so as K_P tends to ever higher values then so y_{ss} tends to 1. The steady-state offset is the difference between the input and the steady-state value and thus, for the unit step input, the offset when $K_P = 0$ is 1, when K_P is 1 it is 0.5, when $K_P = 4$ it is 0.2, when K_P is 10 it is 0.09 and so as K_P tends to ever higher values then so the offset tends to 0.

> Increasing the proportional gain decreases the steady-state offset.

13.2.2 Disturbance rejection

In Section 9.6.2 there was a consideration of the effect of a disturbance on the performance of a closed-loop control system and how it was better at minimising disturbances than an open-loop system. Now consider how the effect of a disturbance is affected by the gain of a proportional controller.

Consider the system shown in Figure 13.10 when there are two possible sources of disturbances, one being a disturbance affecting the input to the process and the other affecting its output.

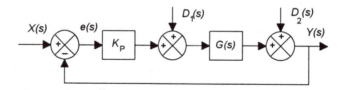

Figure 13.10 *A closed-loop control system with disturbances*

The output $Y(s)$ is $K_PG(s)e(s) + G(s)D_1(s) + D_2(s)$ with the error $e(s)$ being $X(s) - Y(s)$. Thus:

$$Y(s) = K_PG(s)[X(s) - Y(s)] + G(s)D_1(s) + D_2(s)$$

$$Y(s)[1 + K_PG(s)] = K_PG(s)X(s) + G(s)D_1(s) + D_2(s)$$

Hence:

$$Y(s) = \frac{K_PG(s)}{1+K_PG(s)}X(s) + \frac{G(s)}{1+K_PG(s)}D_1(s) + \frac{1}{1+K_PG(s)}D_2(s)$$

The first term is the normal expression for the closed-loop system with no disturbances. The other terms are the terms arising from the two disturbances. The factor $1/[1 + K_P G(s)]$ is thus a measure of how much the effects of the disturbances are modified by the closed-loop. Thus:

> Increasing the proportional gain reduces the effect of disturbances.

13.2.3 Integral wind-up

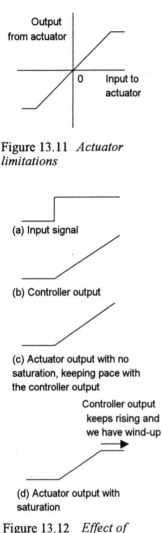

Figure 13.11 *Actuator limitations*

(a) Input signal

(b) Controller output

(c) Actuator output with no saturation, keeping pace with the controller output

Controller output keeps rising and we have wind-up

(d) Actuator output with saturation

Figure 13.12 *Effect of saturation on actuator output*

Real actuators can often cause problems as a consequence of their having input–output characteristics with limitations of the range of actuator output that is possible (Figure 13.11), this often being termed *saturation*. Thus a valve might have a linear characteristic between the fully open and fully closed positions but cannot operate for inputs which would endeavour to cause outputs outside this range. Such a limitation leads to an effect called *integral wind-up* and can result in a deterioration of control action with PI or PID control.

Consider a step input (Figure 13.12(a)) to a system having integral action, the output then being as in Figure 13.12(b). Figure 13.12(c) shows the initial response of a system with unrestricted integral action for a step input. Figure 13.12(d) shows how this response would be modified as a result of saturation. The actuator output rises to its saturation level where it becomes constant. The integrator is said to be *winding-up* during this saturation region because the controller output keeps on rising. The output from the controller drops when the error signal between the input and the fed-back value of the variable decreases as a result of the control action. However, because the controller output has been wound-up to a higher level than the actuator can respond to, it takes some time for the controller output drop to reach the point where the actuator output begins to drop. The integrator is said to be *unwinding* during the time is takes to reach the point where the output begins to drop.

The effect of winding-up and unwinding delays the action of the control signal. Anti-wind-up circuits are used to switch off and on the integral term when saturation of the actuator output occurs.

13.3 Frequency response

The frequency response of controllers can be represented by Bode and Nyquist plots. Figure 13.13 shows the basic forms of the Bode plots for P, PI and PD controllers.

1   *Proportional control* (Figure 13.13(a)) increases the overall system gain but does not affect the phase plot.

2   *PI control* (Figure 13.13(b)) decreases the high frequency gain of a system and thus decreases the phase margin. Because noise tends to be high frequency, PI control decreases the effect of noise on a system.

3   *PD control* (Figure 13.13(c)) decreases the low-frequency gain and reduces the high frequency gain, thus increasing the phase margin.

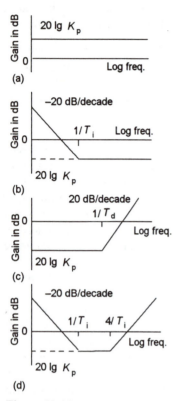

Figure 13.13 *Bode gain plot for (a) P, (b) PI, (c) PD,(d) PID*

Because noise tends to be high frequency, PD control thus tends to increase the effect of noise on a system.

4   There is no standard form of Bode plot for a *PID control* since the shape depends on the relative values of T_i and T_d. Generally the values are chosen to give $T_i = 4T_d$. The result is then as shown in Figure 13.13(d).

Example

Figure 13.14 shows a system with proportional gain of 12. What should the proportional gain be changed to if there is to be gain margin of 3 dB?

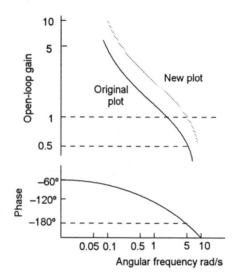

Figure 13.14 *Example*

Proportional gain will move the entire gain plot up or down by a constant amount but will not change the phase plot. The gain margin is the factor by which the gain must be multiplied at the phase crossover to have the value 1, the phase crossover being the frequency at which the phase reaches −180°. A 3 dB gain margin means the gain should be 0.5 at the phase crossover. Thus the new proportional gain for the system should be $0.5 \times 12 = 6$.

13.4 Systems with dead time

Section 5.1.2 included a preliminary discussion of dead time. Most systems have some *time delays* as a result of time being necessary for the transportation of mass or energy along paths and so have dead time. We can think of *dead time* as an element in a control system where the output signal is identical in form to the input signal except for a time delay. Figure 13.15 illustrates for when there is a step input to a dead time element.

Figure 13.16 shows how the magnitude of the output from a first-order system for a unit step input is affected by there being dead time.

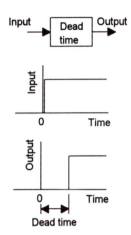

Figure 13.15 *Dead time*

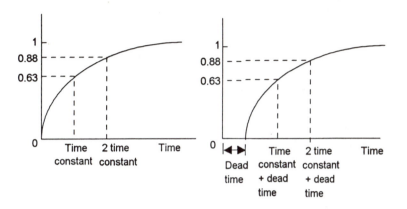

Figure 13.16 *The effect of dead time on the response of a first-order system to a step input, (a) being the output with no dead time and (b) with dead time*

Application

The process reaction controller tuning method of Ziegler and Nichols (Section 5.7.1) is based on a model of an open-loop system being a first-order system with a dead-time element.

Application

An example of a control system with a dead-time element is where a heater immersed in a liquid is controlled so that the liquid emerging from an outlet pipe is at the required temperature. The temperature is monitored in the exit pipe and is some distance from the heater element and so there is a delay as a result of this transport lag.

Another example of a transport lag is in the control system used to control the thickness of steel sheet in a steel rolling mill. There is some distance *d* between the rollers controlling the thickness and the point at which the thickness is monitored. If the sheet is moving at a velocity *v* then the lag time is *d*/*v*.

If a signal is delayed by a time T then its Laplace transform is multiplied by e^{-sT}.

Thus a unit step input with its Laplace transform of $1/s$ has a Laplace transform when delayed by T of $(1/s)e^{-sT}$. We can think of a dead-time element as thus having a transfer function of :

transfer function of dead time element $= e^{-sT}$

It will thus have a frequency response function of $e^{-j\omega T}$. Because a dead-time process delays a signal but does not change the size of a signal, it adds a phase shift to the frequency response of a system without altering the magnitude plot:

magnitude unchanged   added phase $\phi = -\omega T$

Figure 13.17 shows the Bode plot for a dead-time element. The magnitude is constant at 1 for all frequencies and the phase angle increases as the frequency increases. The example shows an element where the dead time has been taken as 20 s. Thus, at $\omega = 0.1$ rad/s we have $\phi = -0.1 \times 20 = -2$ rad or $-2(360/2\pi) = -115°$ and at $\omega = 1$ rad/s we have $\phi = -1 \times 20 = -20$ rad or $-20(360/2\pi) = -1146°$. The phase lag becomes very large as the frequency increases.

Figure 13.18 is a Bode plot which illustrates the effect of a dead time on the performance of a system.

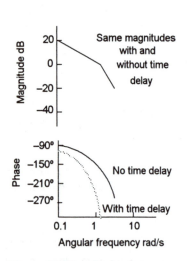

Figure 13.18   *Bode plot showing effect of time delay*

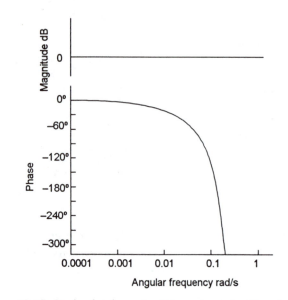

Figure 13.17 *Bode plot for a dead-time element with a dead time of 20 s*

In designing a control system, wherever possible, time delays should be avoided. Thus, feedback paths should have the sensor placed as close as possible to the source of the variable being controlled.

Example

A hopper feeds salt into water flowing along a pipe, the concentration being monitored at a point 2 m from the hopper feed. If the fluid velocity is 1.2 m/s, what will be the dead-time lag and the transfer function of the dead-time element?

The dead-time lag = d/v = 2/1.2 = 1.7 s and the dead-time element will have a transfer function of $e^{-1.7s}$.

Example

A control system has a proportional controller with a gain of 5, this controlling an actuator with a transfer function of $4/(s + 1)$ and a dead time of 2 s. The process has a transfer function of $20/(7s + 1)$ and the negative feedback path a transfer function of $1/(s^2 + 2s + 5)$. Determine the forward path transfer function of the system.

The forward path transfer function is the product of the transfer functions of the controller, the actuator, the dead-time element and the process and is thus:

$$5 \times \frac{4}{s+1} \times e^{-2s} \times \frac{20}{7s+1}$$

13.5 Cascade control

Cascade control involves the use of two controllers with the output of the first controller providing the setpoint for the second controller, the feedback loop for one controller nestling inside the other (Figure 13.19). Such a system can give a improved response to disturbances.

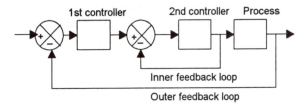

Figure 13.19 *Cascade control*

Figure 13.20 illustrates how cascade control might be used in controlling t27 March 2004he level of a liquid in a container. With the single loop control system, the level sensor provides the feedback which gives the error signal to the controller and hence initiates the controlling action on the rate at which fluid enters the container. We have a situation where changes in the level occur rather slowly, because of the large cross-sectional area of the container. Changes in flow can, however, occur very rapidly. Thus if we have a disturbance which causes a change in the input flow rate, there can be considerable delay before the level changes enough to initiate a correction. With the cascade system, the level sensor provides the feedback to the outer loop controller which then gives an output which provides the set point input to the second controller, this being used to control the rate of flow of the liquid. The flow level loop responds quickly to flow disturbances and so considerably reduces the level fluctuations that would have occurred with the single loop control system.

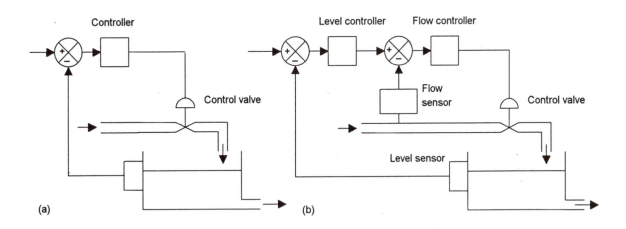

Figure 13.20 *Level control system: (a) single loop, (b) cascade control with a flow control loop inside a level control loop*

13.6 Feedforward control   In the basic closed-loop control system, a disturbance has to propagate through the process and show up as an error signal input to the controller before action can be taken to correct it. This means that there will be some deviation from the set point while this propagation through the system takes place and the control action is taken. With some control systems, this is unacceptable. To overcome this problem, feedforward control is used. With this form of control, the disturbance is measured and a signal added to the controller output to enhance the control law. Thus corrective action is initiated without waiting for the effect of the disturbance to show up in the error signal input to the controller. Figure 13.21 shows such a feedforward system.

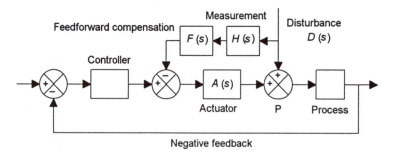

Figure 13.21 *Feedforward control*

At the summation P we have the disturbance input $D(s)$ which we want to cancel out by the feedforward loop signal. The feedforward loop has an input of $D(s)$ and reaches P via $H(s)$, $F(s)$ and $A(s)$. Thus, for cancellation we must have:

$$H(s)F(s)A(s)D(s) + D(s) = 0$$

Suppose we have the transfer function of the disturbance measurement system element as 1, then:

$$F(s) = -\frac{1}{A(s)}$$

If we have a first-order actuator with a lagging transfer function of $1/(\tau s + 1)$ (see Section 10.4), then the feedforward compensation element must have a leading transfer function of $(\tau s + 1)$. A PD controller element has a transfer function of this form and so a derivative time of τ is required.

Problems   *Questions 1 to 6 have four answer options: A, B, C or D. Choose the correct answer from the answer options.*

1 Decide whether each of these statements is True (T) or False (F).

Increasing the gain of a proportional control system will:

(i) Increase the offset error.
(ii) Increase the disturbance rejection.

A   (i) T (ii) T
B   (i) T (ii) F
C   (i) F (ii) T
D   (i) F (ii) F

2 Decide whether each of these statements is True (T) or False (F).

Adding derivative action to a proportional control system will:
(i) Give a faster response to an error signal.
(ii) Eliminate the proportional offset error.

A   (i) T (ii) T
B   (i) T (ii) F
C   (i) F (ii) T
D   (i) F (ii) F

3 Decide whether each of these statements is True (T) or False (F).

Adding integral action to a proportional control system will:
(i) Introduce a $1/s$ term into the Laplace transform of the output.
(ii) Eliminate the proportional offset error.

A   (i) T (ii) T
B   (i) T (ii) F
C   (i) F (ii) T
D   (i) F (ii) F

4 Decide whether each of these statements is True (T) or False (F).

In a control system, a dead-time element with a dead time of 5 s
will have a Bode plot which has:
(i) A constant magnitude of 0 dB.
(ii) A phase lag of $-573\omega°$.

A   (i) T (ii) T
B   (i) T (ii) F
C   (i) F (ii) T
D   (i) F (ii) F

5 The steady-state error for a step input with a proportional control
system is, when steady-state conditions occur:

A   The magnitude of the final output
B   The error input to the controller
C   The initial size of the step input
D   The difference between the final output and the step input.

6 A proportional controller of gain K is used with a system with a
transfer function $4/(2s + 1)$ and a unity negative feedback loop. The
closed-loop transfer function is:

A $4K/(2s + 1)$
B $4K/(2s + 5)$
C $K(2s + 1)/4$
D $4K (2s - 3)$

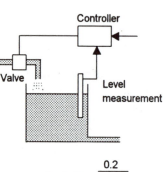

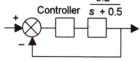

Figure 13.22   *Problem 9*

7 A closed-loop control system has a PID controller with transfer function $K_p + (K_i/s) + K_ds$ and is used with a process having a transfer function of $10/(s + 5)(s + 10)$. If the system has unity feedback, what is the transfer function of the closed-loop system?

8 A closed-loop control system has unity feedback and a plant with transfer function $100/[s(s + 0.1s)(1 + 0.2s)]$. By drawing the Bode diagrams, determine the phase margin when the following controllers are used: (a) a proportional controller with transfer function 1, (b) a PD controller with transfer function $1 + 0.5s$.

9 Figure 13.22 shows a liquid level control system and its representation by a block diagram. Determine the way the output will vary with time if the controller is (a) proportional only with a proportional gain of 2, (b) integral only with an integral gain of 2.

10 For the system shown in Figure 13.22, what will be the steady-state error for a unit step input if the controller is (a) only proportional with a gain 1, (b) PI with a gain 1?

Appendix A
Errors

Measurement errors

In Section 1.3 the idea of accuracy and errors associated with instrumentation were discussed. Here we discus other sources of error and how all such errors affect the accuracy of measurements.

When a physical quantity is measured, the value obtained should not be expected to be exactly what the quantity actually is. With every measured quantity there is associated some error. These errors can arise from:

1 *Instrumentation errors*

Such errors are an inherent feature of an instrument and can arise in the manufacture of the instrument from such causes as tolerances on the dimensions of mechanical components and the values of electrical components used in the construction of the instrument. In addition there can be errors due to other factors such as the accuracy with which the instrument has been calibrated or readings being taken under different conditions to which the instrument was calibrated, e.g. at a different temperature, non-linearity and hysteresis.

2 *Reading errors*

Such errors arise due to the limited accuracy with which scales can be read. When the pointer of an analogue instrument falls between two scale markings (Figure ApA.1) there is some uncertainty as to what the reading should be quoted as. Thus analogue instrument readings should not be quoted as precise numbers but some indication given of the uncertainty with which a reading is made and hence the extent to which the reading could be in error. The worse the reading error could be is that the value indicated by a pointer is somewhere between two successive markings on the scale. In such circumstances the reading error can be quoted as ± half the scale interval. Thus with a rule with scale readings every millimetre, the reading error might be quoted as ± 0.5 mm. Thus when using such a rule to make a measurement, if the nearest mark on the rule is, say, 65 mm, then the reading would be quoted as 65 ± 0.5 mm.

With digital displays there is no uncertainty regarding the value displayed but there is still an error associated with a reading. This is because the reading of a digital instrument goes in jumps, a whole digit at a time. It is not possible to tell where the value is between two successive digits. Thus the reading error can be quoted as ± the

Figure ApA.1 *Reading error*

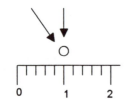

Figure ApA.2 *The effect of parallax: viewed directly from above the reading is 1.0 while viewed at an angle to the vertical the reading can be 1.2*

smallest digit. For example a digital display giving a reading of, say, 6789 would have an error of ±1 and be quoted as 6789 ± 1.

3. *Human errors*

These include errors such as misreading of the position of a pointer on a scale or, where there are multiple scales, the misreading of scales. If the scale and the pointer of an instrument are not in the same plane (Figure ApA.2), then the reading obtained depends on the angle at which the pointer is viewed against the scale; such errors are termed *parallax errors*. To reduce the chance of such errors occurring, some instruments incorporate a mirror alongside the scale so that the scale is read when the pointer and its image are superimposed, thus ensuring that the pointer is being viewed at right angles to the scale. Other potential sources of human errors are the varying reaction times of individuals in timing events or applying varying pressures when using a micrometer screw gauge.

4   *Insertion errors*

In some measurements, the insertion of the instrument into the position to measure a quantity can affect the value obtained (see Section 1.1.3). For example, inserting an ammeter into a circuit to measure the current can change the value of the current in the circuit due to the ammeter's own resistance. Similarly, putting a cold thermometer into a hot liquid to measure its temperature can cool the liquid and so change the temperature being measured.

Random errors

All errors, whatever their source, can be described as being either random or systematic. *Random errors* are ones which can vary in a random manner between successive readings of the same quantity. These may be due to personal errors by the person making the measurements or perhaps due to random electronic fluctuations (termed noise) in the instruments or circuits used, or perhaps varying frictional effects. *Systematic errors* are errors which do not vary from one reading to another. These may be due to some defect in the instrument such as a wrongly set zero so that it always gives a high or low reading, or perhaps incorrect calibration, or perhaps an instrument is temperature dependent and the measurement is made under conditions which differ from those for which it was calibrated, or there is an insertion error.

> Random errors can be minimised by taking a number of readings and obtaining a mean value, systematic errors require the use of a different instrument or measurement technique to establish them.

Mean values

Random errors give a reading that is sometimes too high, sometimes too low. The mean (average) value of a set of results will thus give a more accurate value than just a single value. The *mean* or *average* of a set of readings is given by:

$$\bar{x} = \frac{x_1 + x_2 + \dots x_n}{n}$$

where x_1 is the first reading, x_2 the second reading, ... x_n the nth reading. The more readings we take the more likely it will be that we can cancel out the random variations that occur between readings.

> The *true value* of a measurement might be regarded as the value given by the mean of a very large number of readings of the particular variable concerned.

Standard deviation

Consider the two following sets of time readings:

20.1, 20.0, 20.2, 20.1, 20.1 and 19.5, 20.5, 19.7, 20.6, 20.2 s

Both sets of readings have the same average of 20.1 s, but the second set of readings is more spread out than the first and thus shows more random fluctuations. The *deviation* of any one reading from the mean is the difference between its value and the mean value. Thus the first set of readings has deviations of 0.0, –0.1, +0.1, 0.0 and 0.0; the second set has deviations of –0.6, +0.4, –0.4, +0.5 and + 0.1. The second set of readings has greater deviations than the first set and thus if we had only considered one reading of the less spread out set of readings it would have had a greater chance of being closer to the mean value than any one reading in the more spread out set.

> The spread of a set of readings is taken as a measure of the certainty we can attach to any one reading being close to the mean value, the bigger the spread the greater the uncertainty. The spread of the readings is specified by a quantity termed the *standard deviation*.

The standard deviation is given by:

$$\text{standard deviation} = \sqrt{\frac{(d_1^2 + d_2^2 + \dots d_n^2)}{n - 1}}$$

where d_1 is the deviation of the first result from its average, d_2 the deviation of the second reading, ... d_n the deviation of the nth reading from the average. Note that sometimes the above equation is written with just n instead of $(n - 1)$ on the bottom line. With just n it is assumed that the deviations are all from the true value, i.e. the mean when there are very large numbers of readings. With smaller numbers of readings, the deviations are taken from the mean value without assuming that it is necessarily the true value. To allow for this, $(n-1)$ is used. In fact, with more than a very few readings, the results using n and $(n-1)$ are the same, to the accuracy with which the standard deviations are usually quoted. Table ApA.1 illustrates the calculation of the standard deviation for the two sets of results.

Table ApA. 1 *Calculation of standard deviations*

1st set of readings

Reading (s)	Deviation (s)	(Deviation)^2 s^2
20.1	0.0	0.00
20.0	−0.1	0.01
20.2	+0.1	0.01
20.1	0.0	0.00
20.1	0.0	0.00

Sum of (deviation)^2 = 0.02, hence standard deviation = $\sqrt{(0.02/4)}$ = 0.07.

2nd set of readings

Reading (s)	Deviation (s)	(Deviation)^2 s^2
19.5	−0.6	0.36
20.5	+0.4	0.16
19.7	−0.4	0.16
20.6	+0.5	0.25
20.2	+0.1	0.01

Sum of (deviation)^2 = 0.94, hence standard deviation = $\sqrt{(0.94/4)}$ = 0.48.

The first set of readings has a standard deviation of 0.07 and the second set 0.48. The second set of readings has thus a much greater standard deviation than the first set, indicating the greater spread of those results. A consideration of the statistics involved shows that we can reasonably expect about 68.3% of the readings will lie within plus or minus one standard deviation of the mean, 95.45% within plus or minus two standard deviations and 99.7% within plus or minus three standard deviations.

Error of a mean

If we take a set of readings and obtain a mean, how far from the true value might we expect the mean to be? Essentially what we do is consider the mean value of our set of results to be one of the many mean values which can be obtained from the very large number of results and calculate its standard deviation from that true value mean. To avoid confusion with the standard deviation of a single result from a mean of a set of results, we use the term *standard error* for the standard deviation of the mean from the true value.

> The extent to which we might expect the mean of a set of readings to depart from the true value is given by the *standard error of the mean*.

The standard error is given by:

$$\text{standard error} = \frac{\text{standard deviation of the set of results}}{\sqrt{n}}$$

We can reasonably expect that there is a 68.3% chance that a particular mean value lies within plus or minus one standard error of the true value, a 95.45% chance within plus or minus two standard errors and a 99.7% chance within plus or minus three standard errors. The greater the number of measurements made, the smaller will be the standard error. Note that such a reduction in error is only a reduction in the random error, there may still be an unaffected systematic error.

Thus, for the data in Table ApA.1, the first set of measurements had a standard deviation of 0.07 s and so a standard error of $0.07/\sqrt{5} = 0.03$ s. Thus, with the mean value of 20.1 s, the chance of the true value being within ±0.03 s of 20.1 is about 68%. The chance of the true value being within ±0.06 s is about 95%. Mean values are generally quoted with the 95% chance and thus would be quoted, for the first set of measurements, as 20.1 ± 0.06 s. For the second set of measurements, the standard deviation was 0.48 s and so there is a standard error of $0.48/\sqrt{5} = 0.21$ s. The mean is 20.1 s and the chance of the true value being within ±0.21 s is 68%, within ±0.42 s about 95%. The two sets of measurements are thus likely to be written as: 20.1 ± 0.06 s and 20.1 ± 0.42 s.

Combination of errors The determination of the value of some quantity might require several measurements to be made and their values inserted into an equation. For example, in a determination of the density ρ of a solid, measurements might be made of the mass m of the body and its volume V and the density calculated from m/V. The mass and volume measurements will each have errors associated with them. How then do we determine the consequential error in the density? This type of problem is very common. The following illustrates how we can determine the worst possible error in such situations.

Errors when adding quantities

Consider the calculation of the quantity Z from two measured quantities A and B where $Z = A + B$. If the measured quantity A has an error $\pm\Delta A$ and the quantity B an error $\pm\Delta B$ then the worst possible error we could have in Z is if the quantities are at the extremes of their error bands and the two errors ΔZ are both positive or both negative. Then we have:

$$Z + \Delta Z = (A + \Delta A) + (B + \Delta B)$$

$$Z - \Delta Z = (A - \Delta A) + (B - \Delta B)$$

Subtracting one equation from the other gives the worst possible error as:

$$\Delta Z = \Delta A + \Delta B$$

> When we add two measured quantities the worst possible error in the calculated quantity is the sum of the errors in the measured quantities.

Errors when subtracting quantities

If we have the calculated quantity Z as the difference between two measured quantities, i.e. $Z = A - B$, then, in a similar way, we can show that the worst possible error is given by

$$Z + \Delta Z = (A + \Delta A) - (B + \Delta B)$$

$$Z - \Delta Z = (A - \Delta A) - (B - \Delta B)$$

and so subtracting the two equations gives the worst possible error as:

$$\Delta Z = \Delta A + \Delta B$$

> When we subtract two measured quantities the worst possible error in the calculated quantity is the sum of the errors in the measured quantities.

Errors when multiplying quantities

If we have the calculated quantity Z as the product of two measured quantities A and B, i.e. $Z = AB$, then we can calculate the worst error in Z as being when the quantities are both at the extremes of their error bands and the errors in A and B are both positive or both negative:

$$Z + \Delta Z = (A + \Delta A)(B + \Delta B) = AB + B\Delta A + A\Delta B + \Delta A\Delta B$$

The errors in A and B are small in comparison with the values of A and B so we can neglect the quantity $\Delta A\Delta B$ as being insignificant. Then:

$$\Delta Z = B\Delta A + A\Delta B$$

Dividing through by Z gives:

$$\frac{\Delta Z}{Z} = \frac{B\Delta A + A\Delta B}{Z} = \frac{B\Delta A + A\Delta B}{AB} = \frac{\Delta A}{A} + \frac{\Delta B}{B}$$

Thus, when we have the product of measured quantities, the worst possible fractional error in the calculated quantity is the sum of the fractional errors in the measured quantities. If we multiply the above equation by 100 then:

> The percentage error in the product of two measurements is equal to the sum of the percentage errors in each of the measured quantities.

If we have the square of a measured quantity, then all we have is the quantity multiplied by itself and so the error in the squared quantity is just twice that in the measured quantity. If the quantity is cubed then the error is three times that in the measured quantity.

Errors when dividing quantities

If the calculated quantity is obtained by dividing one measured quantity by another, i.e. $Z = A/B$, then the worst possible error is given when we have the quantities at the extremes of their error bands and the error in A positive and the error in B negative, or vice versa. Then:

$$Z + \Delta Z = \frac{A + \Delta A}{B - \Delta B} = \frac{A\left(1 + \frac{\Delta A}{A}\right)}{B\left(1 - \frac{\Delta B}{B}\right)}$$

Using the binomial series we can write this as:

$$Z + \Delta Z = \frac{A}{B}\left(1 + \frac{\Delta A}{A}\right)\left(1 + \frac{\Delta B}{B} - \ldots\right)$$

Neglecting products of ΔA and ΔB and writing A/B as Z, gives:

$$Z + \Delta Z = Z\left(1 + \frac{\Delta A}{A} + \frac{\Delta B}{B}\right)$$

Hence:

$$\frac{\Delta Z}{Z} = \frac{\Delta A}{A} + \frac{\Delta B}{B}$$

The worst possible fractional error in the calculated quantity is the sum of the fractional errors in the measured quantities or, if expressed in percentages:

> The percentage error in the result of the division of two measurements is equal to the sum of the percentage errors in each of the measured quantities.

Example of error calculation

The above rules for determining errors when measurements are combined can be summed up:

1 When measurements are added or subtracted, the resulting worst error is the sum of the errors.

2 When measurements are multiplied or divided, the resulting worst percentage error is the sum of the percentage errors.

The following examples illustrate the use of the above to determine the worst errors.

Example

The distance between two points is determined from the difference between two length measurements. If these are 120 ± 0.5 mm and 230 ± 0.5 mm, what will be the error in the distance?

Adding the errors gives the difference as 110 ± 1.0 mm.

Example

The resistance R of a resistor is determined from measurements of the potential difference V across it and the current I through it, the resistance being given by V/I. The potential difference has been measured as 2.1 ± 0.2 V and the current measured as 0.25 ± 0.01 A. What will be the error in the resistance?

The percentage error in the voltage reading is $(0.2/2.1) \times 100\% = 9.5\%$ and in the current reading is $(0.01/0.25) \times 100\% = 4.0\%$. Thus the percentage error in the resistance is $9.5 + 4.0 = 13.5\%$. Since we have $V/I = 8.4 \ \Omega$ and 13.5% of 8.4 is 1.1, then the resistance is $8.4 \pm 1.1 \ \Omega$.

Example

The cross-sectional area A of a wire is to be determined from a measurement of the diameter d, being given by $A = \frac{1}{4}\pi d^2$. The diameter is measured as 2.5 ± 0.1 mm. What will be the error in the area?

The percentage error in d^2 will be twice the percentage error in d. Since the percentage error in d is $\pm 4\%$ then the percentage error in d^2, and hence A since the others are pure numbers, is $\pm 8\%$. Since $\frac{1}{4}\pi d^2 = 4.9$ mm^2 and 8% of this value is 0.4 mm^2, the result can be quoted as 4.9 ± 0.4 mm^2.

Appendix B
Differential equations

Differential equations

As will be evident from the models discussed in Chapter 8, many systems have input–output relationships which have to be described by differential equations.

> A *differential equation* is an equation involving derivatives of a function, i.e. terms such as dy/dt and d^2y/dt^2.

Thus:

$$\tau \frac{dy}{dt} + y = 0$$

is a differential equation. The term *ordinary differential equation* is used when there are only derivatives of one variable, e.g. we only have terms such as dy/dt and d^2y/dt^2 and not additionally dx/dt or dx^2/dt^2.

> The *order* of a differential equation is equal to the order of the highest derivative that appears in the equation.

For example,

$$\tau \frac{dy}{dt} + y = 0$$

and

$$\tau \frac{dy}{dt} + y = kx$$

are first-order ordinary differential equations since the highest derivative is dy/dt and there are derivatives of only one variable. The first of the above two equations is said to be *homogeneous* since it only contains terms involving y. Such an equation is given by a system which has no forcing input; one with a forcing input gives a non-homogeneous equation. For example, an electrical circuit containing just a charged capacitor in series with a resistor will give a homogeneous differential equation describing how the potential difference across the capacitor changes with time as the charge leaks off the capacitor; there is no external source of voltage. However, if we have a voltage source which is switched into the circuit with the capacitor and resistor then the voltage source gives a forcing input and a non-homogeneous equation results.

The equations:

$$m\frac{d^2y}{dt^2} + c\frac{dy}{dt} + ky = 0$$

and

$$m\frac{d^2y}{dt^2} + c\frac{dy}{dt} + ky = F$$

are examples of a second-order differential equation since the highest derivative is d^2y/dt^2. The first of the two is homogeneous since there is no forcing input and the second is non-homogeneous with a forcing input F.

Solving differential equations

The following is a brief discussion of the basic methods used to solve first-order and second-order ordinary differential equations.

Solving a first-order differential equation

With a first-order differential equation, if the variables are separable, i.e. for variable y and t, it is of the form:

$$g(y)\frac{dy}{dt} = f(t)$$

then we can solve such equations by integrating both sides of the equation with respect to x:

$$\int g(y)\frac{dy}{dt}\, dt = \int f(t)\, dt$$

This is equivalent to separating the variables and writing:

$$\int g(y)\, dy = \int f(t)\, dt$$

> A first-order separable differential equation can be solved by separating the terms involving the variables so that on one side of the equals sign are the terms involving one variable, e.g. y, and on the other side the other variable e.g. t, then integrate both sides with respect to their variables.

Consider the response of a first-order system to a step input, e.g. the response of a thermometer when inserted suddenly into a hot liquid. This sudden change is an example of a step input. In Chapter 8 the differential equation for such a change was determined as:

$$RC\frac{dT}{dt} + T = T_L$$

where T is the temperature indicated by the thermometer, T_L the temperature of the hot liquid, R the thermal resistance and C the thermal capacitance. We can solve such an equation by the technique of 'separation of variables'. Separating the variables gives:

$$\frac{1}{T_L - T}\, dT = \frac{1}{RC}\, dt$$

Integrating then gives:

$$-\ln (T_L - T) = (1/RC)t + A$$

where A is a constant. This equation can be written as:

$$T_L - T = e^A\, e^{-t/\tau} = B\, e^{-t/\tau}$$

with B being a constant and $\tau = RC$. τ is termed the *time constant* and can be considered to be the time that makes the exponential term e^{-1}. If we consider the thermometer to have been inserted into the hot liquid at time $t = 0$ and to have been indicating the temperature T_0 at that time, then, since $e^0 = 1$, we have $B = T_L - T$. Hence the equation can be written as:

$$T = (T_0 - T_L)\, e^{-t/\tau} + T_L$$

The exponential term will die away as t increases and so gives the transient part of the response. T_L is the steady-state value that will be attained eventually.

Complementary function and particular integral

Suppose we have the first-order differential equation $dy/dt + y = 0$. Such an equation is homogeneous and, if we apply the technique of separation of the variables, has the solution $y = C\, e^{-t}$. Now suppose we have the non-homogeneous equation $dy/dt + y = 2$. If we now use the separation of variable technique we obtain the solution $y = C\, e^{-t} + 2$. Thus, its solution is the sum of the solution for the homogeneous equation plus another term. The solution of the homogeneous differential equation is called the *complementary function* and, when there is a forcing input with that differential equation, the term added to it for the non-homogeneous solution is called the *particular integral*. We can, easily, obtain a particular integral by assuming it will be of the same form as the forcing input. Thus if this is a constant then we try $y = A$, if of the form $a + bx + cx^2 + ...$ then $y = A + Bx + Cx^2 + ...$ is tried, if an exponential then $y = A\, e^{kx}$ is tried, if a sine or cosine then $y = A \sin \omega x + B \cos \omega x$ is tried.

> Determining the general solution of a non-homogeneous differential equation can be reduced to the problem of finding any solution of the non-homogeneous equation, i.e. the particular integral, and adding to it the general solution of the equivalent homogeneous equation, i.e. the complementary function.

As an illustration, consider the system solved in the previous section of a thermometer being inserted into a hot liquid and the relationship being given by:

$$RC\frac{dT}{dt} + T = T_L$$

The homogeneous form of this equation is:

$$RC\frac{dT}{dt} + T = 0$$

We can solve this equation by the technique of the separation of variables to give the complementary function:

$$\frac{1}{T}\,dT = -\frac{1}{RC}\,dt$$

Integrating then gives:

$$-\ln T = -(1/RC)t + A$$

where A is a constant. This equation can be written as:

$$-T = e^A\,e^{-t/\tau} = B\,e^{-t/\tau}$$

$$T = -B\,e^{-t/\tau}$$

Now consider the particular integral for the non-homogeneous equation and, because the forcing input is a constant, we try $T = C$. Since $dT/dt = 0$ then the substituting values in the differential equation gives:

$$0 + C = T_L$$

Thus the particular solution is $T = T_L$. Hence the full solution is the sum of the complementary solution and the particular integral and so:

$$T = -B\,e^{-t/\tau} + T_L$$

If we consider the thermometer to have been inserted into the hot liquid at time $t = 0$ and to have been indicating the temperature T_0 at that time, then, since $e^0 = 1$, we have $B = T_L - T$. Hence, as before, the equation can be written as:

$$T = (T_0 - T_L) \, e^{-t/\tau} + T_L$$

As a further illustration, consider the thermometer, in equilibrium in a liquid at temperature T_0, when the temperature of the liquid is increased at a constant rate a, i.e. a so-called ramp input. The temperature will vary with time so that after a time t it is $at + T_0$. This is then the forcing input to the system and so we have the differential equation:

$$RC\frac{dT}{dt} + T = at + T_L$$

The homogeneous form of this equation is:

$$RC\frac{dT}{dt} + T = 0$$

and, as before, the complementary solution is:

$$T = -B \, e^{-t/\tau}$$

For the particular integral we try a solution of the form $T = D + Et$. Since $dT/dt = E$, substituting in the non-homogeneous differential equation gives:

$$RCE + D + Et = aT + T_0$$

Equating all the coefficients of the t terms gives $E = a$ and equating all the non-t terms gives $RCE + D = T_0$ and so $D = T_0 - RCa$. Thus the particular integral is:

$$T = T_0 - RCa + at$$

and so the full solution, with $\tau = RC$, is:

$$T = -B \, e^{-t/\tau} + T_0 - \tau a + at$$

When $t = 0$ we have $T = T_0$ and so $B = -\tau a$. Hence we have:

$$T = \tau a \, e^{-t/\tau} + T_0 - \tau a + at$$

Solving a second-order differential equation

We can use the technique of finding the complementary function and the particular integral to obtain the solution of a second-order differential equation. Consider a spring–damper–mass system (see Section 8.3.1) which gives the differential equation for the displacement y of the mass when subject to step input at time $t = 0$ of a force F as:

$$m\frac{d^2y}{dt^2} + c\frac{dy}{dt} + ky = F$$

In the absence of damping and the force F we have the homogeneous differential equation :

$$m\frac{d^2y}{dt^2} + ky = 0$$

This describes an oscillation which has an acceleration d^2y/dt^2 which is proportional to $-y$ and is a description of simple harmonic motion; we have a mass on a spring allowed to freely oscillate without any damping. Simple harmonic motion has a displacement $y = A \sin \omega t$. If we substitute this into the differential equation we obtain:

$$-mA\omega^2 \sin \omega t + kA\omega \sin \omega t = 0$$

and so $\omega = \sqrt{(k/m)}$. This is termed the *natural angular frequency* ω_n. If we define a constant called the *damping ratio* of:

$$\zeta = \frac{c}{2\sqrt{mk}}$$

then we can write the differential equation as:

$$\frac{1}{\omega_n^2}\frac{d^2y}{dt^2} + \frac{2\zeta}{\omega_n}\frac{dy}{dt} + y = \frac{F}{k}$$

This differential equation can be solved by the method of determining the complementary function and the particular integral. For the homogeneous form of the differential equation, i.e. the equation with zero input, we have:

$$\frac{1}{\omega_n^2}\frac{d^2y}{dt^2} + \frac{2\zeta}{\omega_n}\frac{dy}{dt} + y = 0$$

We can try a solution of the form $y = A\,e^{st}$. This, when substituted, gives:

$$\frac{1}{\omega_n^2}s^2 + \frac{2\zeta}{\omega_n}s + 1 = 0$$

$$s^2 + 2\omega_n\zeta s + \omega_n^2 = 0$$

This equation has the roots:

$$s = \frac{-2\omega_n\zeta \pm \sqrt{4\omega_n^2\zeta^2 - 4\omega_n^2}}{2} = -\omega_n\zeta \pm \omega_n\sqrt{\zeta^2 - 1}$$

When we have:

1   *Damping ratio between 0 and 1*
    There are two complex roots:

$$s = -\zeta\omega_n \pm j\omega_n\sqrt{1-\zeta^2}$$

If we let:

$$\omega = \omega_n\sqrt{1-\zeta^2}$$

we can write:

$$s = -\zeta\omega_n \pm j\omega$$

Thus:

$$y = A\,e^{(-\zeta\omega_n + j\omega)t} + B\,e^{-\zeta(\omega_n - j\omega)t}$$

$$= e^{-\zeta\omega_n t}(A\,e^{j\psi t} + B\,e^{-j\omega t})$$

Using Euler's equation, we can write this as:

$$y = e^{-\zeta\omega_n t}(P\cos\omega t + Q\sin\omega t)$$

This can be written in an alternative form. If we consider a right-angled triangle with angle ϕ with P and Q being opposite sides of the triangle (Figure ApB.1) then $\sin\phi = P/\sqrt{(P^2 + Q^2)}$ and $\cos\phi = P/\sqrt{(P^2 + Q^2)}$. Hence, using the relationship $\sin(\omega t + \phi) = \sin\omega t\cos\phi + \cos\omega t\sin\phi$, we can write:

$$y = C\,e^{-\zeta\omega_n t}\sin(\omega t + \phi)$$

where C is a constant and ϕ a phase difference. This describes a damped sinusoidal oscillation. Such a motion is said to be *under damped*.

Figure ApB.1   *Angle ϕ*

2 *Damping ratio equal to 1*
   This gives two equal roots $s_1 = s_2 = -\omega_n$. and the solution:

$$y = (At + B)\,e^{-\omega_n t}$$

where A and B are constants. This describes an exponential decay with no oscillations. Such a motion is said to be *critically damped*.

3 *Damping ratio greater than 1*
   This gives two real roots:

$$s_1 = -\omega_n\zeta + \omega_n\sqrt{\zeta^2 - 1}$$

$$s_2 = -\omega_n\zeta - \omega_n\sqrt{\zeta^2 - 1}$$

and hence:

$$y = A\,e^{s_1 t} + B\,e^{s_2 t}$$

where A and B are constants. This describes an exponential decay taking longer to reach the steady-state value than the critically damped case. It is said to be *over damped*.

The above analysis has given the complementary functions for the second-order differential equation. For the particular integral, in this case where we have a step input of size F, we can try the particular integral $x = A$. Substituting this in the differential equation gives $A = F/k$ and thus the particular integral is $y = F/k$. Thus the solutions to the differential equation are:

1 *Damping ratio between 0 and 1, i.e. under damped*

$$y = C\, e^{-\zeta\omega_n t} \sin(\omega t + \phi) + F/k$$

2 *Damping ratio equal to 1, i.e. critically damped*

$$y = (At + B)\, e^{-\omega_n t} + F/k$$

3 *Damping ratio greater than 1, i.e. over damped*

$$y = A\, e^{s_1 t} + B\, e^{s_2 t} + F/k$$

In all cases, as t tends to infinite then y tends to the value F/k. Thus the steady-state value is F/k.

Appendix C
Laplace transform

The Laplace transform

A quantity which is a function of time can be represented as $f(t)$ and is said to be in the *time domain*, e.g. if we have a voltage v which is a function of time we can write as $v(t)$ to show that it is. In discussing control systems we are only concerned with values of time greater than or equal to 0, i.e. $t \geq 0$, and so to obtain the Laplace transform of $f(t)$ function we multiply it by e^{-st} and then integrate with respect to time from zero to infinity; s is a constant with the unit of 1/time. The result is then said to be in the *s-domain*. The Laplace transform of the function of time $f(t)$, which is written as $\mathcal{L}\{f(t)\}$, is thus:

$$\mathcal{L}\{f(t)\} = \int_0^\infty e^{-st} f(t)\, dt$$

A function of s is written as $F(s)$. It is usual to use a capital letter F for the Laplace transform and a lower-case letter f for the time-varying function $f(t)$. Thus:

$$\mathcal{L}\{f(t)\} = F(s)$$

The procedure for using the Laplace transform is to:

1   Transform functions of time into functions of s.

2   Carry out algebraic manipulations in the s-domain. We can carry out algebraic manipulations on a quantity in the s-domain, i.e. adding, subtracting, dividing and multiplying in the way we do with any algebraic quantities.

3   Transform the functions of s back into functions of time, i.e. the so-termed inverse operation. This involves finding the time domain function that could have given the s-domain expression obtained by the algebraic manipulations.

The *inverse Laplace transformation* is when the function of time is obtained from the Laplace transform.

> The inverse Laplace transformation is the conversion of a Laplace transform $F(s)$ into a function of time $f(t)$. This operation is written as $\mathcal{L}^{-1}\{F(s)\} = f(t)$

Obtaining the transform

The following examples illustrate how we can obtain, from first principles, the Laplace transform of functions of time. However, in practice, engineers use a table of Laplace transforms to avoid having to carry out such calculations; Table 10.1 shows some of the transformations commonly encountered.

The unit step is the type of function encountered when we suddenly apply an input to a system (Figure Ap.C.1). It describes an abrupt change in some quantity from zero to a steady value, e.g. the change in the voltage applied to a circuit when it is suddenly switched on. Thus, at time $t = 0$ it suddenly switches to the value 1 and has the constant value of 1 for all values of time greater than 0, i.e. $f(t) = 1$ for $t \geq 0$. The Laplace transform of the unit-step function is:

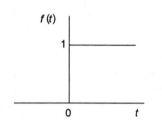

Figure Ap.C.1 *Unit-step function*

$$\mathcal{L}\{f(t)\} = F(s) = \int_0^\infty 1\, e^{-st}\, dt = -\frac{1}{s}[e^{-st}]_0^\infty$$

For $t = \infty$ the value of the exponential term $e^{-\infty}$ is 0 and with $t = 0$ the value of e^{-0} is -1. Hence:

$$F(s) = \frac{1}{s}$$

As a further illustration, consider the Laplace transform of the function $f(t) = e^{at}$, where a is a constant :

$$F(s) = \int_0^\infty e^{at} e^{-st}\, dt$$

$$= \int_0^\infty e^{-(s-a)t}\, dt = -\frac{1}{s-a}[e^{-(s-a)t}]_0^\infty$$

When $t = \infty$ the term in the brackets becomes 0 and when $t = 0$ it becomes -1. Thus:

$$F(s) = \frac{1}{s-a}$$

Properties of the Laplace transform

To use the Laplace transforms of functions given in tables it is necessary to know the properties of such transforms. The following are basic properties:

1 *Linearity property*

   The Laplace transform of the sum of two time functions, e.g. $f(t)$ and $g(t)$, can be obtained by adding the Laplace transforms of the two separate Laplace transforms, i.e.

$$\mathcal{L}\{f(t) + g(t)\} = \mathcal{L}f(t) + \mathcal{L}g(t)$$

Thus the Laplace transform of $1 + t$ is:

$$\mathcal{L}\{1 + t\} = \mathcal{L}1 + \mathcal{L}t = \frac{1}{s} + \frac{1}{s^2}$$

2   *Multiplication by a constant*

As a consequence of the linearity property, suppose we want the Laplace transform of $2t$. This can be considered to be $t + t$ and so the Laplace transform of $2t$ is the sum of the transforms of t and t and so $2 \times$ the Laplace transform of t. Since the Laplace transform of t is $1/s$ then the Laplace transform of $2t$ is $2/s$. The Laplace transform $4t^2$ is given by 4 times the Laplace transform of t^2 and so, since the Laplace transform of t^2 is $2/s^2$, is $8/s^2$. In general:

$$\mathcal{L}\{af(t)\} = a\mathcal{L}f(t)$$

The Laplace transform of $1 + 2t + 4t^2$ is given by the sum of the transforms of the individual terms in the expression and so is:

$$F(s) = \frac{1}{s} + \frac{2}{s^2} + \frac{8}{s^3}$$

3   *Derivatives*

The Laplace transform of a derivative of a function $f(t)$ is given by:

$$\mathcal{L}\left\{\frac{\mathrm{d}}{\mathrm{d}t}f(t)\right\} = sF(s) - f(0)$$

where $f(0)$ is the value of the function when $t = 0$. For a second derivative:

$$\mathcal{L}\left\{\frac{\mathrm{d}^2}{\mathrm{d}t^2}f(t)\right\} = s^2F(s) - sf(0) - \frac{\mathrm{d}}{\mathrm{d}t}f(0)$$

where $\mathrm{d}f(0)/\mathrm{d}t$ is the value of the first derivative at $t = 0$.

For example, if we have a function of time with an initial zero value, i.e. $f(0) = 2$, then the Laplace transform of $\mathrm{d}f(t)/\mathrm{d}t$ is $sF(s)$. If the function had $f(0) = 2$ then the Laplace transform of $\mathrm{d}f(t)/\mathrm{d}t$ is $sF(s) - f(0) = sF(s) - 2$.

As another illustration, if we have $\mathrm{d}^2f(t)/\mathrm{d}t^2$ with $f(0) = 0$ and $\mathrm{d}f(0)/\mathrm{d}t = 0$ then the Laplace transform of $\mathrm{d}^2f(t)/\mathrm{d}t^2$ is $s^2F(s)$. If the function had $f(0) = 2$ and $\mathrm{d}f(0)/\mathrm{d}t = 1$ then the Laplace transform of $\mathrm{d}^2f(t)/\mathrm{d}t^2$ is $s^2F(s) - 2s - 1$.

4   *Integrals*

The Laplace transform of the integral of a function $f(t)$ which has a Laplace transform $F(s)$ is given by:

$$\mathcal{L}\left\{\int_0^t f(t)\,\mathrm{d}t\right\} = \frac{1}{s}F(s)$$

For example, the Laplace transform of the integral of the function e^{-t} between the limits 0 and t is:

$$\mathcal{L}\left\{\int_0^t e^{-t}\, dt\right\} = \frac{1}{s}\, \mathcal{L}\{e^{-t}\} = \frac{1}{s(s+1)}$$

5 *The first shift theorem, factor e^{-at}*
   This theorem states that if $\mathcal{L}\{f(t)\} = F(s)$ then:

$$\mathcal{L}\{e^{-at}f(t)\} = F(s + a)$$

Thus the substitution of $s + a$ for s corresponds to multiplying a time function by e^{-at}. This can be demonstrated as follows:

$$\mathcal{L}\{e^{-at}f(t)\} = \int_0^\infty e^{-at}f(t)\, e^{-st}\, dt = \int_0^\infty f(t)\, e^{-(s+a)}\, dt$$

6 *The second shift theorem, time shifting*
   The second shift theorem states that if a signal is delayed by a time T then its Laplace transform is multiplied by e^{-sT}. A function $u(t)$ which is delayed is represented by $u(t - T)$, where T is the delay. Thus if $F(s)$ is the Laplace transform of $f(t)$ then:

$$\mathcal{L}\{f(t - T)u(t - T)\} = e^{-sT}F(s)$$

This can be demonstrated by considering a unit step function which is delayed by a time T (Figure ApC.2). For such a function:

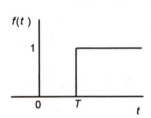

$$\mathcal{L}\{u(t - T)\} = \int_0^\infty u(t - T)\, e^{-st}\, dt = \int_0^T 0\, dt + \int_T^\infty 1\, e^{-st}\, dt$$

$$= \left[\frac{e^{-st}}{-s}\right]_T^\infty = e^{-sT}\frac{1}{s}$$

Figure ApC.2 *Delayed unit step*

Initial and final values

The *initial value* of a function of time is its value at zero time, the *final value* being the value at infinite time. Often there is a need to determine the initial value and final values of systems, e.g. for an electrical circuit when there is, say, a step input. The final value in such a situation is often referred to as the *steady-state value*. The initial and final value theorems enable the initial and final values to be determined from a Laplace transform without the need to find the inverse transform.

1 *The initial value theorem*
   The Laplace transform of $f(t)$ is given by:

$$\mathcal{L}\{f(t)\} = \int_0^\infty e^{-st}f(t)\, dt$$

and so:

$$\mathcal{L}\left\{\frac{\mathrm{d}}{\mathrm{d}t}f(t)\right\} = \int_0^\infty e^{-st}\frac{\mathrm{d}}{\mathrm{d}t}f(t)\ \mathrm{d}t$$

Integration by parts then gives:

$$\mathcal{L}\left\{\frac{\mathrm{d}}{\mathrm{d}t}f(t)\right\} = [e^{-st}f(t)]_0^\infty - \int_0^\infty (-s\ e^{-st})f(t)\ \mathrm{d}t = -f(0) + sF(s)$$

As s tends to infinity then e^{-st} tends to 0. Thus we must have, as a result of equation [18], $\mathcal{L}\{\mathrm{d}f(t)/\mathrm{d}t\}$ tending to 0 as s tends to infinity. Hence:

$$\lim_{s\to\infty}[-f(0) + sF(s)] = 0$$

$$\lim_{s\to\infty}sF(s) = f(0)$$

But $f(0)$ is the initial value of the function at $t = 0$. Thus, provided a limit exists:

$$\lim_{t\to0}f(t) = \lim_{s\to\infty}sF(s)$$

This is known as the *initial value theorem*.

2   *The final value theorem*
As with the initial value theorem, for a function $f(t)$ having a Laplace transform $F(s)$ we can write:

$$\mathcal{L}\left\{\frac{\mathrm{d}}{\mathrm{d}t}f(t)\right\} = \int_0^\infty e^{-st}\frac{\mathrm{d}}{\mathrm{d}t}f(t)\ \mathrm{d}t = -f(0) + sF(s)$$

As s tends to zero then e^{-st} tends to 1 and so:

$$\lim_{s\to0}\left[\int_0^\infty e^{-st}\frac{\mathrm{d}}{\mathrm{d}t}f(t)\ \mathrm{d}t\right] = \int_0^\infty \frac{\mathrm{d}}{\mathrm{d}t}f(t)\ \mathrm{d}t$$

We can write this integral as:

$$\int_0^\infty \frac{\mathrm{d}}{\mathrm{d}t}f(t)\ \mathrm{d}t = \lim_{t\to\infty}\int_0^t \frac{\mathrm{d}}{\mathrm{d}t}f(t)\ \mathrm{d}t = \lim_{t\to\infty}[f(t) - f(0)]$$

Hence, we obtain:

$$\lim_{s\to0}[-f(0) + sF(s)] = \lim_{t\to\infty}[f(t) - f(0)]$$

and so, provided a limit exists:

$$\lim_{t\to\infty}f(t) = \lim_{s\to0}sF(s)$$

This is termed the *final value theorem*.

For example, consider the final value of e^{-3t} as t tends to infinity. Using Table 10.1, we have $e^{-at} = 1/(s + a)$ and so $e^{-3t} = 1/(s + 3)$. The final-value theorem thus gives:

$$\lim_{t \to \infty} f(t) = \lim_{s \to 0} sF(s) = \lim_{s \to \infty} \frac{s}{s+3} = 0$$

and so the steady-state value is 0.

The inverse transform   The linearity property of Laplace transforms means that if we have a transform as the sum of two separate terms then we can take the inverse of each separately and the sum of the two inverse transforms is the required inverse transform.

$$\mathcal{L}^{-1}\{aF(s) + bG(s)\} = a\mathcal{L}^{-1}F(s) + b\mathcal{L}^{-1}G(s)$$

To illustrate how rearrangement of a function can often put it into the standard form shown in Table 10.1, the inverse transform of $3/(2s + 1)$ can be obtained by rearranging it as:

$$\frac{3(1/2)}{s + (1/2)}$$

Table 10.1 contains the transform $1/(s + a)$ with the inverse of e^{-at}. Thus the inverse transformation is just this transform with $a = \frac{1}{2}$ and multiplied by the constant $(3/2)$ and so is $(3/2)\, e^{-t/2}$.

Expressions often have to be put into simpler standard forms of Table 10.1 by the use of partial fractions. For example, the inverse Laplace transform of $(3s - 1)/s(s - 1)$ is obtained by first simplifying it using partial fractions:

$$\frac{3s - 1}{s(s - 1)} = \frac{A}{s} + \frac{B}{s - 1}$$

and so we must have $3s - 1 = A(s - 1) + Bs$. Equating coefficients of s gives $3 = A + B$ and equating numerical terms gives $-1 = -A$. Hence:

$$\frac{3s - 1}{s(s - 1)} = \frac{1}{s} + \frac{2}{s - 1}$$

The inverse transform of $1/s$ is 1 and the inverse of $1/(s - 1)$ is e^t. Thus:

$$\mathcal{L}^{-1}\left\{ \frac{3s - 1}{s(s - 1)} \right\} = 1 + 2\, e^t$$

Solving differential equations   Laplace transforms offer a method of solving differential equations. The procedure adopted is:

1   Replace each term in the differential equation by its Laplace transform, inserting the given initial conditions.

2 Algebraically rearrange the equation to give the transform of the solution.

3 Invert the resulting Laplace transform to obtain the answer as a function of time.

As an illustration, the first-order differential equation:

$$3\frac{dx}{dt} + 2x = 4$$

has the Laplace transform:

$$3[sX(s) - x(0)] + 2X(s) = \frac{4}{s}$$

Given that $x = 0$ at $t = 0$:

$$3sX(s) + 2X(s) = \frac{4}{s}$$

Hence:

$$X(s) = \frac{4}{s(3s + 2)}$$

Simplifying by the use of partial fractions:

$$\frac{4}{s(3s + 2)} = \frac{A}{s} + \frac{B}{3s + 2}$$

Hence $A(3s + 2) + Bs = 4$ and so $A = 2$ and $B = -2/3$. Thus:

$$X(s) = \frac{2}{s} - \frac{2}{3(3s + 2)} = \frac{2}{s} - 2\frac{\frac{2}{3}}{\frac{2}{3}\left(s + \frac{2}{3}\right)}$$

and so:

$$x(t) = 2 - 2\,e^{-2t/3}$$

As a further illustration, the second-order differential equation:

$$\frac{d^2x}{dt^2} - 5\frac{dx}{dt} + 6x = 2\,e^{-t}$$

has the Laplace transform:

$$s^2X(s) - sx(0) - \frac{d}{dt}x(0) - 5[sX(s) - x(0)] + 6X(s) = \frac{2}{s + 1}$$

Given that $x = 0$ and $dx/dt = 1$ at $t = 0$:

$$s^2 X(s) - 1 - 5sX(s) + 6X(s) = \frac{2}{s+1}$$

Hence:

$$X(s) = \frac{\frac{2}{s+1} + 1}{s^2 - 5s + 6} = \frac{2}{(s+1)(s-2)(s-3)} + \frac{1}{(s-2)(s-3)}$$

We can simplify the above expression by the use of partial fractions:

$$\frac{2}{(s+1)(s-2)(s-3)} = \frac{A}{s+1} + \frac{B}{s-2} + \frac{C}{s-3}$$

Hence $A(s-2)(s-3) + B(s+1)(s-3) + C(s+1)(s-2) = 2$ and so $A = 1/6$, $B = -2/3$ and $C = \frac{1}{2}$. For the other fraction:

$$\frac{1}{(s-2)(s-3)} = \frac{D}{s-2} + \frac{E}{s-3}$$

Hence $D(s-3) + E(s-2) = 1$ and so $D = -1$ and $E = 1$. Thus:

$$X(s) = \frac{\frac{1}{6}}{s+1} + \frac{-\frac{2}{3}}{s-2} + \frac{\frac{1}{2}}{s-3} + \frac{-1}{s-2} + \frac{1}{s-3}$$

$$= \frac{\frac{1}{6}}{s+1} - \frac{\frac{5}{3}}{s-2} + \frac{\frac{3}{2}}{s-3}$$

The inverse transform is thus:

$$x(t) = \frac{1}{6} e^{-t} - \frac{5}{3} e^{2t} + \frac{3}{2} e^{3t}$$

Answers

The following are answers to the numerical and multiple-choice problems and brief clues as to the answers to other problems.

Chapter 1   1 A   2 B   3 B   4 C   5 B

6 Sensor, signal processor, data presentation, see section 1.2.
7 See sections 1.4 and 1.3.3. Reliability: the probability that a system will operate to an agreed level of performance for a specified period, subject to specified environmental conditions. Repeatability: the ability of a system to give the same value for repeated measurements of the same variable.
8 See section 1.6.3.
9 See Section 1.6.
10 Only give required accuracy 6 times in 10.
11 In 1 year, 1 in 100 will be found to fail.
12 (a) 9 mm/kg, (b) 0.060 mV/°C, (c) 3 pC/N
13 5%

Chapter 2   1 C   2 B   3 A   4 C   5 B   6 A

7 B   8 B   9 D   10 B   11 C   12 C
13 A   14 C   15 A   16 C   17 C   18 A
19 For example: (a) turbine meter, (b) bourdon gauge, (c) LVDT, (d) thermocouple.
20 For example: (a) Wheatstone bridge, (b) ADC.
21 $454 \, \Omega$
22 $117 \, \Omega$
23 $0.13 \, \Omega$
24 $0.126 \, \Omega$
25 -29.7 pF/mm
26 55.3 pF
27 12.227 mV
28 -0.89%
29 As in the question
30 0.059 V
31 5.25×10^{-5} V
32 See Section 2.9.3, $R_2 = 19.9 \, \Omega$
33 50
34 $10 \, \text{k}\Omega$
35 $490 \, \text{k}\Omega$
36 $100 \, \text{k}\Omega$

37 2.33 kΩ
38 2.442×10^{-4}
39 9
40 19.6 mV

Chapter 3   1 A   2 A   3 C   4 C   5 D   6 A

7 For example: painted stripes on floor and optical sensor; optical encoder to monitor rotation of wheel.
8 For example: (a) weight-operated electrical switch, (b) thermistor, potential divider circuit, voltmeter, (c) load cell using strain gauges, Wheatstone bridge, amplifier.
9 244 μV
10 About 200 to 400 gives the range.

Chapter 4   1 D   2 A   3 D   4 C

5 (a) Closed-loop with a sensor, e.g. LVDT, giving a negative feedback signal which is compared with the required value, (b) discrete-time control system with perhaps limit switches being used to sense when a package is on the belt and at the pick-up point, (c) discrete-time control system where a sensor provides a signal which results in an output when the sensor signal ceases, i.e. a NOT gate, (d) closed-loop with a sensor perhaps monitoring the weight of a sack as it is filled, or perhaps open-loop with the amount of chemical supplied being for a fixed length of time, (e) closed-loop with the input of water being controlled by the feedback signal from a level sensor, (f) open-loop, there being no feedback, (g) closed-loop.
6 (a) Closed-loop with a signal from the controller, the differential amplifier, being used to switch on or off, by means of the relay, the supply to the heater, (b) The sensor, the thermistor, provides an analogue feedback signal which is converted into a digital signal for comparison by the microcontroller with a digital set point signal, the output of the microcontroller being used to switch a heater on or off.
7 See Figure A.1
8 Measurement – level probe, controller – relay and solenoid valve, correction – flow control valve, process – water tank, measurement – level probe
9 See Figure A.2
10 (a) See Figure A.3(a), (b) see Figure A.3(b)

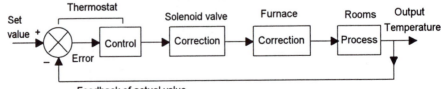

Figure A.1 *Chapter 4, problem 7*

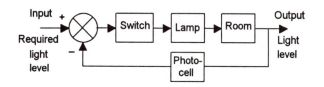

Figure A.2   *Chapter 4, problem 9*

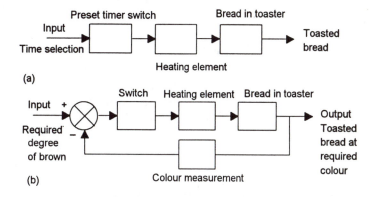

(a)

(b)

Figure A.3   *Chapter 4, problem 10*

Chapter 5   1 C   2 A   3 B   4 D   5 A   6 C
7 B   8 D   9 A   10 B   11 D
12   (a) −10°C to +30°C, (b) 40°C
13   (a) 2°C, (b) 10%
14   0.40 mA/°C
15   (a) 255 rev/min, (b) 324 rev/min
16   50%, 60%
17   See Figure A.4
18   $K_p = 2$, $T_i = 5.5$ min, $T_d = 1.4$ min
19   $K_p = 1.3$, $T_i = 6$ min, $T_d = 1.5$ min
20   $K_p = 0.62$, $T_i = 3$ min, $T_d = 0.75$ min

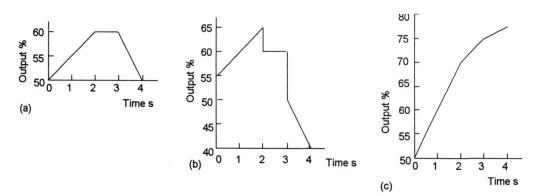

Figure A.4   *Chapter 5, problem 17*

Chapter 6

1 B	2 C	3 A	4 C	5 A	6 A
7 B	8 C	9 D	10 A	11 A	12 C
13 D	14 B	15 A	16 B	17 C	18 C

19 A
20 0.004 m²
21 5.85 m³/s
22 504 pulses/s
23 1260 mm
24 800 mm

Chapter 7

1 D	2 A	3 C	4 A	5 A	6 C
7 D	8 B	9 C	10 A	11 C	12 C
13 C	14 A	15 D	16 A	17 A	18 C

19 See Figure A.5
20 See Figure A.6
21 A+ and B+, C+, A– and B–, C–
22 A+, B+, A–, B–, A+, A–

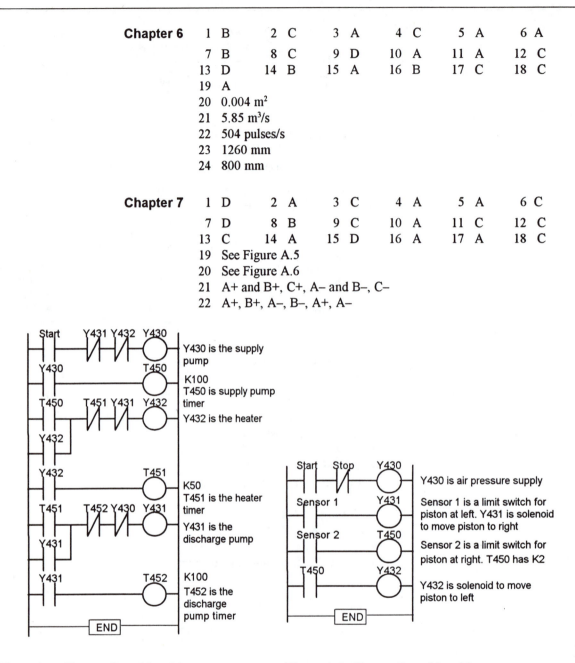

Figure A.5 *Chapter 7, problem 19*   Figure A.6 *Chapter 7, problem 20*

Chapter 8

1 B	2 B	3 C	4 A	5 D	6 D

7 C
8 10 V
9 100
10 (a) 6/11, (b) 6/13
11 (a) $k_1 x = M\dfrac{d^2 y}{dt^2} + c\dfrac{dy}{dt} + (k_1 + k_2)y,$

(b) $T = I\dfrac{d^2\theta}{dt^2} + c\dfrac{d\theta}{dt} + k\theta$,

(c) $v = L\dfrac{di}{dt} + Ri$,

(d) $p = c\dfrac{d\theta}{dt} + cpq\theta$, where $\theta = \theta_o - \theta_i$

Chapter 9 1 D 2 B 3 C 4 A

5 $5/[s(s + 1)]$

6 1.7

7 $2/(s + 1.2)$

8 $1/(1 + RCs)$

9 (a) $G_1G_2G_3$, (b) $G_1G_2/(1 + G_1G_2G_3)$, (c) $G_1G_2G_3/(1 + G_1G_2G_3)$,
   (d) $G_1G_2/(1 + G_1G_2H_1 + G_2H_2)$

10 (a) $1/[s(s + 2) + 2]$, (b) $K/[s(s - 1) + K(s + 2) + Ks]$,
    (c) $2/(s^2 + 9s + 6)$

11 $K_1N/[s(sL + R) + K_1K_2N]$

12 $75/[(s + 1)^2(s + 2) + 7.5]$

13 $Y(s) = [2/(s^2 + 4)]X_1(s) + [s/(s^2 + 4)]X_2(s)$

14 $Y(s) = [4/(s^2 + 16s + 40)]X_1(s) + [s/(s^2 + 16s + 40)]X_2(s)$

Chapter 10 1 C 2 A 3 D 4 B 5 A 6 A

7 D 8 C 9 A 10 B

11 $3/s$

12 2

13 $5/s^2$

14 e^{-5t} V

15 (a) $(5/3)(1 - e^{-3t})$ V, (b) $5e^{-3t}$ V

16 (a) $6(1 - e^{-t})$ V, (b) $6e^{-t}$ V

17 (a) $2(1 - e^{-2t})$ V, (b) $\frac{1}{2}[t - \frac{1}{2}(1 - e^{-2t})]$ V

18 (a) $5/(s - 1) - 4/(s - 2)$, (b) $4/(s + 1) - 3/(s + 2)$,
    (c) $2/(s + 1) - 3/(s + 1)^2$

19 $24 - 12e^{-2t} - 4e^{-4t}$

20 $8e^{-2t} - 8t\,e^{-2t}$

21 (a) Critical, (b) overdamped, (c) underdamped

22 $-1.5 + 3.0t + 1.5e^{-2t}$

23 $0.5 - e^{-t} + 0.5e^{-2t}$

24 $0.5(e^{-t} - e^{-3t})$

25 $0.5(1 - e^{-10t})$

26 10, 0.05

27 Underdamped

28 Critically damped

29 $1/53$ s

30 (a) 9.2 s, 8.8 s, (b) 2.3 s, 2.2 s, (c) 0.77 s, 0.73 s

31 69.1 s, 65.9 s

32 8.7 s, 20 s, 19.1 s

33 (a) 10 rad/s, 0.2, 0.16 s, 53%, 2 s, (b) 7 rad/s, 0.29, 0.23 s, 39%,
    2.0 s

34 4 rad/s, 0.63, 0.51 s, 7.8%, 1.6 s

35 2.6

36 Stable (a) and (c); oscillatory (b) and (d)
37 (a) Stable, (b) unstable, (c) unstable
38 (a) Stable, (b) unstable, (c) stable
39 (a) Stable, (b) unstable

Chapter 11   1 C   2 A   3 D   4 B   5 C   6 D

7 A
8 (a) $3/(5 + j\omega)$, (b) $7/(2 + j\omega)$, (c) $1/(10 + j\omega)(2 + j\omega)$
9 (a) 1.34, $-26.6°$, (b) 1.06, $-45°$
10 See Figure A.7
11 See Figure A.8
12 (a) $1/s$, (b) $3.2/(1 + s)$, (c) $2/(s^2 + 2\zeta s + 1)$,
    (d) $3.2 \times 100/(s^2 + 2\zeta s + 100)$, (e) $10/(s^2 + 0.8s + 4)$
13 $16 \times 15/(s + 15)$
14 1.82, $50°$

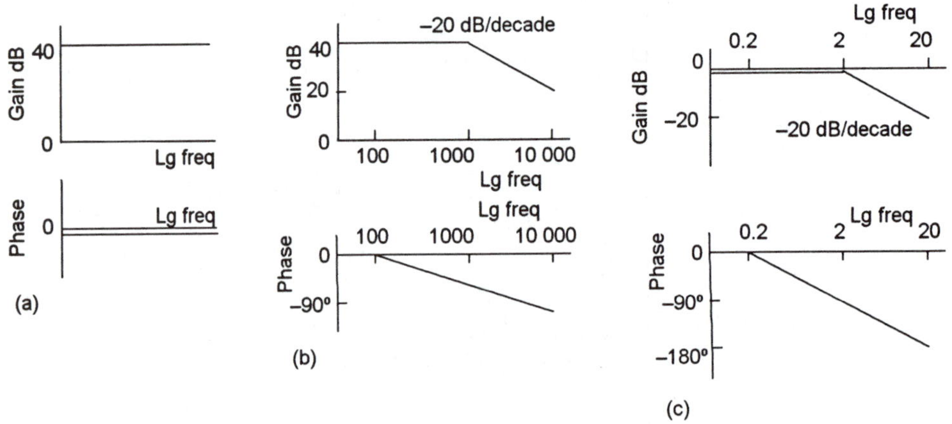

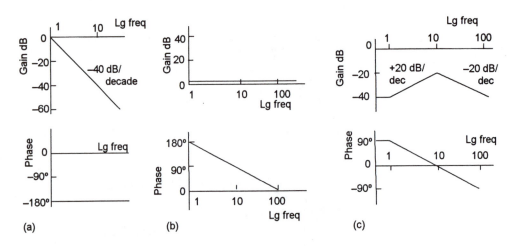

Figure A.7   *Chapter 11, problem 10*

Figure A.8   *Chapter 11, problem 11*

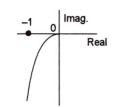

Figure A.9 *Chapter 12, problem 4*

15 (a) Stable, (b) about 36 dB, (c) about 80°
16 (a) Stable, (b) about 15 dB, (c) about 43°
17 (a) Stable, (b) about 15 dB, (c) about 32°
18 About 5.7 dB, reduced by factor of 1.9
19 About 34°, 1.7 Db, reduced by factor of 1.2
20 55°
21 $(0.38s + 1)/(0.13s + 1)$
22 $(10s + 10)/(80s + 1)$

Chapter 12　　1　A　　　　2　C　　　　3　C

4　See Figure A.9
5　$(-1, j0)$ point not to be enclosed
6　8.0 dB, 20°
7　4.4 dB, 10°
8　3.5 dB, 11°
9　28 dB, 76°

Chapter 13　　1　C　　　2　B　　　3　A　　　4　A　　　5　D　　　6　B

7　$$\frac{10(K_d s^2 + K_p s + K_i)}{s(s = 5)(s + 10) + 10(K_d s^2 + K_p s + K_i)}$$

8　(a) –40°, (b) 15°
9　(a) $4 - 4\,e^{-0.5t}$, (b) $5[1 - 0.997\,e^{-0.05t} \sin(0.63t - 1.49)]$
10　(a) 0.71, (b) 0

Index